COMPLICATIONS OF
EXTERNAL SKELETAL FIXATION

COMPLICATIONS OF EXTERNAL SKELETAL FIXATION
Causes, Prevention, and Treatment

By

STUART ALAN GREEN, M.D.

Chief, Osteomyelitis Service
Rancho Los Amigos Hospital
Downey, California

CHARLES C THOMAS • PUBLISHER
Springfield • Illinois • U.S.A.

Published and Distributed Throughout the World by
CHARLES C THOMAS • PUBLISHER
2600 South First Street
Springfield, Illinois 62717, U.S.A.

This book is protected by copyright. No part of it
may be reproduced in any manner without written
permission from the publisher.

© *1981 by* CHARLES C THOMAS • PUBLISHER
ISBN 0-398-04482-1
Library of Congress Catalog Card Number: 81-414

*With THOMAS BOOKS careful attention is given to all details of
manufacturing and design. It is the Publisher's desire to present books that are
satisfactory as to their physical qualities and artistic possibilities and appropriate for their particular use. THOMAS BOOKS will be true to those laws of
quality that assure a good name and good will.*

Library of Congress Cataloging in Publication Data

Green, Stuart Alan.
    Complications of external skeletal fixation.

    Bibliography: p.
    Includes index.
    1. Fracture fixation—Complications and sequelae. I. Title. II. Title:
External skeletal fixation: causes, prevention & treatment. [DNLM: 1. Fracture
fixation—Adverse effects. 2. Orthopedic fixation devices—Adverse effects. WE
185 G798c]
RD101.G67   617'.1501   81-414
ISBN 0-398-04482-1   AACR1

Printed in the United States of America
PS-R-1

PREFACE

This monograph deals with the complications of external fixation that have vexed four generations of orthopaedic surgeons. It was written to focus attention on the causes, prevention, and management of these troublesome problems. It is based on the premise that more is to be learned from problems and failures than from successes.

The work evolved from a scientific exhibit: *Complications of External Fixation: Why They Occur; How to Avoid Them*, which was first displayed at the 1979 meeting of the American Academy of Orthopaedic Surgeons. The organization and structure of the book germinated from the exhibit.

The section dealing with nerve and vessel injury was created especially for this monograph after I realized that some of the most frightful complications of external fixations might be avoidable if the problem was approached in a logical manner.

This monograph is not intended to be a treatise on fracture management utilizing external fixation as compared with other modalities. In fact, the matter has not yet been resolved.

Consideration will be given to various fixator frame configurations as they apply to the problems being discussed. It will become apparent from the illustrations that we utilize the Hoffmann system for external fixation at Rancho Los Amigos Hospital. This should not be construed as an endorsement of the system, neither should it be inferred that the Hoffmann system in any way contributes to, or helps prevent, the complications to be considered.

The opinions and recommendations in this book are my own, as are any errors and omissions. I have tried to convey the current wisdom among authorities in the field of external fixation, but it is not implied that the measures outlined in this book are the only strategies available to reduce the likelihood of complications. Some complications are preventable, while others seem to strike capriciously. I hope this volume will aid the reader in avoiding the former and managing the latter.

TO MY PARENTS

Leo Arthur Green, M.D.
Mary G. Green

ACKNOWLEDGMENTS

Much of the credit for my involvement with external fixation belongs to two progressive orthopaedic surgeons: Vert Mooney, M.D., formerly chief of the Problem Fracture Service at Rancho Los Amigos Hospital; and Douglas Garland, M.D., his successor. Additionally, Dr. Garland read the entire manuscript and made many helpful suggestions.

Appreciation is due to Edward Miller, M.D., my coexhibitor for the original display on complications of external fixation.

Mr. Payne Thomas, the publisher, deserves credit for suggesting this monograph. His encouragement and counsel have been invaluable.

It has been my pleasure to work closely with Michele Predisik, a talented medical illustrator, whose beautiful drawings are evident in this volume.

Special thanks to Georges Deutsch, Richard Wheeler, and Ian Teague who have given me the opportunity to exchange ideas with orthopaedic colleagues at various meetings.

Gratitude is also due to those orthopaedic surgeons in the Long Beach, California area who have been kind enough to share their clinical material with me.

My family deserves so much credit for their patience with me that it is impossible to thank them enough.

CONTENTS

	Page
Preface	v

Chapter

1	Introduction	3
2	Pin Tract Infection	12
3	Nerve and Vessel Injury	31
4	Interference With Limb Function	78
5	Failure to Obtain Union	84
6	Unsuccessful Arthrodesis	107
7	Persistent Wound Infection	124
8	Fixator Problems	136
9	Post Fixation Complications	145
10	Summary and Conclusions	156
Bibliography		159
Index		169

COMPLICATIONS OF EXTERNAL SKELETAL FIXATION

Chapter 1

INTRODUCTION

HISTORY OF EXTERNAL FIXATION

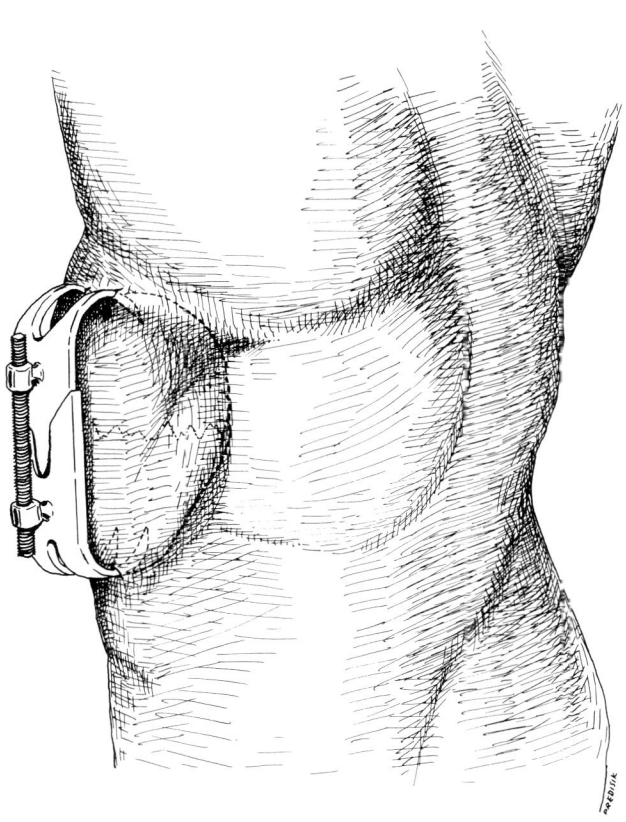

Figure 1: Malgaigne's patellar fixator—1843.

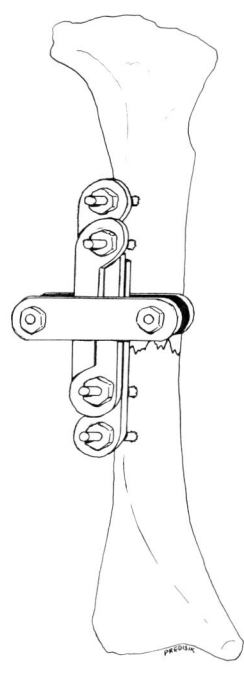

Figure 2: Parkhill's "bone-clamp"—1897.

The external fixator was invented twelve years before the plaster cast. In 1843, Jean Francois Malgaigne[184] devised an ingenious mechanism consisting of a clamp that approximated four transcutaneous metal prongs for use in reducing and maintaining patellar fractures. In the 130 years since Malgaigne's invention, many other external fixation systems have been introduced. Clayton Parkhill,[214] an American surgeon working in Denver, in 1897 devised a system of fracture management utilizing transcutaneous pins connected to a rigid external plate. He published his experiences with the device noting excellent results and an absence of complications when his technique was employed. At about the same time (1902), Albin Lambotte[168] of Belgium publicized a similar system that he invented independently. Both men were stimulated by the observation that metal pins which penetrated bone and protruded through the skin were remarkably well tolerated.

In the United States, other orthopaedic surgeons developed external fixation systems of fracture management in the early part of the twentieth century.[60,7,99] In 1934, Roger Anderson[8] devised an apparatus for the

3

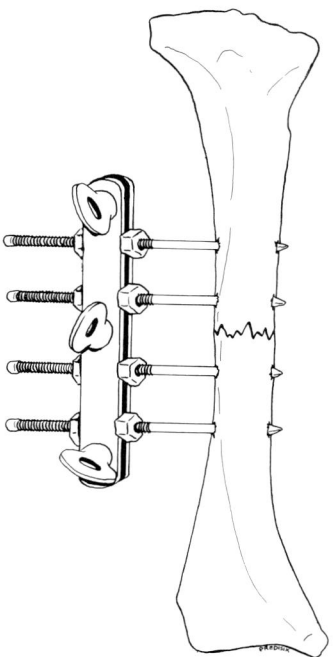

Figure 3: Lambotte's pioneering external fixator—1902.

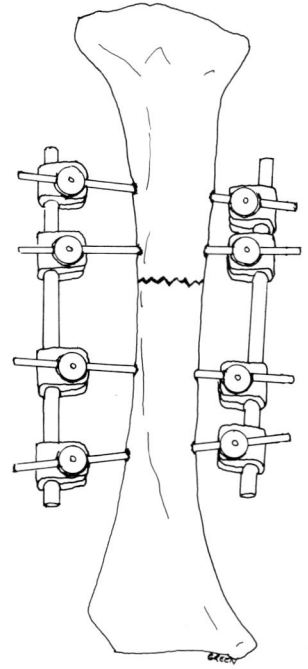

Figure 4: Anderson frame (1938) employing transfixion pins.

mechanical reduction of fractures utilizing transcutaneous pins connected to movable metal yokes. He used full (through-and-through) pins that protruded from both sides of the limb while transfixing bone. Anderson's yokes were part of what he called a "fracture robot," which permitted multiplanar adjustment of the fracture fragments and allowed significant compression at the fracture site prior to cast application. The cast was applied while the limb was still in the fracture robot. Compression was maintained by tension on slightly bent pins. If necessary, the entire apparatus could be autoclaved for use with compound fractures. Once fracture reduction was complete with Anderson's device, a plaster of Paris cast was applied over the pins, and the fracture robot was removed and available for use on another patient. Anderson did not mention significant complications with his technique in his early reports. On the contrary, he stated that "many years' experience of securely incorporating pins, tongs or wires in plaster, warrants the statement that bone infection does not follow a sterile insertion." Later, Anderson extended his concept and devised an external fixation system that connected the transcutaneous pins to bars on the medial and lateral side of the limb by means of articulations, thereby eliminating the plaster cast.[9]

Meanwhile, Otto Stader,[251] a veterinarian, had devised a system of fracture management for dogs con-

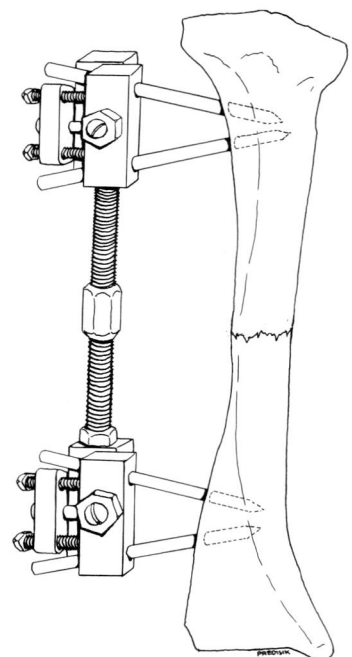

Figure 5: Stader's reduction-fixation apparatus.

sisting of an external skeletal fixator which permitted not only stabilization of the fracture but also reduction of the fracture fragments in three planes independently. The device was developed because dogs tended to destroy casts by biting and tearing at them, and by fouling them with body excretions. In 1937, Lewis and Breidenbach[173] observed a police dog with a fractured femur that had been treated with the Stader apparatus. They were encouraged by both the excellent alignment obtained and the potential for early ambulation. With Stader, they refined and enlarged the device somewhat, reporting their experience with its use for treating fractures of the long bones in 1942. They described three cases of pin sepsis out of twenty patients treated, with one case of chronic pin hole osteomyelitis. The patient died of septicemia subsequent to the infection, which occurred in the days before antibiotics were widely available. They attributed the pin hole sepsis to a galvanic reaction between the bone, the steel pins, and the aluminum fixator, and consequently began to employ pins that had been insulated with a plastic pin-gripper.

Another pioneer in the use of external fixation techniques was H.H. Haynes,[120] of Clarksburg, West Virginia, who first reported on the device he had invented in 1938. He described six cases without any note of pin tract infections or other complications.

Meanwhile, in Europe, Lambotte's original concept of external skeletal fixation was being expanded significantly. Early pioneers included Juvara,[15] Verbrugge,[262] Goosens,[108] and others.[71] In 1938, Raoul Hoffmann,[128] a mechanically minded Swiss surgeon, devised an external fixator that incorporated a universal ball joint connecting the external bar of his fixator to strong pin-gripping clamps. This universal joint permitted fracture reduction in three planes to be carried out while the fixator was in place. It also made the fixator more versatile because the pin-gripping clamps could be placed in any position dictated by the anatomic considerations of the fracture under treatment. Furthermore, Hoffmann could substitute a sliding compression-distraction bar for the rigid bar which connected the pin-gripping clamps. In this manner, interfragmentary compression or limb length restoration could be achieved.

In the United States during World War II, the development of external skeletal fixation techniques was stimulated by the appearance of a treatise on fracture management by Shaar and Kreuz.[245] Their experience with the Stader splint was extensive, and their book included a chapter outlining the technical details to be considered in order to prevent undue complications associated with the use of external skeletal fixation. They also described pin placement positions designed to prevent neurovascular injury. While their discussion (both in their book and in various articles)[243-247] of the strategies for avoiding complications was extensive, Shaar and Kreuz did not

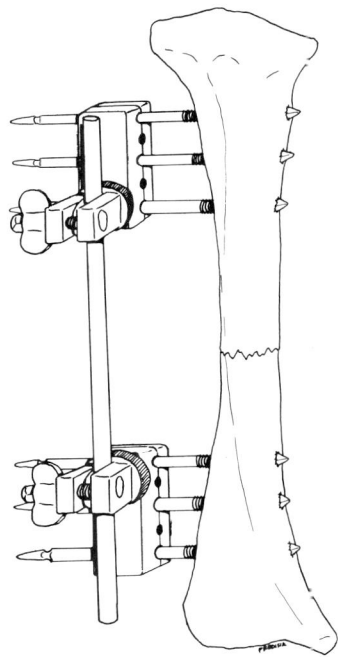

Figure 6: Hoffmann external fixator (1938).

document well the incidence and details of complications in their series.

Toward the end of World War II, however, the high incidence of significant complications associated with external fixation became apparent. This observation resulted in a directive issued to military surgeons of the United States Armed Forces to discontinue the use of external skeletal fixation, retarding development in the use of external fixation in the United States for a time.

Following the war, The Committee on Fracture and Trauma Surgery of the American Academy of Orthopaedic Surgeons began investigating the efficiency, practicality, and rightful place of external skeletal fixation in the armamentarium of fracture management. A study begun by Clay Ray Murray[200] was commissioned, to be followed up later by Mather Cleveland and Herman F. Johnson.[140] Questionnaires were sent to members of the American Academy of Orthopaedic Surgeons, the American Association of Surgery and Trauma, and the Iowa Medical Association. (Iowa was considered to be a representative state and the questionnaire responses from the IMA group were used to assess the use of external fixation by non-orthopaedic surgeons.)

Of the 3,082 questionnaires sent, 768 were returned. Of the returned questionnaires, 48 percent came from practitioners who had no experience with external skeletal fixation. These were discarded. The remaining

395 replies were carefully analyzed by the committee. Of this group, 28 percent felt that external skeletal fixation had a definite place in fracture management, while 29.4% felt that external fixation was inadvisable except in rare, carefully chosen instances. Just over 43 percent of the respondents had used external fixation at one time but had abandoned it completely by the time of the survey.

An interesting observation by the committee was the positive correlation between the number of cases treated and the surgeon's attitude toward external skeletal fixation. Generally speaking, surgeons who had the most experience with external fixation were most likely to recommend its continued use, while surgeons with very limited experience with external fixation tended to be among those who had discarded it completely. Respondents who felt that external fixation was advisable listed a number of advantages offered by the technique. They were impressed by the more secure and adequate immobilization external fixation provided, as well as the ease of application, the prospect of early joint motion, and the reduced period of hospitalization. The major disadvantages noted by the group who had discarded external fixation included the presence of soft tissue infections at the pin sites, the possibility of ring sequestra and osteomyelitis, and the danger of delayed union or nonunion. Other surgeons were distressed by the mechanical difficulty associated with external fixators, as well as by the prospect of converting a closed fracture to an open fracture. They were also concerned by the difficulties they had in obtaining and maintaining a reduction and by the dangers of distraction at the fracture site.

In view of the above considerations, the committee concluded that any physician who contemplated the use of external skeletal fixation required special training under the supervision of a surgeon who had treated at least 200 cases by this method. They recommended that physicians without adequate training not attempt to use the technique. As a consequence, by 1950, the majority of American orthopaedic surgeons were not using this modality. From 1950 to 1970, mechanical fixators were generally unpopular with American orthopaedists, although the pins-in-plaster technique was used for special problems, such as unstable wrist fractures[62,111] and displaced fractures of the tibia and fibula.[7,247]

In Europe, on the other hand, clinical research on external skeletal fixation continued throughout the years during and following World War II. Raoul Hoffmann improved his device,[132] providing a stronger universal joint and an enlarged pin-gripper that held the pins more securely. A pin guide was developed; this permitted application of the pins through the center of the bone using short guide pins to determine the extent of the cortex. Judet,[142-148] working in France, devised an external skeletal fixator that was elegant in its simplicity. Charnley,[52] of England, presented his concept of compression arthrodesis of the major joints, utilizing a rather simple external skeletal fixator which provided continuous compression across cut cancellous surfaces of the joint to be fused. In time, the AO group[270] of Switzerland modified Charnley's device in order to permit the addition of several more pins to his frame configuration.

In France during the 1960s, Jacques Vidal[269] began recognizing the importance of extremely rigid fixation in dealing with problems associated with septic nonunions of long bone. He utilized Hoffmann's equip-

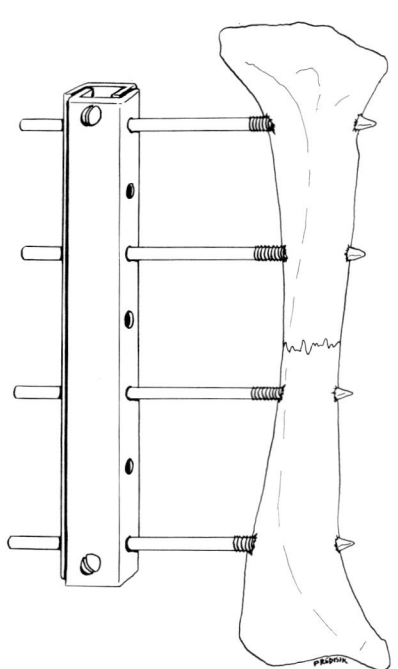

Figure 7: Judet's fixator.

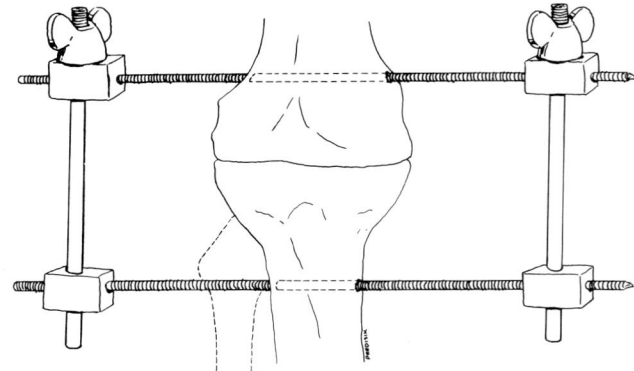

Figure 8: Charnley compression clamp.

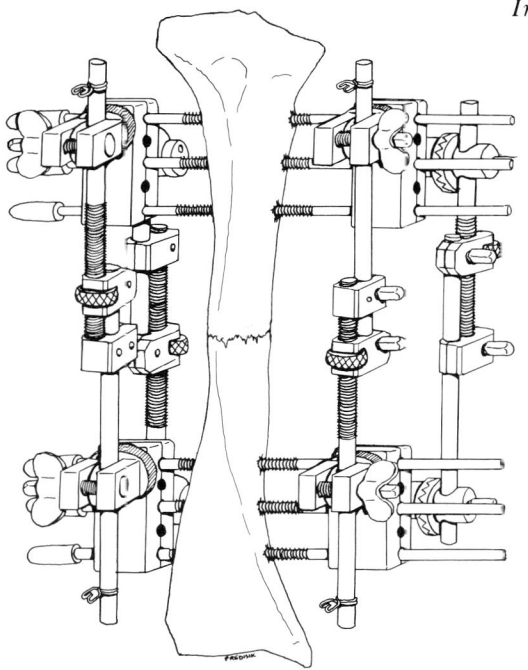

Figure 9: Vidal-Adrey quadrilateral frame utilizing Hoffmann system components.

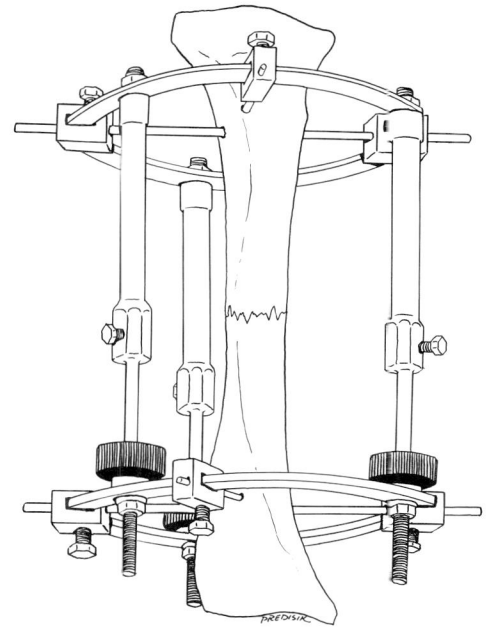

Figure 10: Ilisarov full-ring fixator.

ment but designed a quadrilateral frame to provide rigid stabilization of complex fracture problems under treatment (Fig. 9). Jose Adrey[2], working with Vidal, did the appropriate biomechanical studies on the device and determined that the quadilateral configuration was quite stable. Meanwhile, in Belgium, a continuation of Dr. Hoffmann's original concept of a unilateral frame, utilizing a single connecting bar and half-pins, was carried out at the University of Brussels by Franz Burny[35] and others.[40,41] Burny's immeasurable contribution to the field of external skeletal fixation has been the accurate and comprehensive statistical analysis of several large series of fracture problems. In England, Day and Freeman[100] have reported their results with an elegantly simple fixator they developed, similar to Hoffman's in design in that the pin-gripping clamps are connected to rigid bars by universal ball joints. Following Anderson's formula, they employ full transfixion pins.

In Russia, external fixation as a modality for fracture treatment remained viable in the period subsequent to World War II. Work by Ilisarov[137] and others[278] focused attention on rigid ring-type fixators that were connected to the bone by very thin transfixion wires. The wires were maintained in tension by special wire-gripping clamps. (In order to maintain fixation on a bone segment, at least two non-coplanar wires had to be inserted.) While these fixators are quite cumbersome, some contain ingenious geared articulations that permit precise displacement of the rings in any of three planes independently. Kronner,[167] an American surgeon, modified the Russian design by employing plastic components and transfixion pins in place of the thin wires used by Soviet surgeons. He also eliminated the geared articulations, substituting universal ball joints from Hoffmann's original design. Fischer, of Minneapolis, has recently developed a half-ring fixator that has several unique features.

External fixators specifically designed for limb lengthening began to appear after W.V. Anderson[10] developed an apparatus that employed full transcutaneous pins connected to threaded distraction bars on both sides of the limb. The device permitted gradual controlled distraction of an osteotomized bone. His apparatus has been modified slightly by Coleman and Noonan,[63] and by Kawamura[155] of Japan. Heinz Wagner,[283,284] working in Germany, modified Anderson's concept even further, substituting half-pins (actually Schanz screws) for Anderson's full-pins, while employing a universal distraction bar that the patient could lengthen himself. These pioneers have accurately recorded the incidence of complications with their techniques, some of which are unique to limb lengthening.

Worldwide interest in external skeletal fixation is on the rise again, as it was before World War II. A reason for this is the increased incidence of serious bone and

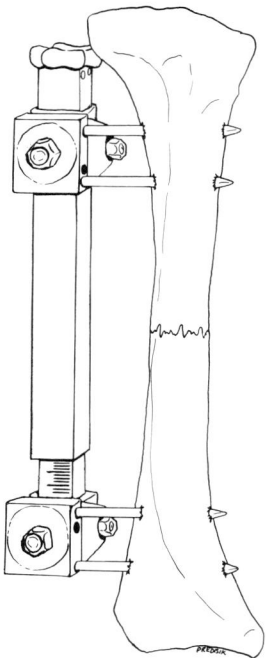

Figure 11: Wagner limb-lengthening device.

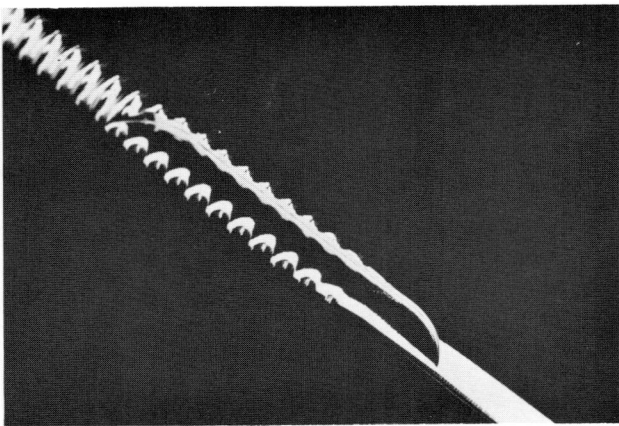

Figure 12A: Bonnel central threaded (self-tapping) full pin.

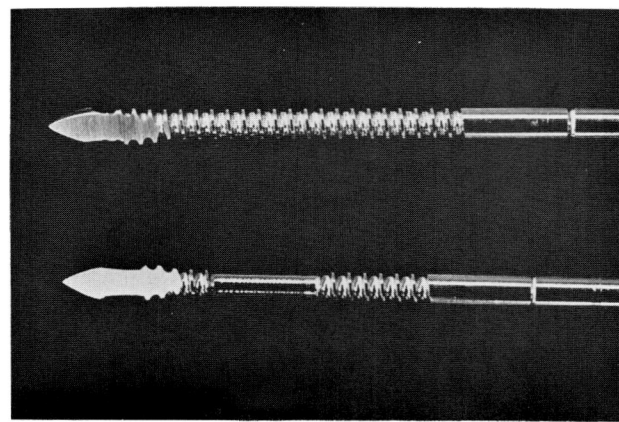

Figure 12B: Half-pins: end threaded (above) and interrupted thread (below).

soft tissue injury associated with high-speed road accidents, especially among motorcyclists.

A wide variety of fixators possessing ingenious articulations and pin-grippers are currently available. Surgical appliance manufacturers continue to add new components and fixator frames to the marketplace at a bewildering pace. The devices vary considerably in configuration and in the technique of their frame assembly. The feature common to all fixators, however, is that they are attached to the human body with pins that penetrate the skin and affix to bone. The complications associated with transcutaneous pins are thus common to all past, present, and future fixators, regardless of design or construction.

Terminology

Components

PIN: The term pin refers to that portion of the fixator which penetrates the skin and soft tissues and attaches to bone. In the European literature pins are sometimes referred to as screws or nails (the distinction resting perhaps on the presence or absence of threads). In this book, the term is used to refer to both the threaded and nonthreaded (smooth pin) varieties.

FULL-PIN: A full-pin is one that protrudes through the skin and soft tissues on both sides of the limb. Such pins are sometimes referred to as transfixion pins or through-and-through pins.

HALF-PINS: A half-pin is one that penetrates the skin and soft tissues on one side of the limb only and that penetrates bone but does not emerge on the other side of the limb. Such pins have been referred to elsewhere as *end-threaded pins* or *bicortical pins*. When inserted, such pins are supposed to penetrate both cortices of the bone but not much beyond the second cortex.

PIN-GRIPPER: The apparatus that holds the pin is called a pin-gripper (or pin-gripping clamp). Needless to say, the pin-gripper must have a means of attachment to the rest of the fixator.

BAR: The bar is the portion of the apparatus that connects the pin-grippers. Bars may be solid or hollow, smooth or threaded, and they may incorporate a compresson-distraction apparatus in their structure.

RING: A ring is a circular bar (or modified bar) that attaches to pin-grippers in a plane that is usually perpendicular to the long axis of the limb. The rings

Figure 13A: Multiple position pin-gripper (Day system).

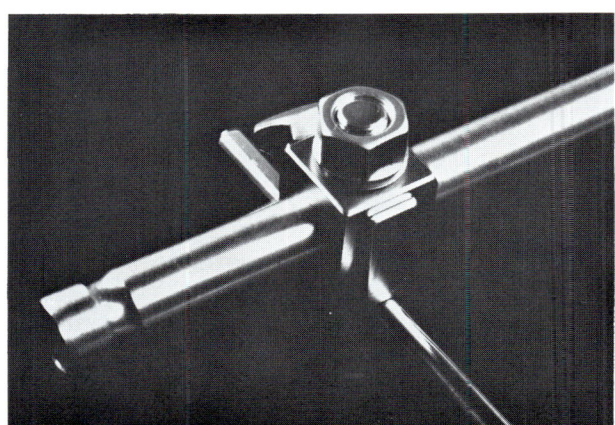

Figure 13B: Single position pin gripper (ASIF tubular system).

may or may not completely encircle the limb. (Incomplete circles are called *half-rings*.) The rings must be connected to each other by bars in order to create a fixator configuration.

ARTICULATIONS: A device that connects one bar to another (or a bar to a ring) is referred to as an articulation. Some articulations consist of universal joints or hinges, but most do not.

Frame Configuration

Throughout this book, the frame configuration terminology employed by Chao and coworkers at the Mayo Clinic will be used.

UNILATERAL: The unilateral frame is one that employs one bar connecting two or more pin-gripping clamps, which are attached to half-pins. It is the simplest configuration. This category includes Parkhill's original bone clamp, Lambotte's external fixator,

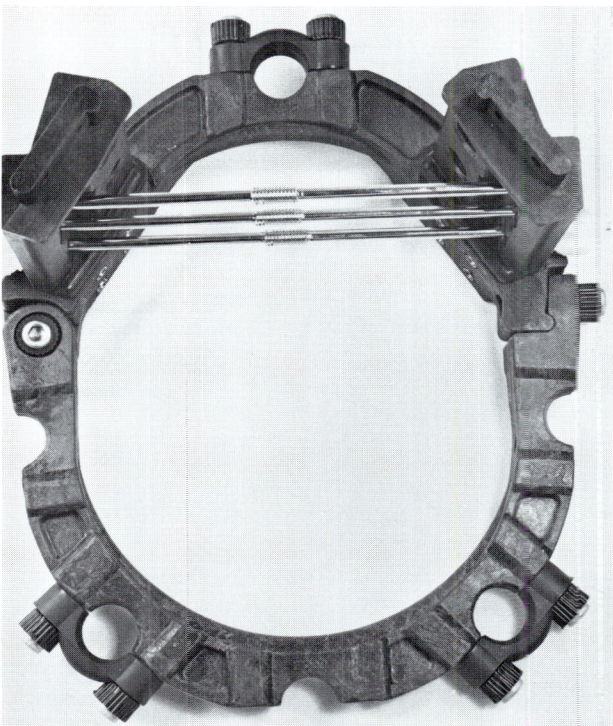

Figure 14: Full ring (Kronner system).

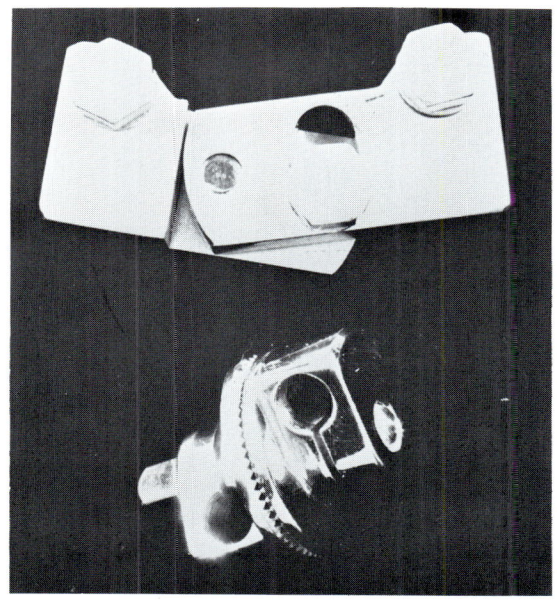

Figure 15: Articulation: Hinged ASIF (above); Hoffmann (below).

and the apparatuses devised by Stader, Hoffmann, and Wagner.

BILATERAL: A bilateral frame is one that employs a rigid bar on both sides of the limb, connected to full-pins that transfix the bone. Roger Anderson's external skeletal fixator was of bilateral design, as is the Day frame.

QUADRILATERAL: A quadrilateral frame is one that has four bars, two on each side of the limb, connected to pins that transfix the bone. The Vidal-Adrey frame is the typical example.

BIPLANAR: A biplanar frame is one that employs pins placed in two (or more) planes for increased stability. The ASIF tubular fixator is frequently used in this configuration, and Vidal now recommends biplanar stabilization for the management of septic nonunions. In this situation, the quadrilateral frame design is modified with the addition of a fifth bar and additional half-pins.

RING: A ring fixator is one that uses rounded bars that completely encircle the limb in a plane that is usually transverse to the axis of the limb. Pins transfix the limb and connect to the rings in various locations. The rings, as noted earlier, are connected to each other by additional bars. Russian investigators have been utilizing these fixators for many years.

HALF-RING: A half-ring fixator is one employing bars that incompletely encircle the limb in a manner similar to the ring fixator. Fischer of Minneapolis has devised such a fixator.

Fixation Systems

Prefabricated Fixators

External fixation systems in which the components are prefabricated by a manufacturer can be divided into two broad categories: those with fixed configurations and those with variable configurations.

FIXED CONFIGURATION: These external fixation frames are characterized by a relatively fixed, but usually adjustable, spatial configuration that dictates the position, direction, or number of transcutaneous pins. Examples of the fixed configuration design include the Russian ring-type fixators and the Kronner ring frame developed in the United States. This category also includes several simple frame configurations, including the pioneering Stader apparatus, as well as the system developed by Judet of France. The Wagner, Anderson, and Kawamura limb-lengthening frames, Charnley compression clamps, and the original ASIF threaded external fixation system also have fixed configuration.

VARIABLE CONFIGURATION: The variable configuration fixator systems are similar to each other in that they consist of separate components that can be assembled into any spatial configuration as dictated by the nature of the musculoskeletal problem. Precise pin position is generally required only with the individual

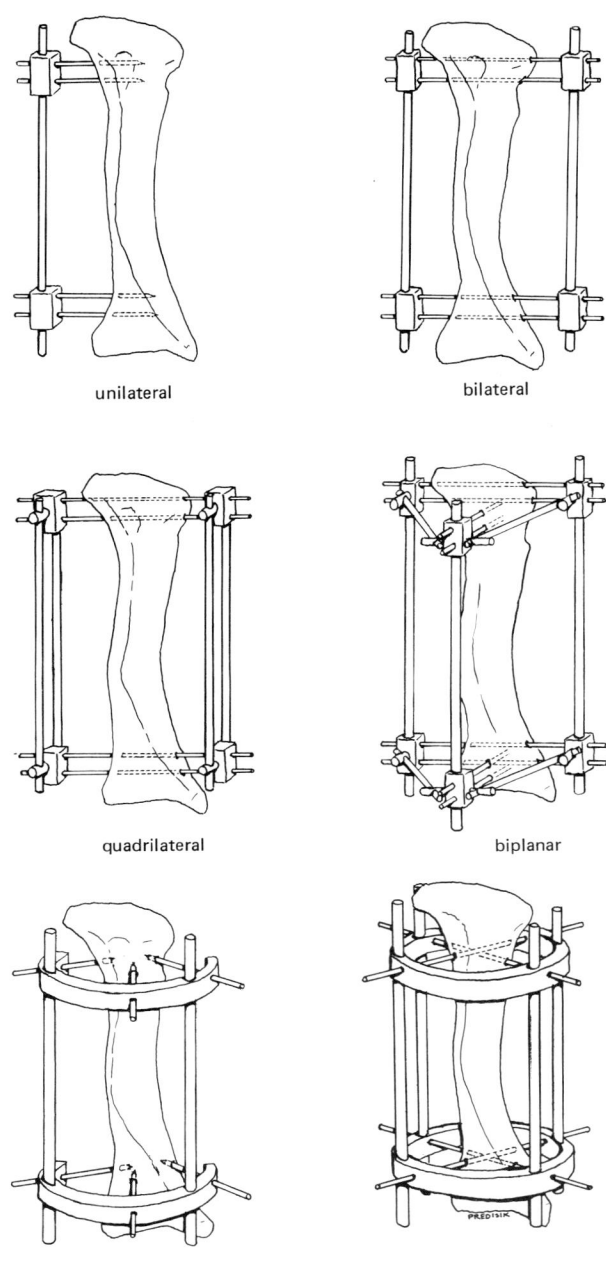

Figure 16: Fixator configuration. (From Chao, personal communication.)

pins within a cluster (those held by the same pin-gripping clamp). Currently available fixators of this type include the Anderson, Hoffmann, Day, Fischer, and ASIF tubular system. Undoubtedly, additional systems will become available in the future.

Improvised Systems

This category consists of systems of fracture management where transcutaneous pins are connected by an unsolidified substance that hardens within a few minutes after being applied. The classic pins-in-plaster technique, methyl methacrylate external pin fixation, and the Murray[182] epoxy-filled tube system belong in this group. These systems permit unlimited pin positions, but they lack adjustability.

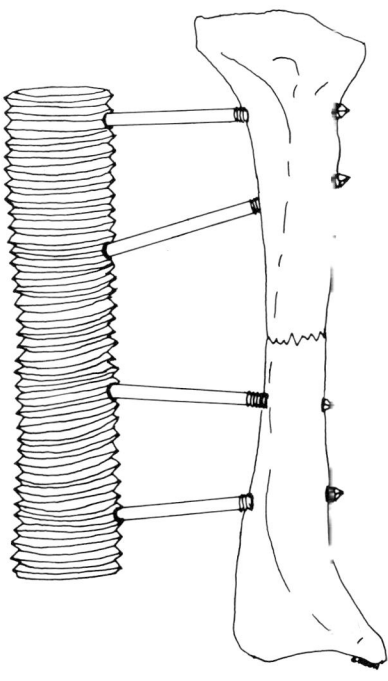

Figure 17: Murray epoxy-filled tube fixator.

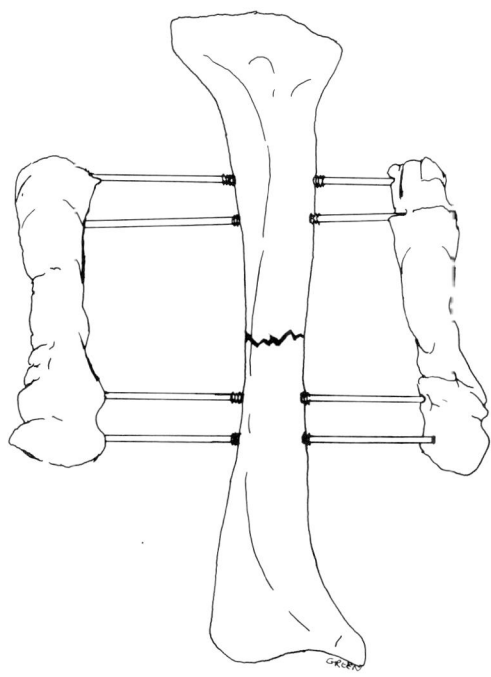

Figure 18: Methyl methacrylate fixator.

Chapter 2

PIN TRACT INFECTION

Introduction

Pin tract infection has always been the principal drawback to the use of external fixation. Unfortunately, preliminary reports announcing the development of new fixators during the 1930s rarely took note of this complication.[8,120,128] By the 1940s, however, the problem of pin sepsis had become apparent to many workers in the field. For example, Lewis, Breidenbach, and Stader,[173] in applying the Stader apparatus to an initial series of twenty human patients, noted pin "seepage" in ten patients and described abscess formation (requiring incision and drainage) in three patients. They distinguished between benign pin seepage (described as a serous discharge) and true infection (infection requiring additional treatment). The Stader group felt that the high proportion of pin problems was due to "galvanism" resulting from the use of steel pins and a duralumin fixator, noting the absence of pin seepage when they used plastic to insulate the pins from the fixator.

In 1939, Naden,[202] a Canadian, described his experience with the Roger Anderson apparatus, which he applied to 950 fractures and orthopaedic conditions. His series included the observation of 4,800 pins. He noted only 5 percent minor local skin irritation about the pins in the entire series. Naden also described five patients who required hospitalization for skin infections not involving bone, and four patients who had small sequestra that healed after the pins were removed.

In 1944, Siris[248] reported his experience with the Roger Anderson and Haynes devices, which he utilized to treat long bone fractures in eighty patients. He noted clear or purulent discharge from the pin sites in 36 percent of the cases; 22.5 percent of the patients developed a pin-site osteomyelitis; 10 percent of the patients developed systemic sepsis associated with pin tract infections. Siris concluded: ". . . infection of the pin sites is the most serious potential and real complication of using external pin transfixion of fractures. Infection far transcends the danger of distraction, delayed union or non-union."

The American experience with external fixation during World War II was generally unfavorable, due to the high incidence of pin hole sepsis that followed employment of the device by physicians who were not knowledgeable about the appropriate techniques. In 1950, as noted in Chapter 1, Johnson and Stovall[140] surveyed the American orthopaedic community to assess the extent of use and the role of external skeletal fixation in fracture management. They found that 63 percent of the orthopaedic surgeons who had tried the external fixation approach had discarded it because of the danger of soft tissue infection at the pin sites, the formation of ring sequestra, and the occurrence of pin hole osteomyelitis. By 1950, the majority of American orthopaedic surgeons were not using external fixation because of the problem of pin hole sepsis.

In Europe, on the other hand, continued use of external skeletal fixation following World War II resulted in refinements of both the device and application techniques, but never in a complete solution to the problem of pin tract infection. Fellander,[93] for example, working in Stockholm, reported his experience with thirty-nine applications of the Hoffmann fixator in 1963. He recorded that eight (16.3%) of the patients developed slight pin infection at one or more pin sites; another eight of the patients had evidence of more marked infection with manifestations of osteitis (osteomyelitis); one patient had long-lasting suppuration from a pin tract.

The Belgian surgeon Burny has reported a detailed statistical analysis of a large number of patients treated with external skeletal fixation.[37] Burny utilizes Hoffmann's original concept of external skeletal fixation: two groups of half-pins connected by a single bar. Burny calls pin sepsis "reactions of pin intolerance"— these include three levels: redness; loosening of the pins; and osteolysis. For reporting the location and extent of pin hole sepsis in his cases, Burny employs a standardized form that is in use in eight trauma centers (in Europe and Algeria). His observation of 5,125 pins revealed that 10 percent of them developed the described signs of pin hole sepsis within the first 150 days of fixator application. After 200 days, approximately 25 percent of the pins showed evidence of sepsis. By day 250, the rate of pin reactions had approached 30 percent.

It is noteworthy that, after the first 150 days, there is a distinct difference in the rate of pin tract infection between the proximal and distal pin groups. The proximal pins showed a 20 percent reaction rate at 250 days while the distal group showed a reaction rate of 50 percent.

Burny also analyzed those of his cases where external skeletal fixation had been applied to humeral fractures,[41] using the detailed assessment procedure described above for tibial pins. He found that the percentage of pin tract reactions rose steadily, to 34 percent by the seventieth day after application. Thereafter, the rate diminished somewhat—probably because the pins likely to become septic had already been removed. Altogether, Burny recorded thirteen patients with osteolysis around one or more pins, out of 100 cases of humeral fractures treated with external skeletal fixation.

Karlstrom and Olerud[153] (Sweden), reporting their experience with the Vidal quadrilateral frame, reviewed fifty-one cases of open tibial shaft fractures that were managed with external fixation. Although they described several minor pin reactions, they noted no significant pin tract infections requiring care after the fixator was removed.

External skeletal fixators have been used for limb-lengthening for many years. Employing the device for this purpose with 141 polio victims, Ahmadi and coworkers[4] compared the rate of infection with three different types of fixators. They noted that the pin tract infections were all mild and resolved shortly after removal of the transcutaneous pins. The Anderson lengthening apparatus was associated with a 12 percent pin infection rate, while the Rezaian[228] frame (designed in Iran) was associated with a 20 percent rate of infection. The Wagner leg-lengthening fixator, used with fifty-one of the patients, was associated with a 22 percent infection rate; nevertheless, the researchers felt that the Wagner device was superior to the other two, for a number of technical reasons. (Wagner,[284] reporting his own experience with his first eighty-eight limb-lengthening cases, noted a rate of 5.6 percent soft tissue infection—which resolved after removal of the fixator—and one instance of bone infection that required sequestrectomy.) Kawamura[155] (Japan) noted thirteen pin hole infections in the course of 291 lower extremity limb-lengthening cases. He employed a half-ring fixator that possesses considerable intrinsic stability, and none of his cases developed chronic infections.

Recent reports of external fixator application provide evidence that the problem of pin tract infection has not yet been solved. In 1979, Lawyer[171] reviewed thirty-one cases of compound tibial fractures treated with the Hoffmann device and noted several minor pin tract infections, which were controlled with antibiotics, and four significant pin tract infections that required surgical debridement. Recently, Klemm[161] (also employing the Hoffmann device) described his experience with fifty infected pseudarthroses of the tibia, which were treated with gentamicin-impregnated methyl methacrylate beads. He noted a 14 percent rate of pin tract infection.

Work reported by Krempen, Silver, and Sotelo[154,165] involved treatment of thirty-three open fractures and twenty-three previously infected pseudarthroses with the Vidal-Adrey frame, utilizing Hoffmann components. They recorded four cases of pin loosening, ten patients with "secretion," and two with "copious drainage" from pin holes. Weis and coworkers[285] have reported on nine complex tibial fractures treated with external fixators, noting one significant pin tract infection in their series.

Edwards' work at the University of Maryland has involved the employment of external fixation techniques developed in Europe (utilizing Hoffmann components) to a group of polytrauma victims.[87] His experience included use of the device with sixty tibias, twenty-one femurs, and six upper-extremity injuries. Edwards noted that 37 percent of his group developed

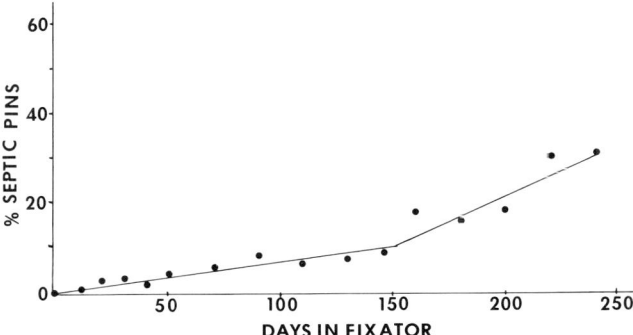

Graph 1. The percentage of septic pins increases steadily while the patient is in a fixator, especially after 150 days (from Burny).[40]

at least one pin tract infection; the rate of pin hole sepsis was 10 percent of the total number of pins. Speaking as cochairman at the 1979 International Hoffmann Fixation meeting, Edwards pointed out that "pin tract infection remains a common problem, even with such enlightened external skeletal fixation systems as the Hoffmann apparatus."[87]

Pins-in-plaster fixators appear to have a much lower incidence of pin tract infections than is true with the other types of devices. Anderson and Hutchins[7] reported their experience with this system in treating tibia-fibula fractures in 1966. They noted only three insignificant pin tract infections in a group of 107 patients. The lower infection rate may be due to the cast's contributing to stabilization of the pin hole.

One problem in determining the overall incidence of pin tract infections is that different authors use different sets of criteria to define pin tract infection. This variance is present even within a single institution, making a review of patients' charts at one hospital sometimes an inaccurate procedure for determining the incidence of pin tract sepsis. (I am inclined to agree with Kretzler's[166] observation that "it is difficult in a retrospective review to distinguish with certainty between scant clear drainage and that which is associated with some redness and swelling.")

Because of this problem, the concept of "major" and "minor" pin tract infection has been introduced. A major pin tract infection is defined as one that is treated with admission to the hospital (in order to control sepsis), remove one or more pins, or remove the entire fixator. A minor pin tract infection is defined as any apparently benign pin reaction. This distinction was introduced to simplify chart review and analysis: a major pin infection will be accompanied by a clinical note indicating the treatment—hospital admission, pin removal, frame removal—while lesser pin reactions are frequently not recorded on the chart. Utilizing these criteria, we recently analyzed our experience with fifty-one applications of external fixation as an adjunct to the treatment of chronic bone and joint infections at Rancho Los Amigos Hospital.[116] We noted instances of major pin tract infection in 30 percent of our patients; the incidence of minor infections was almost 100 percent. (In this group of patients, the average time in the fixator was six months.)

PATHOPHYSIOLOGY OF PIN HOLE SEPSIS

A metallic pin—or any hard foreign substance for that matter—when inserted into the body's tissues will provoke the development of a membrane separating the foreign material from the adjacent tissues. If relative motion is present between the foreign material and the local tissues, a bursal membrane will usually form; the bursal fluid secreted by the membrane acts as a lubricant between the foreign body and the adjacent tissues. Orthopaedic surgeons are familiar with the bursa that develops over stainless steel wires utilized to replace the greater trochanter following total hip replacement arthroplasty. In the case of the wires around the greater trochanter, the bursal fluid is sterile because the bursal sac is isolated from the exterior environment of the patient. With a transcutaneous pin, on the other hand, the bursal fluid becomes contaminated with microorganisms through the external pin hole. Nevertheless, the contamination presents no special problem as long as the pin hole drains freely to the outside, thereby maintaining a relatively low bacterial concentration within the bursa.[95]

Pin holes become infected when the delicate balance between the patient's natural defenses and the bacteria's infective capability is altered. This alteration can result from (1) the development of an *abscess* (closed space) around the pin; (2) the presence of *necrotic tissue* in the pin hole, which can become the focus of sepsis; (3) the presence of *excessive motion* between the pin and the adjacent tissues, which diminishes host resistance.

Abscess Formation

As noted above, the fluid formed around the pin by the local tissues drains to the external surface and is contaminated with microorganisms in the process. The amount of fluid may be extremely limited, especially when there is very little relative motion between the soft tissues and the pin. This occurs most frequently in areas where the pin is inserted into a bone with little overlying soft tissue, such as over the anterior tibia. When the fluid reaches the surface, it dries, forming a crust around the pin-skin interface. If this crust restricts free drainage of the contaminated bursal fluid by sealing the pin hole, deep abscess formation will result. This process, postulated by Fischer,[95] leads to the observation that frequent pin care directed toward removal of the crust from the pin-skin interface is important in reducing the incidence of pin tract sepsis.

Necrosis

Skin Necrosis

Necrosis of the skin will occur if the tension (or compression) produced by the presence of the pin is sufficient to interfere with the circulation of the local subdermal capillary plexus. Plastic surgeons are mindful of this basic principle when transposing skin flaps; trauma surgeons utilizing transcutaneous pins for external skeletal fixation must also keep it in mind. Skin tension can occur immediately following pin insertion, or at any time when a change in alignment or length is made. Skin can also be pinched between two or more pins if the pins are close together within a cluster. Thermal damage to skin and soft tissues related to high-speed bone drilling has also been reported.[175]

Soft Tissue Necrosis

Necrosis of the subcutaneous and deep soft tissues will develop if undue tension (or compression) is created by a transcutaneous pin. This can occur if the soft tissues are displaced by the pin after it has been inserted. Such tension is likely to occur in the anterior compartment of the lower leg, if a pin is used to push the anterior compartment musculature posteriorly while probing for the tibia. (It is far wiser to transfix the muscle by pushing the pin straight in, thereby avoiding undue tension.) Necrosis can also be produced if soft tissue is permitted to "wind up" around the pin while it is being inserted. (This can best be prevented by the use of a drill sleeve.)

Bone Necrosis

Necrosis of bone can occur with the heat generated from drilling. Matthews and Hirsch[187] have reported results of experiments they performed, studying the heat generated from drilling bone with a power drill. The experiments consisted of measuring the temperature in bone adjacent to a drill hole by using tiny thermocouples. They employed 3.2 mm drills, and measured temperatures 0.5 mm to 2.0 mm from the drill hole. They studied the effects of irrigation, dull drills, drill speed, and drill pressure. They observed temperatures as high as 140° C at 0.5 mm from the drill hole. It is noteworthy that damage to osteocytes can occur if they are exposed to temperatures of 55° C for one minute or more.[232] Alterations in the mechanical properties of cortical bone can occur if the bone is exposed to temperatures of 50° C or more.[17] Interestingly, they did not compare hand drilling to power drilling, although they did note very little difference in the temperatures obtained when using different drill speeds. However, they noted that predrilling the bone hole with a 2.2 mm drill-bit, followed by enlarging the drill hole with a 3.2 mm drill-bit, reduced the highest temperature obtained from 105° C to 45° C.

A hand drill (or hand brace) will help prevent thermal damage to bone, periosteum, and soft tissues.

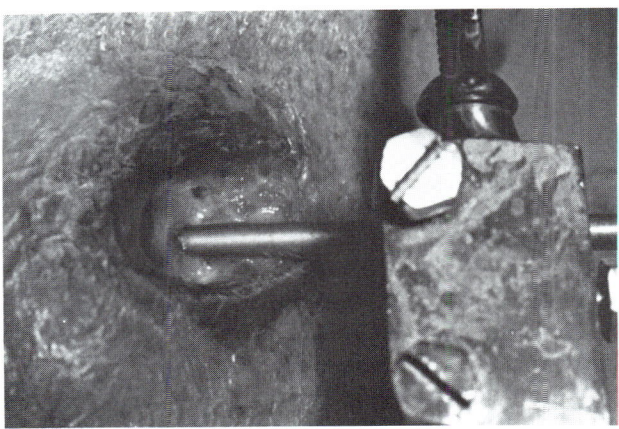

Figure 19: Full thickness thermal burn produced by an overheated pin. The burn was evident to the surgeons immediately after pin insertion. If a pin emerges from a limb too hot to be comfortably held in the surgeon's fingertips, it should be removed, cooled off, and reinserted elsewhere into a predrilled hole. From M. A. Linson and R. A. Scott, Thermal Burns Associated with High Speed Cortical Drilling. *Orthopedics*, 1:394, 1978. Reproduced by permission.

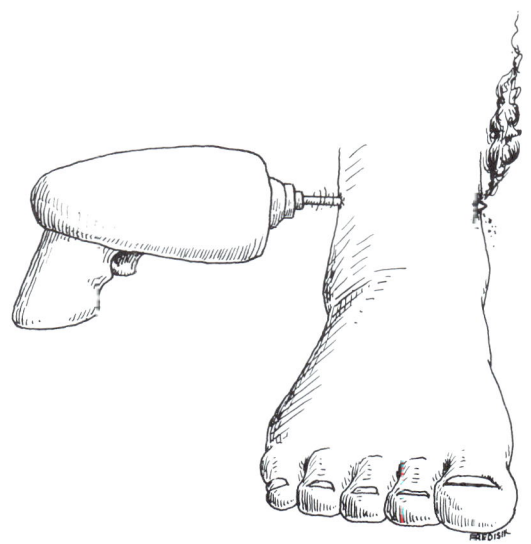

Figure 20: High speed power drills should not be used for pin insertion

One might question the need for a hand brace or hand drill when, after all, power drills are used in many surgical situations without problems. The critical difference between these surgical procedures and the application of an external fixator is, of course, the presence of transcutaneous pins. The pin holes provide a continuous portal of entry for bacteria into the bone. Heat-damaged bone is more likely to become a focus of chronic infection than is normal bone, because the thermal necrosis can result in the formation of a sequestrum. In this case, the sequestrum is dead bone in continuity with the microorganisms of the pin hole microflora. As such, it may become the focus of a chronic pin hole infection.

Excessive pressure—due to compression of the frame—is thought to produce necrosis of bone at the pin-bone interface.[49] Presumably, the pressure reduces local bone circulation, resulting in death of osteocytes. Bone pressure in excess of 100 kg can be developed in such cases with the compression of a Vidal quadrilateral frame, utilizing the compression bar provided by the manufacturer. Finger tightening alone can produce pressures in excess of 45 kg.[49] The bone necrosis that occurs at such high pressures may become the focus of chronic infection.

Motion

Relative motion between the pin and the adjacent tissues contributes to pin hole sepsis. As far as the microenvironment of the pin hole is concerned, it makes little difference whether the pin is moving with respect to the tissue or the tissue is sliding back and forth along the pin. The effect is the same: relative motion between soft tissue and a contaminated foreign body (the pin).

In order to fully appreciate the role motion plays in the development of infection, consider the inflammatory process. The phenomenon by which phagocytic cells (polymorphonuclear leukocytes and macrophages) migrate toward microorganisms is called chemotaxis.[289] The phagocytic cells ingest the microorganisms; once ingested, the microorganisms are destroyed by the action of substances, such as hydrogen peroxide and iodine, produced by the phagocytic cells.[160]

Chemotaxis is a fascinating process. The phagocytic cells migrate in the direction of an increasing gradient of a chemotactic substance, the focus of which is the bacteria itself. The chemotactic substances may be products of the bacteria or, more commonly, protein fragments from the complement system that have become attached to the surface of the bacteria.[289] The complement system consists of a series of proteins that bind sequentially to specific receptor sites on the antibody molecules, which have attached themselves to surface antigens of the microorganism. As the complement molecules bind to the appropriate sites, they split into fragments. These fragments diffuse away from the bacteria with a concentration gradient that is greater immediately adjacent to the bacteria, and decreasing with increasing distance from the bacteria.

Phagocytic white blood cells are very sensitive to the concentration of gradient of chemotactic substances along their surfaces.[290] By a mechanism not fully understood at present, the phagocytic cells stream in an amoebic manner towards the increasing concentration of chemotactic substance.[290] The process involves activation of actinlike and myocinlike proteins in the microfilament system of the white blood cells.[172] Ramsey[224] has observed that neutrophils migrate

Figure 21: Multiple pin tract infections produced, in this patient, by poor pin placement (see Fig. 151) and unprotected weight bearing.

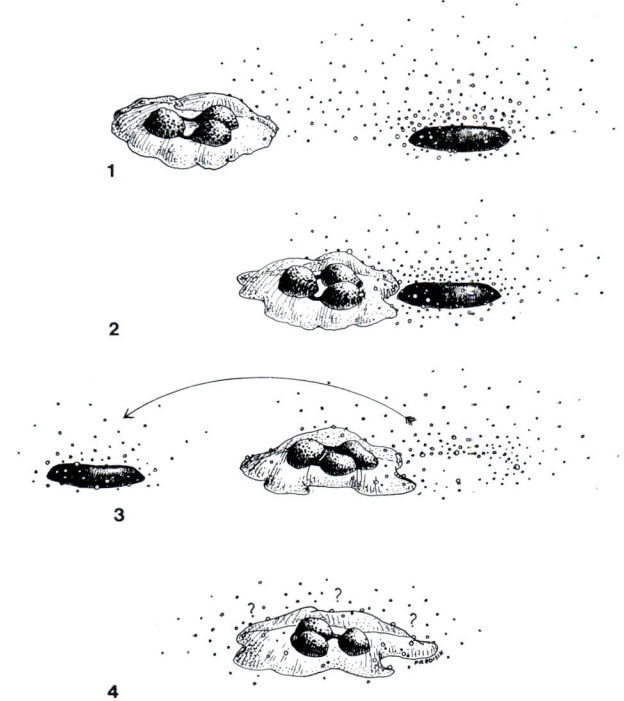

Figure 22: Motion and chemotaxis: (1) and (2) phagocytic white blood cell follows chemotactic gradient to bacterium; (3) white blood cell continues toward focus of gradient even if bacteria is moved away, until gradient becomes reestablished in the opposite direction; (4) constant movement of milieu mixes gradient and reduces chemotactic efficiency.

towards the chemotactic focus at the rate of 10 μm/minute. The direction of movement can be changed by moving the source of the chemotactic substance. If a bacterium is removed from the focus of the chemotactic gradient, the phagocytic cell takes sixty seconds or more to reorient itself to the new gradient.

The above discussion, based on *in vitro* studies, indicates that the phagocytic neutrophils (and macrophages) migrate towards the bacteria by following the chemotactic gradient. Furthermore, movement of the source of the gradient (bacteria) forces the phagocytic cells to change direction. One conclusion that can be drawn from this is that a constantly moving environment is disadvantageous to phagocytic cells in competition with bacteria. A moving environment mixes the chemotactic gradient, making the task of precise location of bacteria difficult.

There is also clinical evidence to support this conclusion. The gratifying response of soft tissue and musculoskeletal infections to immobilization is well known. Our experience at Rancho Los Amigos Hospital fully supports the conclusion that reduction of motion at the pin-tissue interface reduces the incidence of pin tract infections. The low incidence of pin tract infections in reported series of fractures managed with pins-in-plaster [7, 241] is due, no doubt, to skin immobilization by the plaster cast.

PIN-SOFT TISSUE INTERFACE: Motion between the pin and soft tissues can be reduced (but not eliminated) by selecting areas for pin insertion that have a minimal amount of subcutaneous tissue. Further reduction in soft tissue motion relative to the pin can be accomplished by wrapping the pins with a bulky wad of gauze dressing between the skin and the fixator.

PIN-BONE INTERFACE: Pin loosening within the bone is associated with pin hole sepsis. In fact, the association of pin sepsis and pin loosening is so consistent that loosening is considered the most significant feature contributing to the development of pin hole sepsis.[36, 95, 191, 265] The pathophysiology of loosening hardware within the bone has been studied by Perren,[218] Schatzker et al.,[238, 239] and others. They have noted that bone resorption and subsequent implant loosening results from cyclic (rather than constant) pressure at the bone-metal interface. Forces that frequently change direction and magnitude result in local bone surface resorption in areas that have contact with metal. Cortical bone is gradually replaced by fibrous granulation tissue, which is more tolerant of the fluctuating forces that produce strain (deformation) in the tissues around the implant.

Loose hardware tends to become looser with the passage of time as osteolysis around the metal increases.[238] Vidal has observed that "osteolysis around the pins is always the consequence of movement—it should not be found in an external fixation device mounted correctly."[268] Once the pin becomes loose, pin-tissue interface motion will promote sepsis in a manner consistent with mechanisms already described.

Motion at the pin-bone interface can be eliminated in several ways. A preventive measure is to *employ only threaded pins*. If properly inserted, they will not slip back and forth in the bone as will smooth pins. Pins should not be "backed out" once they are inserted, as they tend to loosen more quickly thereafter. It is better to insert them slightly too shallow before checking the depth with fluoroscopy or roentgenograms than to insert them too deep and have to back them out.

Another measure to reduce cyclic pin motion is to increase the stability (stiffness) of the fixator configuration. Chao and coworkers[49] have determined that the stiffness of a fixator can be increased by (1) increasing the number of pins; (2) increasing the distance between the pins within each pin cluster; (3) applying pins closer to the fracture site; and (4) incorporating pins that are mechanically stiff into the fixator. It should be noted that these considerations are theoretical, based on mathematical models and laboratory analysis. I know of no studies comparing fixators of different stability in a clinical setting where other factors are kept constant.

The problem of fixator stability becomes critical if one or more loose septic pins must be removed. Loosening of the remaining pins may occur because the overall stiffness of the frame configuration decreases. The problems can be avoided in the first place if a sufficient number of pins are inserted to allow removal of one or more pins without affecting the integrity of the fixation. As Naden[202] puts it, "'tis better to add a pin than to have one too few."

PREVENTION OF PIN TRACT INFECTION

Preoperative Planning

The objective of preoperative planning is to reduce or eliminate those variables that increase the likelihood of pin tract infections. Considerations include proper patient selection; selection of the appropriate fixator or frame configuration; selection of appropriate pins; and selection of the proper tools for application of the fixator. Of equal importance are technical details of fixator application developed to decrease the likelihood of pin tract infection.

Patient Selection

It is hoped that the surgeon will select a patient for external fixation only after careful consideration of the alternative modalities available. I agree with the position of Kretzler[166] who states that "pins should not be used when simpler methods will suffice." Certain orthopaedic problems are considered well suited for external skeletal fixation. These include the following:
- Fractures associated with extensive skin and soft tissue loss.
- Fractures associated with marked comminution or absolute bone loss.
- Complex pelvic ring fractures and symphysis pubis separation.
- Nonunions of long bone (infected and otherwise).
- Arthrodesis of joints, especially with chronic infection.
- Reduction of bedrest in polytrauma victims.
- Stabilization of limbs for cross-leg skin flaps.
- Immobilization of joints requiring extensive skin coverage.

Fixator Selection

The selection of the appropriate fixator configuration for a specific function is important. In general, the fixator configuration should be sufficiently rigid to prevent pin loosening before the frame is removed. This feature alone will do much to prevent pin sepsis.

When dealing with a chronic bone infection or an extensively contaminated wound, the fixation frame should be capable of extraordinary rigidity. If the orthopaedic problem is less complex and the application short-term, a less rigid configuration will do. If a cross-leg flap or other secondary surgery is to be performed while the fixator is in place, the frame configuration must be selected with the proposed procedure in mind. If compression is required, the frame should be compressible. If comminution of fracture fragments is present, the frame should have the capability to control intermediate fragments. Because it is better to insert pins through intact skin, the fixator should permit pin placement to be dictated by the nature of the injury rather than by the configuration of the frame.

Pin Selection

SMOOTH PINS: Smooth pins should not be used for external skeletal fixation. They create two holes in the bone but do not offer the advantage of "screwed-in" bone fixation. The unfortunate experience with the Stader apparatus in the 1940s was due, I believe, to insufficient fixation resulting from the utilization of smooth pins.

THREADED PINS: Several types of threaded pins are available for use with external skeletal fixation; most such pins are usually fully threaded. The surgeon utilizing them should be aware of certain problems inherent to their use. When drilling a threaded pin into dense cortical bone, the first cortex is drilled and tapped by the threads of the pin itself. The point of the pin then easily traverses the medullary canal, encountering resistance only when it reaches the inner side of the far cortex. When the pin tip meets this resistance, it will not easily advance in the bone, resulting in damage to the threads already tapped in the near cortex, thereby reducing fixation.

Bonnel-type pins, which are smooth at both ends and threaded in the center, will drill the bone holes without damaging them. The pin is designed to allow both cortices to be completely penetrated by the smooth part of the pin before the threaded portion taps the first, and

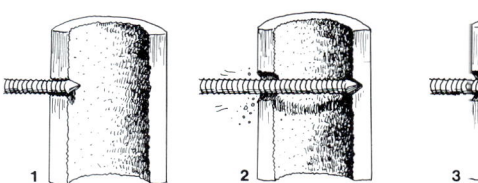

Figure 23: Drilling threaded pins into cortical bone: (1) threaded Steinmann pin drills and taps hole in near cortex; (2) resistance from the inner wall of far cortex impedes forward progress of the pin, stripping the hole in the first cortex; (3) final fixation—secure in far cortex, but loose in near cortex.

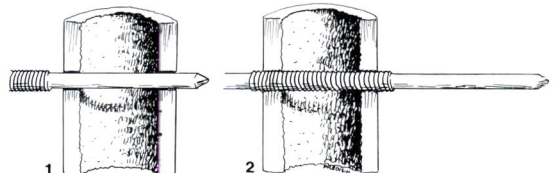

Figure 24: Bonnel-type pins predrill both cortices before the near cortex is tapped.

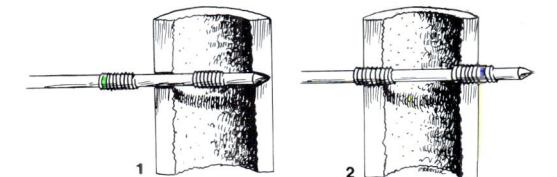

Figure 25: Interrupted threads on half-pin prevent damage to tapped hole in the near cortex while penetrating far cortex.

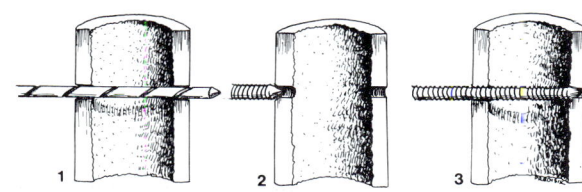

Figure 26: Predrilling cortical bone with a drill bit 1.0 mm smaller than pin will result in secure fixation.

then the second, cortex. In this manner, the pin predrills both bone holes before tapping and engaging the holes with the threaded section.

End-threaded half-pins are similar to fully threaded Steinmann pins in that there is danger of damage to the tapped hole in the near cortex when the pin meets the resistance of the inner wall of the far cortex. Special half-pins, with two threaded sections separated by a smooth section, will protect the tapped hole in the near cortex while the far cortex is being drilled. If a proper size pin is selected, the smooth portion of the pin will turn freely in the first cortex, while the point of the pin drills the second cortex. As the pin begins to advance in the far cortex, the second group of threads engages the near cortex. In practice, however, double-threaded half-pins are difficult to use because of the considerable difference in diameter of the various bones of the body. A large assortment of such pins must be available in order to use them properly.

Pin Insertion Considerations

Predrilling

Inserting a stainless steel pin into the tibia of a young healthy adult male can be an exercise in frustration for the surgeon. After drilling for a while and making no headway, there is a tendency to push harder and turn the drill faster. Heat (from drilling) increases the microhardness of bone[91], making progress difficult. To make matters worse, the cuttings (chaff) from the drilling of bone have no place to go because the pin contains no fluting (groove). The chaff also increases friction, making the drilling even more difficult.[187] Friction created by these factors increases the temperature of the pin point until it is too hot to touch when it emerges from the opposite side of the limb. It is easier (and safer) if the surgeon predrills cortical bone before pin insertion.

A sharp "twist" drill bit will penetrate bone more easily than will a pointed pin, because the fluting of the drill bit permits the chaff to be carried away from the worksite, reducing friction and also the amount of effort required of the surgeon. One might think it reasonable to select a drill bit size that closely matches the diameter of the pin.[187] A smaller drill bit, however, cuts more easily into bone. (A similar phenomenon can be observed with drilling sheet metal. It is much easier to start a hole with a very small drill bit—and then to enlarge it with a second drill bit—than to cut the hole with a larger drill bit alone.) For this reason, I prefer to use a drill bit that is at least 1.0 mm smaller than the diameter of the pin. The drill bits should be *extra long*, because they must pass through soft tissue before encountering bone.

The technical problem with predrilling is to find the hole with the pin after the drill bit is removed. There are three solutions available: (1) utilizing pins that are fabricated with a fluted twist at one end; (2) predrilling through a pin alignment guide; and (3) predrilling through a hollow drill sleeve. Crowe pins, which incorporate a twist drill at the cutting end, certainly seem to be the simplest and best solution to the problem. The surgeon can, with such a pin, drill into bone and insert the pin in one step. However, the drill bit portion of these pins is the same diameter as the basic (core) diameter of the pin itself. The advantages of utilizing a *smaller* drill bit for predrilling have already been noted.

The second solution to the problem of finding the predrilled hole for the pin is to utilize a pin alignment guide if one is available. The technique is quite simple. After making the initial skin incision, predrill the bone hole with an extra long drill bit, which has been passed through one of the holes in the alignment guide. (It is sometimes necessary to slide the drill bit along the surface of the bone in order to determine its width and center.) Have an assistant press the pin alignment guide against the skin, and drill through both cortices. As soon as the second cortex is penetrated, carefully withdraw the drill bit, but maintain the guide against the skin. Insert the pin in the same hole in the alignment guide. Search out the bone hole with the tip of the pin, feeling around the area of the bone until the pin engages the hole. Wiggle the pin slightly to make sure it is seated properly and proceed to drill the pin into the bone. Repeat this procedure until all the pins are inserted.

Another reliable way to find the bone hole after predrilling is to use a *drilling sleeve*. The sleeve (a hollow metal tube) is pressed against the bone while the drill hole is being made by the drill bit. The transcutaneous pin of the external fixator is inserted through the same drill sleeve after the drill bit is removed, thereby solving the problem of locating the bone hole. This concept is hardly new. In the 1940s, the Reduction-Retention apparatus marketed by the Zimmer Corporation used the drill sleeve with its external fixation system.[286] Judet's[149] external fixator also employs a drill sleeve, as does the ASIF external fixator. A drill sleeve is also available for the Hoffmann fixator and for the Fischer external fixation system. Unfortunately, the pins of the ASIF tubular system do not fit into the drill sleeve of that system. (The sleeve is intended to prevent the soft tissues from wrapping around the drill bit and for use as a guide to bone-hole placement.) Both the sleeve of the Fischer system and the drill sleeve available for the Hoffmann system are designed to permit the pin to fit through the sleeve.

Pretapping

Most pins available for external skeletal fixation systems tap the bone hole as the threads of the pin are inserted. Most modern *internal* fixation systems, on the other hand, employ separate taps to prepare the bone hole for screw insertion. The tap not only carries away cuttings from the bone hole but also permits better fixation of the end of the screw, which does not need self-tapping flutes. There is a difference of opinion among investigators as to whether elimination of the flutes at the end of a screw insures a better fit within the bone hole.[198, 239]

Among external skeletal fixation systems, only the ASIF system and the Fischer system possess separate taps for the bone holes. If pretapping the bone hole proves to be as important for external fixation as it has been for internal fixation, other fixator manufacturers will soon include a bone tap to be used before pin insertion. A tap seems especially desirable when an end-threaded half-pin is inserted. A pretapped hole will permit the use of a half-pin with a blunt tip that is threaded to the end, thereby reducing the amount of pin protruding from the far cortex.

Drill Selection

HAND DRILL: Use a hand drill for pin insertion. The hand drill reduces, but does not completely eliminate, the danger of thermal necrosis of bone. A hand brace is even better than a hand drill, because there is less tendency to wobble when inserting the pin.

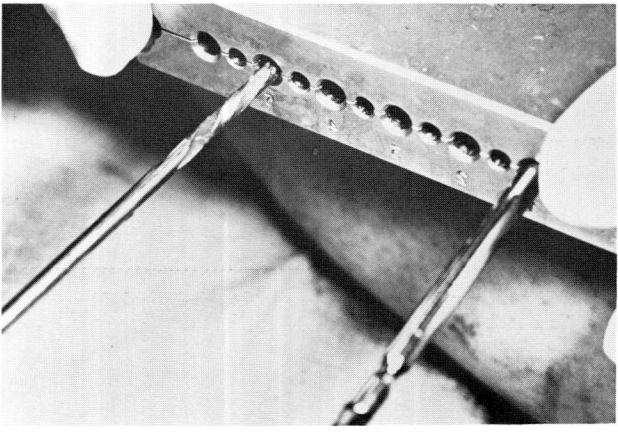

Figure 27: Finding the predrill hole (pin guide): assistant pushes guide firmly against the skin while the drill is removed and a pin inserted into the same hole.

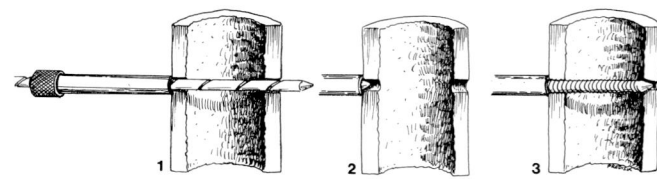

Figure 28: Finding the predrill hole (drill sleeve): sleeve is pressed firmly against the bone while drill is removed.

Pin Tract Infection

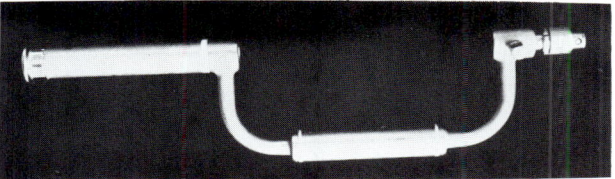

Figure 29: Hand brace reduces wobble during pin insertion.

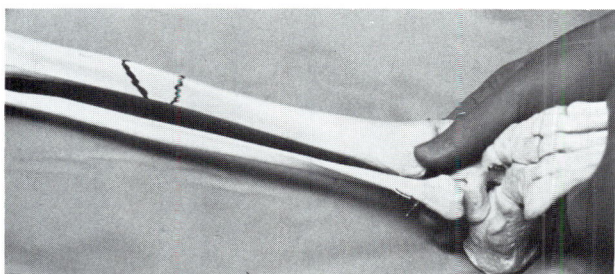

Figure 30: Align the fracture before pin insertion with longitudinal traction.

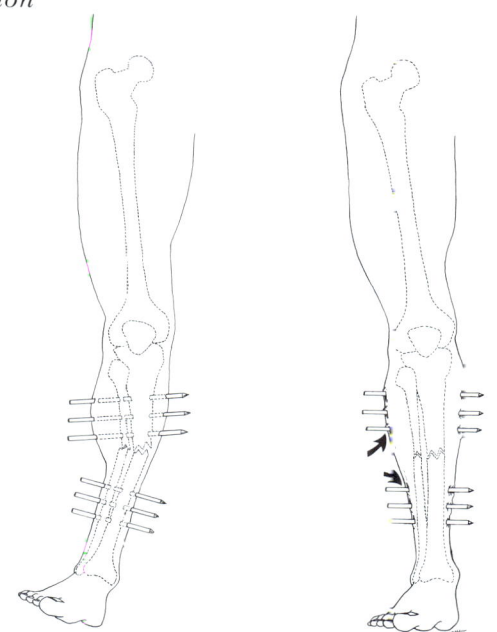

Figure 31: Alignment of fracture *after* pin insertion produces tissue tension on the concave side of the fracture deformity and pinching on the convex side.

POWER DRILL: Power drills should not be used to insert transcutaneous pins. Low r.p.m. drills are available and some orthopaedic surgeons have been using them to insert pins for skeletal traction and external fixation without experiencing undue problems. It is advisable, however, to use such power drills only for predrilling the bone hole with a small drill bit, after which the pin is inserted with a hand drill or hand brace.

Sterile Technique

Needless to say, when applying an external fixator, it is important to observe the usual principles of sterility that pertain to any surgical procedure. A fixator should be applied in the operating room, with the patient anesthetized. When the fixator is applied to a contaminated or infected fracture, debride the involved tissues thoroughly prior to application of the fixator. A separate sterile set-up should be used when the transcutaneous pins are inserted. If the wound area was infected, the limb can be reprepped and redraped after covering the operative site with a sterile gauze wrap.

Fracture Alignment

The fracture should be aligned as precisely as possible prior to pin insertion. This can be accomplished with lower extremity injuries by having an assistant apply traction to the limb by grasping the ankle. If the injury is open, manipulative reduction can be accomplished by the surgeon through the wound. However, if the injury is closed, it may be necessary to make a small skin incision to permit accurate reduction of the fracture fragments.

If the fracture is not aligned before pin insertion, undue skin tension will be created (a source of possible necrosis) by the skin on the concave side of the fracture deformity when the fracture is reduced. The pins may also pinch the skin on the convex side of the fracture deformity, creating additional skin necrosis. Alignment of the fracture prior to the pin insertion insures that the pin groups will be parallel to each other, giving the frame a neat, squared-off appearance.

Some fixator frames require precise alignment of rotation at the fracture site before pin insertion because they do not permit correction of axial malalignment once the pins are in place.

TECHNIQUE OF PIN INSERTION

Proper pin technique serves to (1) prevent undue necrosis of skin, soft tissues, and bone; (2) eliminate unnecessary motion between the pin and the tissues of the pin hole; and (3) avoid the prospect of neurovascular injury. As already noted, a sufficient number of pins should be inserted so that one or more may be removed should they become loose or septic.

Pin Cluster Selection

Divergent Pins

Several fixator frames permit the insertion of pins that are not parallel to each other. Freeman[100] believes that nonparallel pins are better than parallel ones because they prevent the bone from moving from side-to-side along the pins. The Roger Anderson system, for example, allows each pin to be individually clamped to the sidebars by a separate pin-gripper. This configuration not only makes possible considerable variability in pin placement but also permits divergent pins to be placed close to each other without bending. The pins must be placed in the same frontal plane if the frame is to be used in its simplest form. If the pins are not in the same frontal plane, they must be bent to connect with the frame. An alternative strategy is to assemble outriggers, consisting of separate bars that connect the pin-grippers to the principal side-bars. The Day frame employed by Freeman utilizes a rather thin pin-gripper, which allows nonparallel pins to be bent when they are fitted into the gripper.

I find divergent pins somewhat difficult to insert because the pins, once started obliquely, tend to "walk" or slide on the cortical surface of the bone. Although this tendency can be overcome by starting the pin perpendicular to the bone and shifting obliquely after the initial hole is made, I am not quite certain where the pin will emerge on the opposite side of the limb. Furthermore, any change in the position of the pin once it is within the soft tissues may produce undue soft tissue tension, possibly resulting in necrosis and subsequent pin hole sepsis.

Parallel Pins

Several fixator frames have multiposition pin-grippers, which are grooved to hold the pins. The Hoffmann system, for example, permits up to five pins to be held by one pin-gripper. Supplementary pin-grippers can increase to nine the number of pins within one cluster. When such a system is used, it is necessary to employ a pin alignment guide for pin insertion. The guide is flat and broad and permits the pins to be inserted parallel to each other. Many times, in spite of diligent efforts by the surgeon, one or more of the pins in a group will not be parallel to the others. An effort to correct the problem by redirecting a nonparallel pin through the bone usually fails. It is almost impossible to redirect the pin once it has started a hole through the bone in the wrong direction. A far wiser solution is to remove the nonparallel pin and insert another pin in a different direction, rather than forcing the unsatisfactory pin into the pin-gripper. A slight amount of nonparallelism will occur at times and is acceptable. I generally remove any pin with more than five degrees divergence from the other pins within a group.

Skin Incision

All skin flaps should be replaced in their original positions prior to pin insertion. This avoids unnecessary tension when the skin edges are finally approximated. Make each pin incision at least 1.0 cm. The incision can be either longitudinal or transverse. Longitudinal incisions are safer because they are less likely to transect neurovascular structures. If multiple pins are to be placed close together, transverse skin incisions—perpendicular to the row of pins—are useful because wider skin islands are left between the pins. This strategy is especially useful in the upper thigh where skin tension is likely to develop when the patient is supine in bed.

If the initial pin hole is not large enough, the skin will have a tendency to twist around the pin. If this happens, enlarge the skin hole with a sharp pointed scalpel. As the pin emerges from the opposite side of the limb, make an exit hole with the scalpel as soon as the pin begins to tent the skin. After all pins in one group are in place, release any remaining skin tension. Evidence of tension is a ridge of skin on one side of the pin. Place a sharp pointed scalpel parallel to the pin and push the blade into the ridge of skin to release the

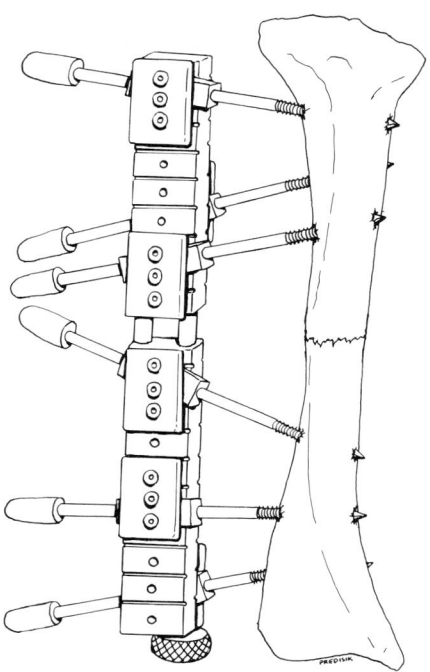

Figure 32: Sukhthian-Hughes unilateral frame: a divergent pin fixator.

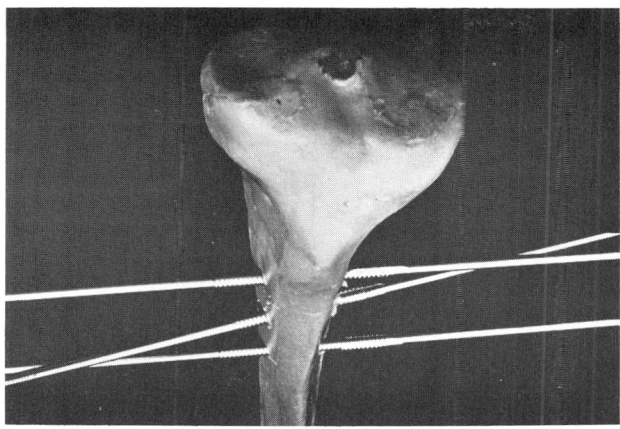

Figure 33: Parallel pin insertion: nonparallel pins occur in spite of diligent effort.

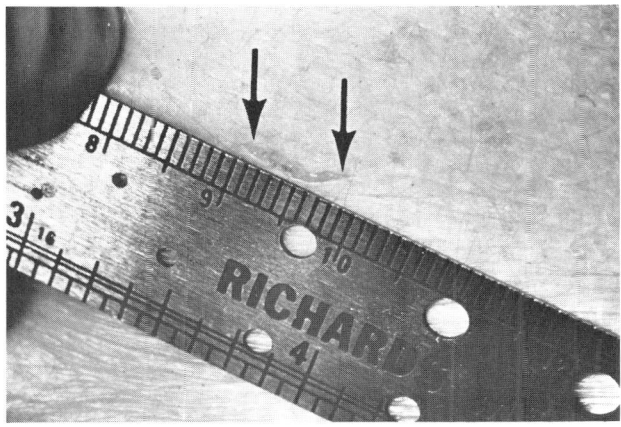

Figure 36: Initial skin incision: 1 cm long.

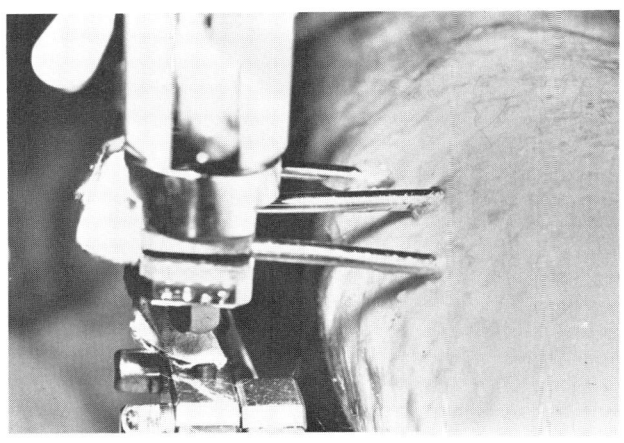

Figure 34: Incorrect technique: forcing a nonparallel pin into a pin gripper can produce bone necrosis and sepsis.

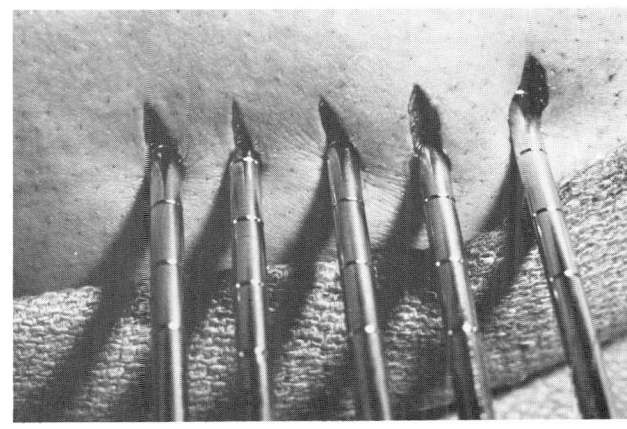

Figure 37: Transverse incisions in lateral thigh reduce skin tension when patient is supine.

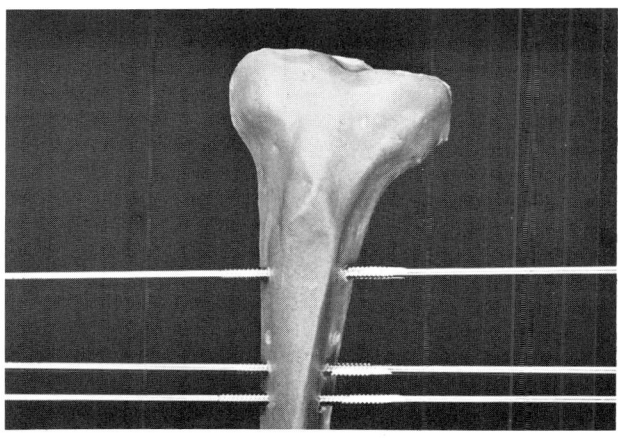

Figure 35: Nonparallel pin removed and inserted elsewhere.

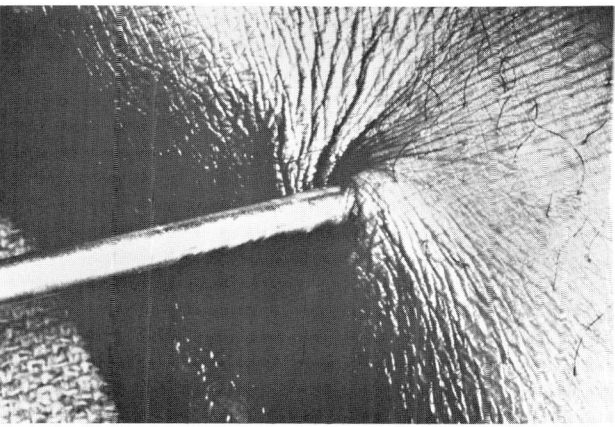

Figure 38: Incorrect technique: the "twirl" requires enlargement of the skin incision.

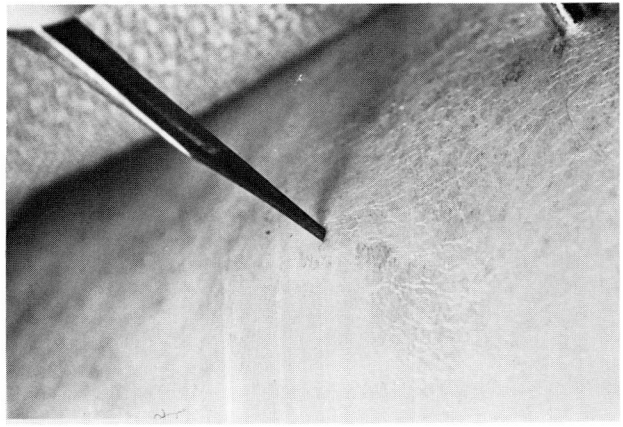

Figure 39: Cut skin before pin emerges.

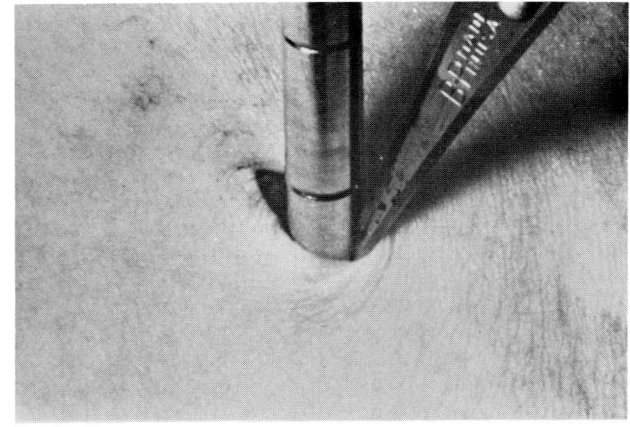

Figure 42: No. 11 scalpel blade pushed into area of tension.

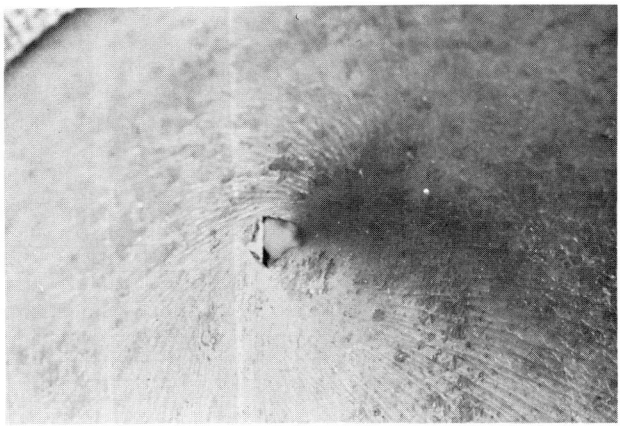

Figure 40: Incorrect technique: pin point penetrating skin.

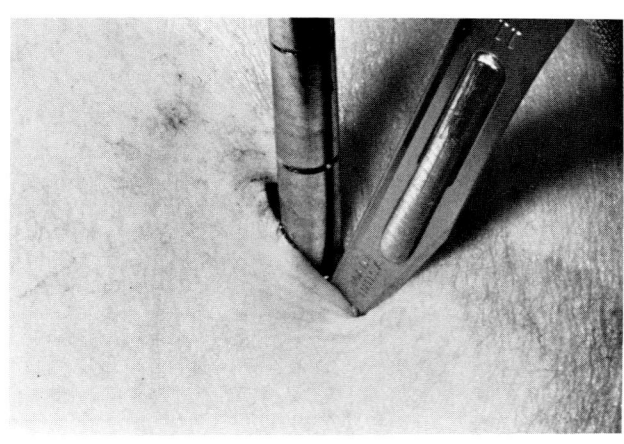

Figure 43: Release of skin tension.

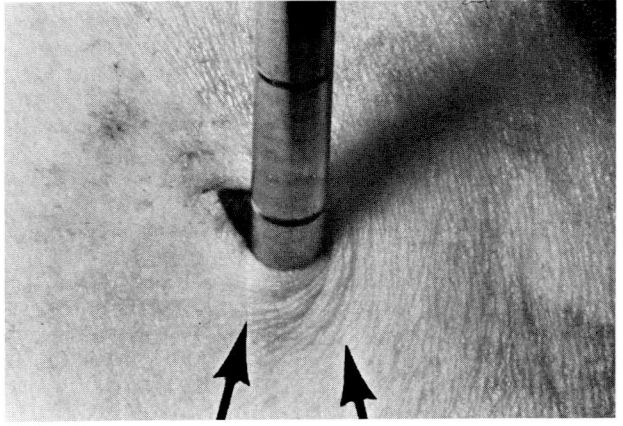

Figure 41: Tension on skin after pin insertion.

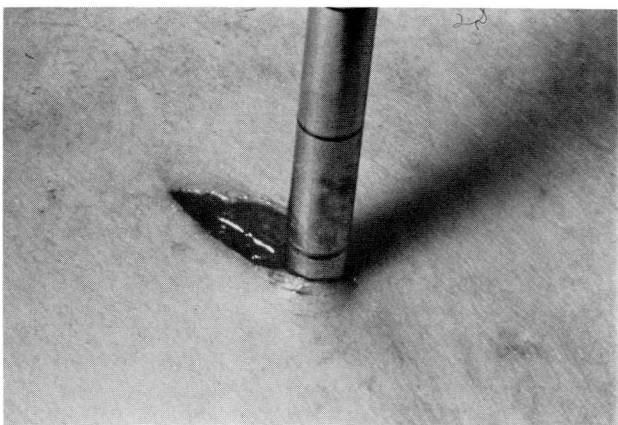

Figure 44: Enlarged skin hole appears on opposite side of released tension.

pressure. The size of the pin hole is of no great concern. A large skin hole will heal without trouble, but tension at the pin-skin interface will lead to necrosis and secondary infection.

After the remaining pins are inserted, repeat the procedure until all tension is released around all pins in each group. After the frame is completed and final fracture reduction is accomplished, it is usually necessary to check each pin again for any remaining evidence of tension. If compression or distraction are performed while the frame is in place, it may be necessary to release skin tension again after the maneuver is completed. Follow the procedure outlined above.

Occasionally, after making the initial skin incision, considerable bleeding will be encountered. Usually the bleeding will stop if light pressure is applied to the pin hole. If the skin incision continues to bleed, the postoperative dressing applied around the pins will act as a tamponade.

Pin Positioning

Fluoroscopic Control

Image intensification fluoroscopy is helpful when inserting pins. The image intensifier permits not only the localization of pin placement but also the depth of pin insertion. Considerable ingenuity is sometimes required by the surgeon to obtain x-ray projections that accurately reflect the position and depth of transcutaneous pins.

Soft Tissue Considerations

Pins should be inserted into bone through intact skin, if possible. This may require an unusual frame

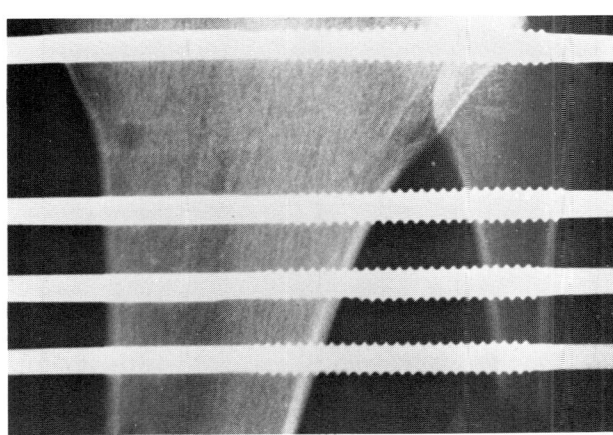

Figure 45A: Incorrect technique: threads do not engage both cortices (fluoroscopy not employed).

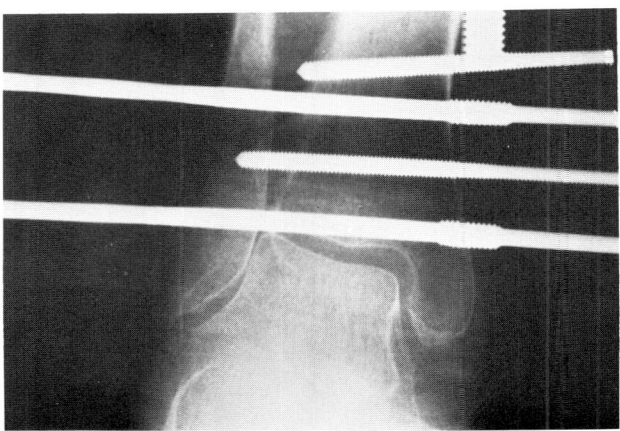

Figure 45B: Incorrect technique: pin penetrates joint (fluoroscopy not employed).

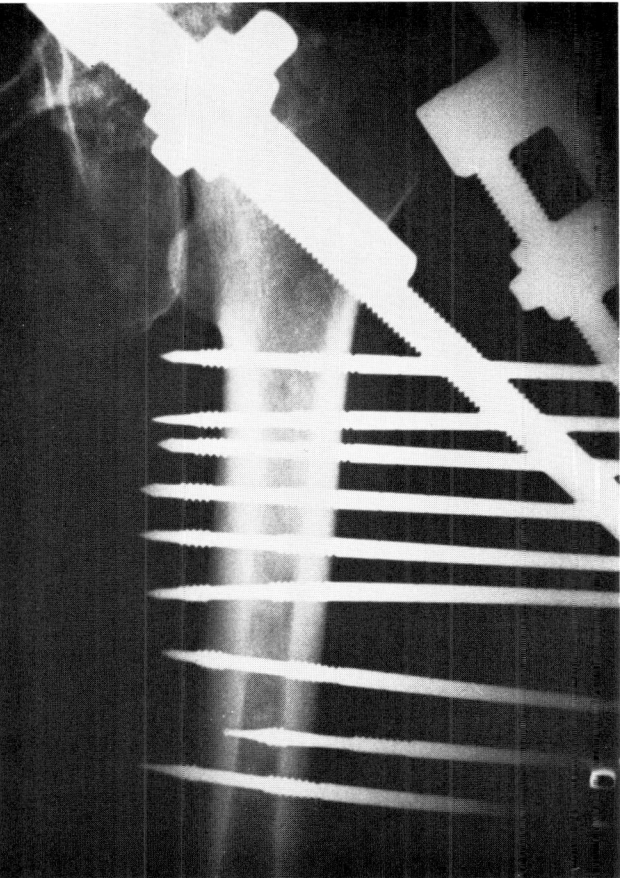

Figure 46: Incorrect technique: pin insertion too deep (no fluoroscopy; incorrect pin length with interrupted threads). Deep femoral artery was not damaged because the pins were aimed anteriorly (See Chapter 3).

design, tailored to fit the situation at hand. If this cannot be accomplished, pin insertion into exposed bone is acceptable, but it may interfere with plans for future soft tissue coverage.

Fracture Considerations

It is best to avoid penetration of the fracture hematoma, which may occur by placing pins too close to the fracture site, lest the fracture hematoma become contaminated with microorganisms entering through the pin hole. Although greater fracture rigidity can be achieved if at least one pin is close to either side of the fracture line, avoid placing pins within one inch of the fracture site, if possible.

Anatomic Considerations

The best pin-to-bone fixation will occur with the maximum separation between the two cortices engaged by the pin. In a circular bone, this means that the pin should be inserted through the center of the bone. In a triangular shaped bone such as the tibia, the pin should be inserted more posteriorly. It is important not to place the pins into extremely dense ridges of cortical bone. The most serious chronic pin hole infections follow pin placement into the dense anterior ridge of the tibia.

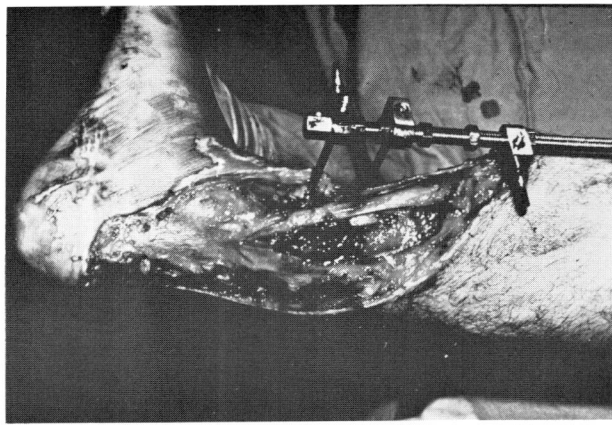

Figure 47: Pin insertion into denuded tissue is acceptable but should not interfere with subsequent plans for skin coverage.

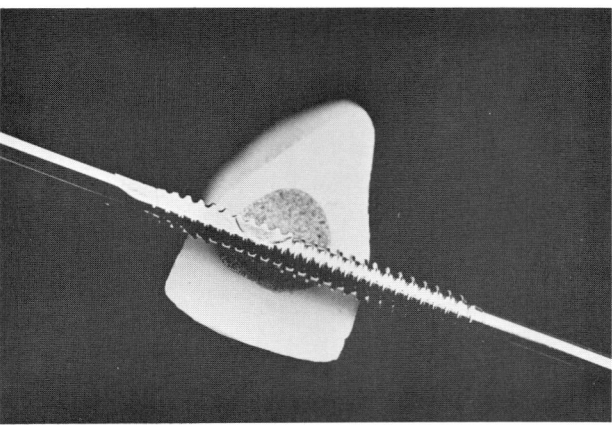

Figure 49: Correct placement in the tibia: two-thirds toward posterior cortex.

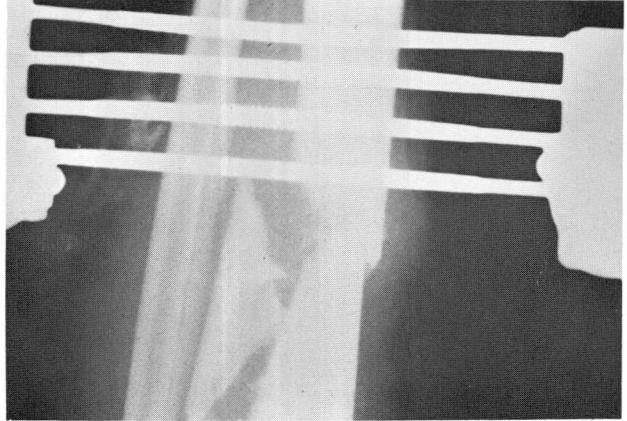

Figure 48: Incorrect technique: pin into fracture site.

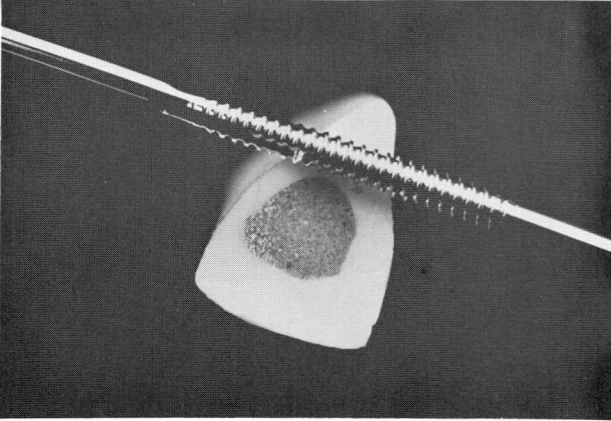

Figure 50: Incorrect technique: pin placed in anterior ridge of tibia.

MANAGEMENT AFTER PIN INSERTION

The strategy of treatment after the pins have been inserted is to apply the frame as quickly as possible and to avoid any action that will cause tissue necrosis around the pin holes. Furthermore, provisions must be made following pin insertion to prevent unnecessary skin motion relative to the pins. Finally, consideration must be given to the selection of a pin care routine that will minimize the risk of pin tract infection.

Frame Assembly

Alignment

Frame assembly can be extremely time-consuming if the surgeon is not familiar with the technical details necessary for constructing the proper spatial configuration of the fixator. It is important to practice frame assembly prior to surgery. A piece of wood or synthetic bone can be used. It is also helpful to learn the correct names for the components, asking for them as one would ask for any surgical instrument. The operating room personnel will quickly learn the names of the components if they are expected to hand them to the surgeon.

Once the frame is assembled, skeletal alignment should be evaluated with roentgenograms or fluoroscopy. Some projections will be difficult to interpret because of the presence of radio-opaque components of the fixator. If this occurs, oblique projections can be obtained of both limbs (for purposes of comparison). If alignment is unsatisfactory, the entire frame should be loosened, and a manual correction of the limb carried out. The frame should not be used to correct malalignment of a fracture by compressing the convex, and distracting the concave, side of the fracture deformity.[95] Figures 51-55 illustrate a case where the fixator frame was used to correct a valgus deformity by compressing the fracture medially and distracting it laterally. Undue pressures were produced at the pin-bone interface because the pins were fixed at right angles by the configuration of the frame itself. Multiple ring sequestra developed. The procedure also resulted in the frame being bent but did not correct the deformity.

Compression

Excessive pressure at the pin-bone interface may also occur when the frame is used for compression arthrodesis of joints, or when compressing stable cortical fractures. After applying several external fixators, one can develop a certain "feel" for the correct amount of compression that can be tolerated by a patient. A valuable way to learn what constitutes excessive compression is to practice on conscious patients in the office or clinic. A good opportunity for this is following compression arthrodesis or compression stabilization of fractures. The frame should be compressed slowly by turning each compression device one-quarter turn in succession. At some point, the patient will describe discomfort. Initially, the discomfort will subside if the limb is rested for a few seconds. There will come a level of compression, however, that will be followed by steady unremitting pain. At this point, the compression device should be distracted one-quarter of a turn at a time until the pain diminishes. This is the compression limit that, in my opinion, is safe for the patient. Learn the feel of this compression limit, so that it can be safely applied to an anesthetized patient.

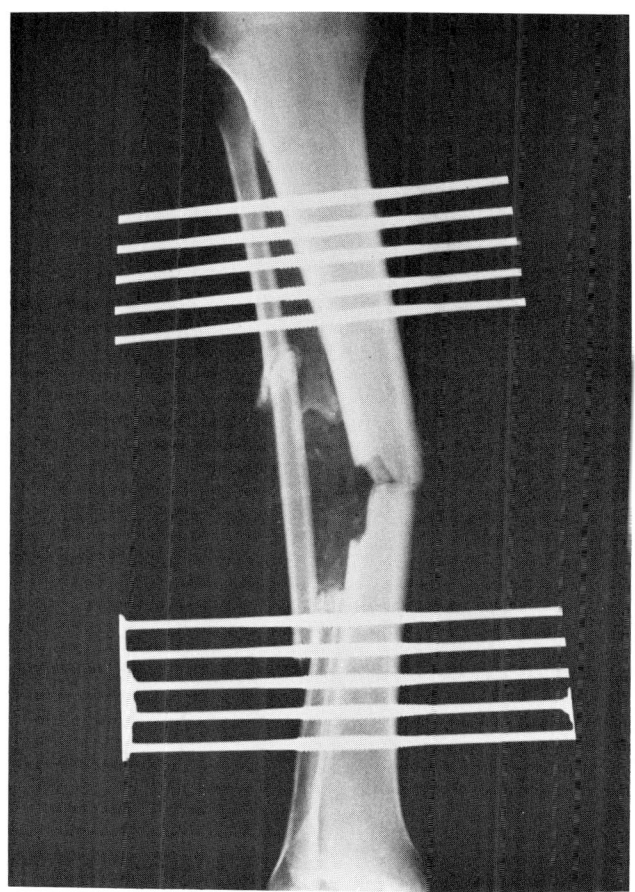

Figure 51: Incorrect technique: valgus deformity in fixator. Correction attempted by distracting concave side of deformity and correcting convex side of deformity (see figures 52-55 for results).

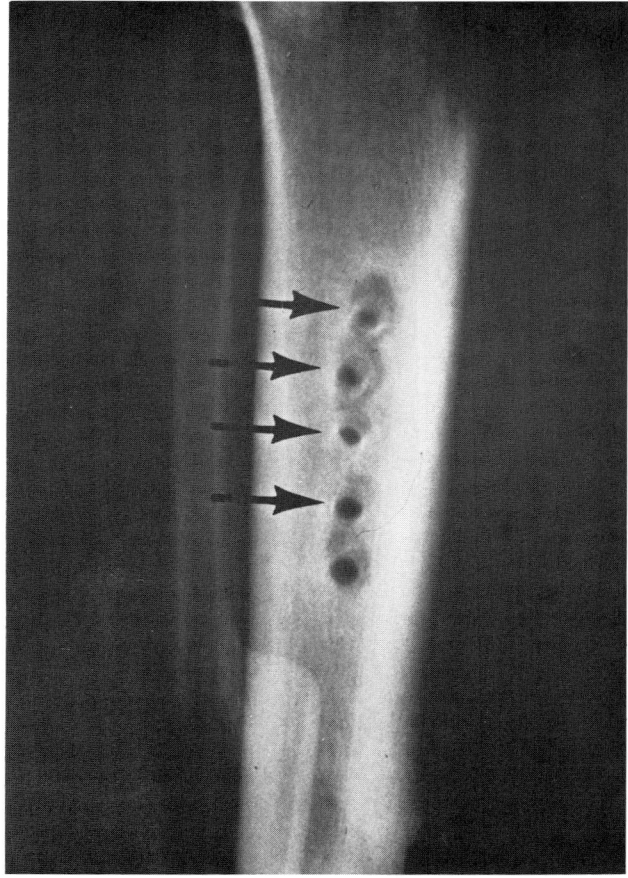

Figure 52: Multiple ring sequestra (proximal). Same patient as Figure 51.

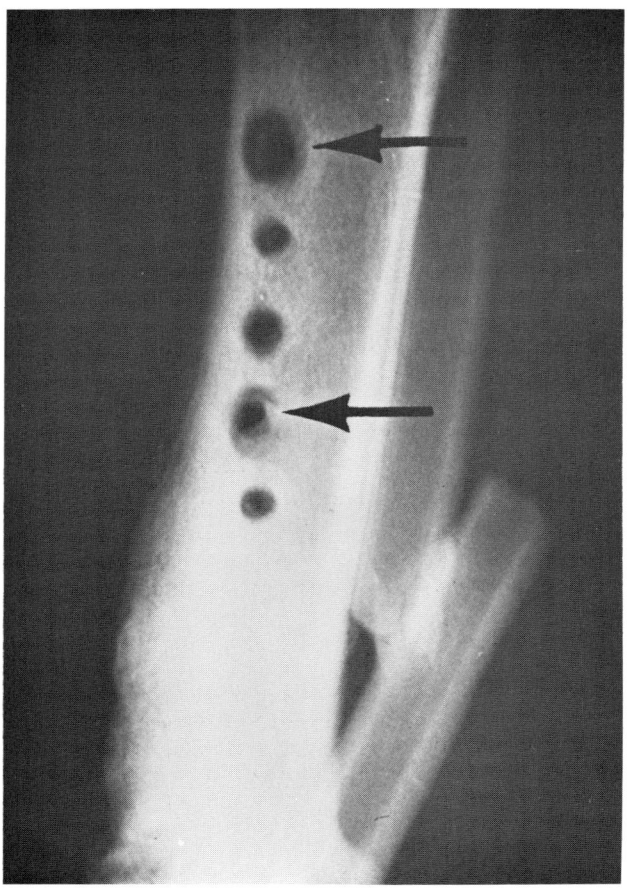

Figure 53: Multiple ring sequestra (distal).

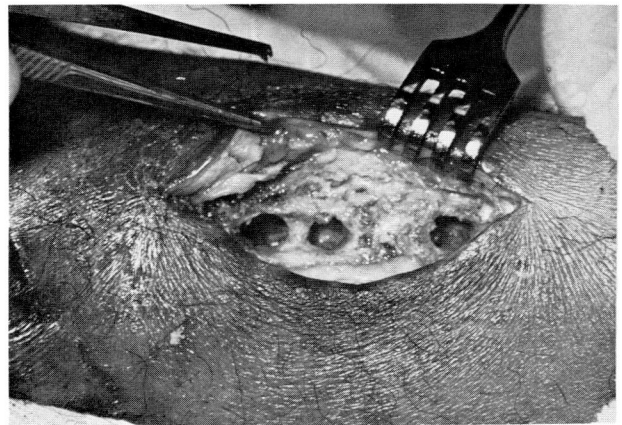

Figure 54: Appearance after sequestrectomy.

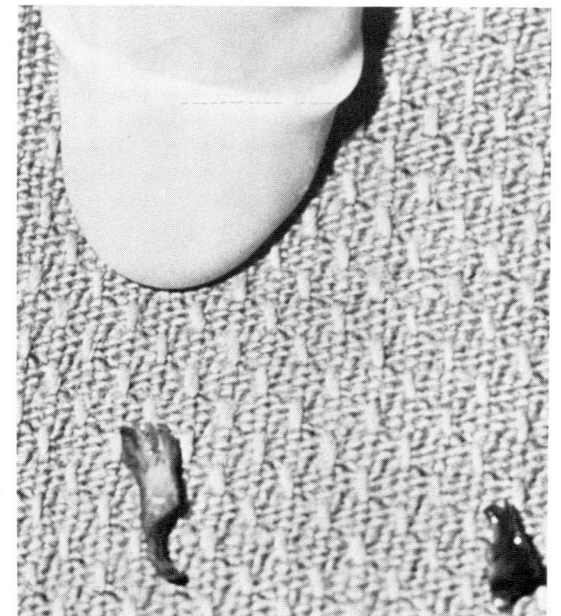

Figure 55: Incomplete ring sequestra.

Distraction

When applying an external skeletal fixator for distraction, it is important to avoid excessive distractive forces at the time of initial frame application. Kawamura[155] recommends that bones not be lengthened more than 1.5 percent following frame application.

Techniques to Prevent Skin Motion

Pin Wrapping

As noted earlier, it makes little difference to the pin hole microflora whether the pin is moving in the soft tissues or the tissue is sliding along the pin. The effect is the same between the tissue and a contaminated foreign body. Reduction of soft tissue motion around the pin hole can be accomplished by forming a bulky wad of gauze dressing wrapped around the pin so as to completely fill the space between the skin and the fixator. This controls sliding of the skin when the limb swells, following ambulation or activity.

Plaster of Paris

Another strategy to prevent skin-pin motion is to incorporate the entire skeletal fixator in plaster of Paris. This technique works well if the pins and frame components are wrapped in Webril® or Softroll® before the plaster is applied. Siris[248] noted that his incidence of pin tract sepsis was greatly reduced by this procedure. I believe that a bulky wrap of soft gauze dressing between the skin and the fixator functions in a manner similar to a plaster of Paris cast in reducing skin mobility along the pins.

Pin Care Routine

The question of daily pin care stirs much controversy among workers in the field of external skeletal fixation. Burny[37] recommends cleansing the pins three times daily with alcohol. Mears[191] utilizes hydrogen peroxide and bacitracin ointment, while Brooker[26] uses hydrogen peroxide and alcohol. Cooney[67] recommends Betadine® ointment, while Naden[202] covers his pins with Zephiran® soaked sponges. Krempen[164] on the other hand, cleanses the pins three times a day with Betadine solution and applies gentamicin ointment. I am inclined to agree with Fischer[95] that manipulation of the pin-skin interface to clean the crust is probably more important than the agent being used.

My routine for pin care consists of daily cleansing of the pins and surrounding skin with hydrogen peroxide, using small swabs or applicator sticks. If the patient is fairly agile, he can wash around the pins with soap and water in the shower. This is followed by

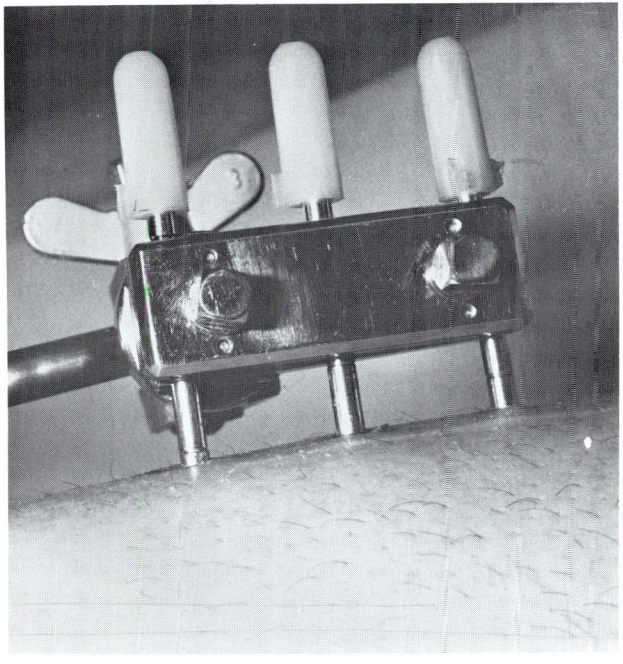

Figure 56A: Pin gripper: one fingerbreadth from skin over subcutaneous portion of tibia.

Figure 56B: Bulky wrap dampens skin motion but should not occlude the pin holes.

application of a bulky wrap—as described above—to control the space between the skin and the fixator. I also expect the patient to keep his fixator scrupulously clean and free of dust and crust.

In spite of diligent efforts, however, some pins will become septic. Furthermore, pin tract infection, at times, seems to occur when least expected. A most carefully placed, thoroughly released, and well-managed pin hole may become infected while others in the same patient do not. Nevertheless, close adherence to the principles outlined in this chapter will do much to control the factors primarily associated with pin hole sepsis.

Ambulatory Aids

In view of the observation that pin hole loosening is associated with pin hole sepsis, efforts should be focused on reducing cyclic stresses at the pin-bone interface. Such stresses occur with unprotected weight-bearing in lower extremity applications. Therefore, do not permit patients to ambulate with a fixator in place without supplementary ambulatory aids such as crutches. The reason for this is obvious. The external fixator serves as an exterior skeleton when there is no continuity of bone following a fracture. The mechanical stresses of weight bearing are transferred from the bone to the fixator at the pin-bone interface. The pins, being flexible, will transmit cyclic pressure associated with ambulation to the bone. This results in bone resorption and subsequent pin loosening. For this reason, unprotected weight bearing with an external skeletal fixator on the lower extremity should be discouraged.

Dealing with Pin Hole Problems

If the patient presents with evidence of pin hole sepsis following application of an external skeletal fixator, the physician should make every effort to resolve the problem. Initial management consists of rest at home, with elevation of the affected limb. The frequency of cleansing around the pin hole should be increased. I routinely enlarge the skin hole by infiltrating the skin around the pin with a local anesthetic, then insert a #11 blade into the skin adjacent to the pin. I also start the patient on oral antistaphylococcal antibiotics.

If these measures fail to promptly relieve the problem, the pin-gripper should be opened slightly and the pin checked for loosening by wiggling it. If the pin is loose, or if the maneuver produces pain, the pin should be removed. If removal of the loose pin affects the stability of the fixator, a new pin can be inserted in another location. The pin hole should be curetted with a small curette after a septic pin is removed (see Chapter 9).

If, on the other hand, the infected pin is securely fastened to the bone, the patient should be admitted to the hospital for a brief course of parenteral antibiotics, bedrest, and a deep incision and drainage of the soft tissues around the pin hole. Antibiotic therapy can be guided by cultures of the pin hole. If the septic process is not resolved by these actions, the involved pin should be removed and replaced with a new one in a different position. If the infection has not extended to involve the bone, drainage should stop in a few days. If draining persists, there is a significant probability that the patient has developed a chronic pin hole infection. Strategy for dealing with this problem will be considered in Chapter 9.

Chapter 3

NERVE AND VESSEL INJURY

Introduction

The vision of transcutaneous pins impaling major nerves or blood vessels is frightening. Yet, reports of serious neurovascular injury from fixator pins are uncommon. In fact, workers reporting large series of external skeletal fixator applications usually note the absence of a significant neurovascular injury. Burny and coworkers, for example, reporting their initial experience with the Hoffmann fixator noted no neurovascular injuries associated with 1,421 tibial[37] and 100 humeral[41] external fixator applications. Fellander[93] recorded no neurovascular injury after forty-nine applications of the Vidal quadrilateral frame configuration, and Krempen[164,165] did not describe any neurovascular injuries in his series of fifty-six patients treated with external skeletal fixation. Because of these reports, one is left with the impression that such injuries are rare. However, they are not unheard of; descriptions of such injuries do appear from time to time in reports dealing with external skeletal fixation.

DESCRIPTIONS OF NERVE AND VESSEL INJURY

Vessel Injuries

Seligson,[242] reporting his experience with the pins-in-plaster technique of external skeletal fixation, describes a young patient who noted bleeding from a pin hole while sitting in class, seven days after pin insertion. The patient was taken to the hospital, where the wound was explored and a bleeding vessel ligated. In a case treated at Rancho Los Amigos Hospital, a pin was placed too close to the superficial femoral artery during fixator application for knee arthrodesis (Figure 57). Three months after the fixator was applied, the patient developed swelling of his thigh. Surgical exploration of the upper medial thigh revealed erosion of the wall of the superficial femoral artery where the vessel was resting against the uppermost pin. Fortunately, sufficient collateral circulation was present to prevent loss of the limb.

Burny[36] reported a case of a patient referred to him who, immediately following pin removal, was noted to have excessive bleeding from one of the pin holes in the upper leg. The patient continued to bleed for seven weeks, although the wound was being packed daily. An arteriogram finally revealed the source of the problem: a false aneurysm of a branch of the anterior tibial artery. Browner, of Baltimore, describes a similar situation occurring after the removal of a Hoffmann external fixator applied to the femur (personal communication). Excessive bleeding from a pin hole eventually led to an arteriogram being taken, which revealed a false aneurysm of the deep femoral artery, adjacent to the medial cortex of the femur. Figure 58 illustrates a case with an arteriovenous fistula of the superficial femoral artery and vein that was discovered after pin removal.

Dwyer,[86] an English surgeon, studied the mechanism of potential neurovascular injury in connection with a new ring-type external fixator that he developed in 1973. He practiced pin insertion on cadavers and amputation specimens of the lower limbs. Having noticed that vessels are pushed to the side by the pin as it penetrates through soft tissue, he concluded that "it is almost impossible to pierce a major vessel with pins." This fact probably accounts for the rather peculiar way vessel injuries become apparent during the course of external skeletal fixation. I suspect that a pin directed at a vessel will usually push it to the side without transecting it. As time passes, the vessel, resting against the pin, develops an erosion in its wall. As a result, the patient may suddenly experience bleeding from the pin hole quite some time after the fixator was applied. Alternatively, the pin may create a hole in the side of a vessel, which does not become apparent until the pin is removed. Excessive bleeding through the pin hole may occur, or a false aneurysm may develop in the

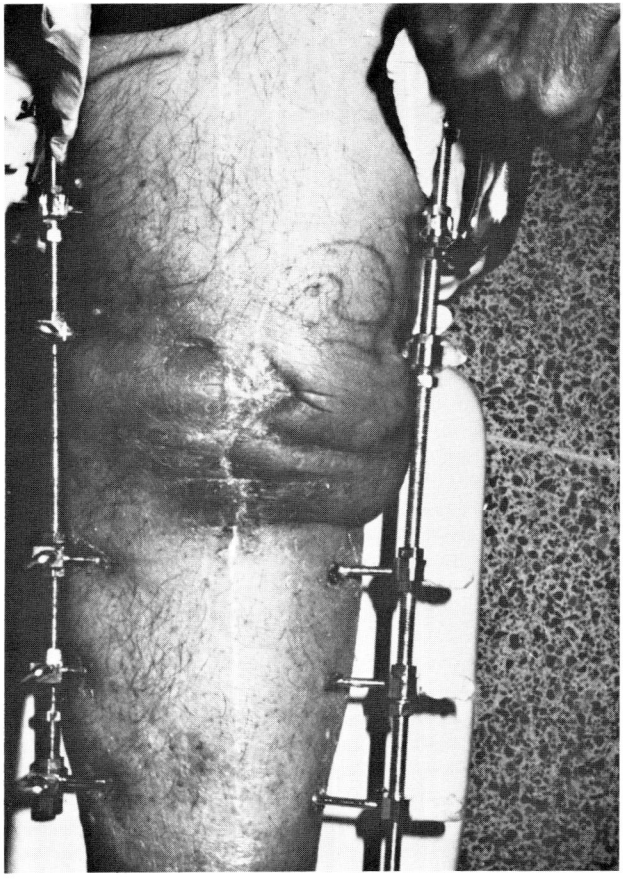

Figure 57: Erosion of superficial femoral artery occurred from upper medial pin in this frame configuration (note scar from emergency exploration).

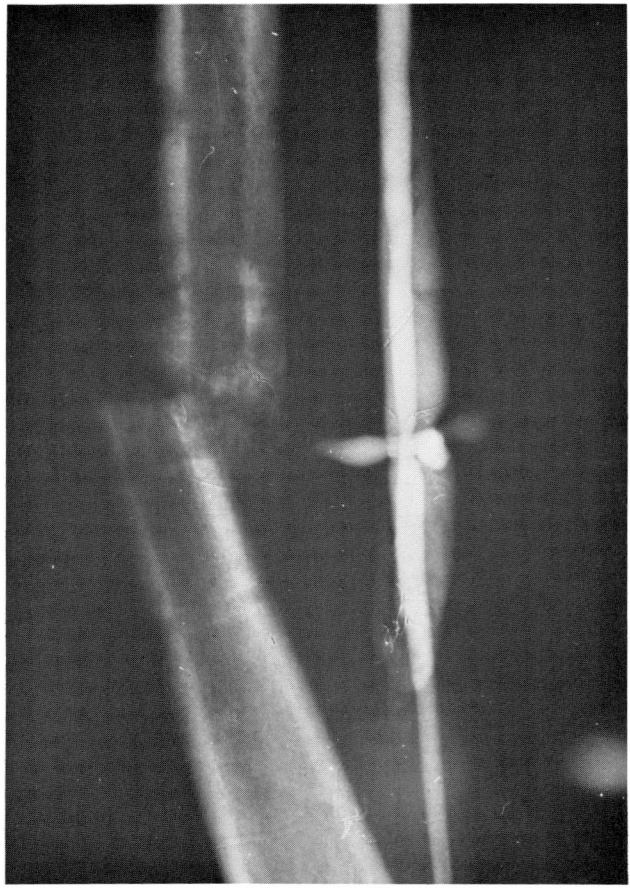

Figure 58A: Femoral arterial venous fistula (with small false aneurysm) discovered after fixator removal.

soft tissues. If the vessel wall necrosis involves an adjacent artery and vein, an arteriovenous fistula may be created shortly after pin removal.

Reports describing serious distal vascular compromise following pin insertion are quire rare. Naden,[202] in reporting his experience with the Roger Anderson apparatus, noted a case where a foot amputation was necessary twenty-four hours after pin insertion, due to damage of the posterior tibial artery. It is noteworthy that the limb had been severely traumatized by the effects of the initial injury, which probably impeded collateral circulation through other major vessels. I am aware of two other such cases, both resulting in amputation shortly following pin insertion. In both situations, one occurring in Florida and one in Minnesota, the presumed site of vessel occlusion was in the distal part of the leg, where the anterior tibial artery crosses the lateral surface of the tibia as it passes from the interosseous membrane to the front of the ankle. In both cases, there was marked traumatic injury to the limb, which probably impeded collateral circulation.

Raimbeau, Chevalier, and Raguin[223] analyzed the problem of damage to the anterior tibial artery caused by transcutaneous pins. They performed arteriograms on cadaver limbs and determined that the region of the tibia consisting of the lower end of the third quarter and the upper end of the fourth quarter is a danger zone for transfixion pin placement. They presented three patients, each of whom presented with a classic anterior compartment syndrome following application of external fixation to the lower leg. Naden[202] noted one case of slough of the anterior compartment musculature without evidence of infection in his series of 950 applications of external skeletal fixation. Vidal[273] describes two patients who came to his attention with severe anterior compartment syndrome problems fol-

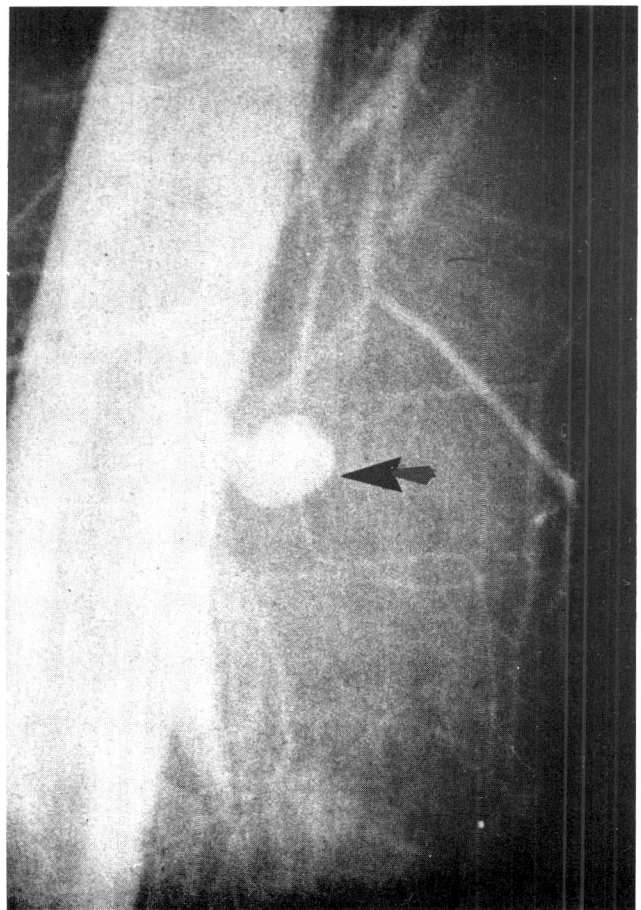

Figure 58B: Aneurysm of deep femoral artery produced by a pin. By permission of Bruce D. Browner, M.D.

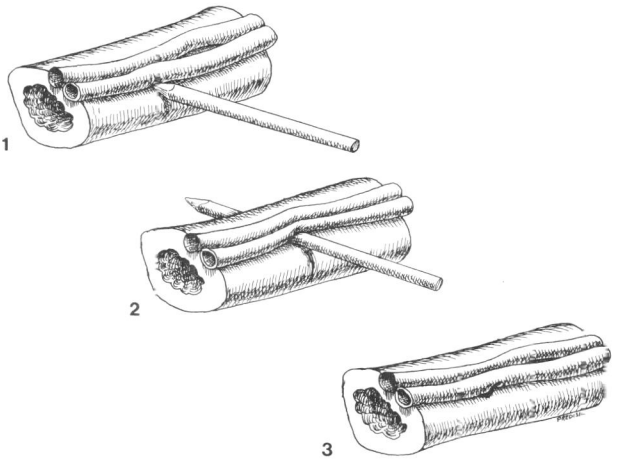

Figure 59: Mechanism of vessel injury. The pin pushes the vessel to the side during insertion; erosion develops later.

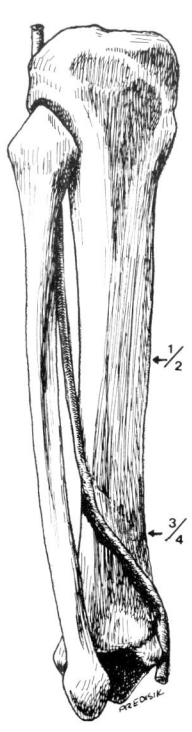

Figure 60: The course of the anterior tibial artery and deep peroneal nerve (not shown) along the lateral surface of the tibia.

lowing application of external fixation. Cabenela[44] also reported a case of anterior compartment syndrome. Edwards[89] evaluated a patient who developed extensive septic necrosis of the musculature of the anterior compartment which may have started with a vascular injury due to pin insertion.

Raimbeau and his associates also measured tissue pressures in the anterior compartment after insertion of transcutaneous pins. They determined that the intracompartmental pressure was not significantly elevated after insertion of one transfixion pin, but it more than doubled when a second pin was inserted. Insertion of a third pin did not significantly raise the pressure any higher. Thus, they identified two vascular syndromes associated with pin fixation of the lower leg. The first, interference with the distal circulation of the anterior tibial artery, is quite rare because adequate collateral circulation is usually present. The second, anterior compartment syndrome, may be due to partial occlusion of the anterior tibial artery combined with the increased compartment pressure associated with transfixion pins.

Nerve Injuries

Acute transection of a major nerve is unlikely with external skeletal fixation. Nerves may, however, be stretched or damaged during the course of pin insertion. Olerud[208] describes transient partial loss of sensation on the dorsum of the foot in a "few cases" in his large series of external skeletal fixation for tibial fractures. Naden[202] likewise describes "several" transient lesions of the peroneal nerve. Cabenela[44] has seen two such cases. In my own experience, a patient developed transient radiating pain and paresthesias following the course of the superficial peroneal nerve to the dorsum of the foot after application of an external skeletal fixator to the tibia. I also have a patient who developed a transient paresthesia of the ulnar nerve following application of an external fixator for stabilization of his elbow. Vidal[268] describes paresthesias of the hand following insertion of pins into the humerus for upper extremity stabilization.

In spite of the relative infrequency of reports of serious neurovascular injury, great care is nevertheless required during pin insertion so that major neurovascular structures are not stretched or damaged during the course of pin insertion.

Some authorities recommend a skin incision, with observation of major neurovascular bundles, when pins are inserted into certain anatomical areas. Vidal[265] recommends identifying the femoral vessels when inserting pins into the second quarter of the femur. He also advises opening and identifying the position of the anterior tibial artery and vein and deep peroneal nerve when inserting pins into the danger area described by Raimbeau. Vidal,[265] Mears,[191] and others[65] recommend a small incision on the dorsum of the forearm to identify the radial sensory nerve and extensor tendons when inserting transcutaneous pins into the forearm bones. Likewise, it is helpful to observe the ulnar nerve if the pins are inserted into the distal humerus from the lateral epicondyle to the medial epicondyle.[67] Instead of exposing these structures surgically, however, one can select pin placement positions that avoid the possibility of damage to these structures.

Two difficulties are encountered during the procedure of pin insertion. First, the surgeon is occasionally unsure of the precise position of a major nerve or vessel with respect to the bone at the level of the limb selected for pin insertion. This confusion arises from the surgeon's orientation to surgical anatomy, which usually considers the position of a nerve or vessel in its longitudinal relationship to surgical exposure. This is because surgical exposures are usually parallel to both the bone and the neurovascular structures in each anatomical region. Second, it is frequently difficult to assess the exact depth to which a pin has penetrated into the bone. This may seem surprising, considering how easy it is to "feel" when a drill bit penetrates the opposite side of a bone during drilling. Nevertheless, because the pin is threaded, there is enough resistance to forward progress to make depth determination difficult.

PIN PLACEMENT TO AVOID NEUROVASCULAR INJURY

An atlas showing pin placement positions appears in this chapter. The atlas is designed to reduce the likelihood of neurovascular injury from transcutaneous pins. By recommending pin placement in certain positions, I do not mean to imply that these are the only acceptably safe positions for pin insertion. At many points in the limb, pins can be safely inserted in several directions that have not been indicated. The descriptions of these positions were omitted for the sake of simplicity and clarity of illustration. With experience (and reference to the atlas), surgeons will find additional pin positions to solve specific clinical problems. In selecting the recommended direction for inserting a pin, I followed several principles designed not only to reduce the incidence of neurovascular injury but also to allow easy, yet solid, pin insertion.

First, whenever possible the pins are inserted *perpendicular to the bone surfaces*. This facilitates the pin insertion process because it reduces the tendency of the pin point to "walk" (slide along the bone surface). The tibia, for example, has a triangular cross-section. When the patient is supine, the lateral surface is vertical and the medial surface is oblique. Full-pins are more easily inserted from lateral to medial because of this anatomic feature.

Second, pin directions should *cross the center of the medullary canal* to engage both cortices. When widely separated cortices are engaged by a pin, the tendency of the pin to wobble and loosen is reduced and maximum stability of pin fixation is achieved.

Third, pin insertion into *dense bony ridges is to be avoided* wherever possible. Drilling into very dense cortical bone with hand tools is tedious and frustrating, tempting the surgeon to try to overcome the resistance by pushing harder and drilling faster, which increases thermal injury to bone and consequently the likelihood of pin hole sepsis.

Fourth, pin positions should have a *margin of safety* on the opposite side of the bone. A pin is considered *safe* if it passes through the bone and emerges from the opposite side of the limb without encountering a major neurovascular structure. Such pins are illustrated as

Nerve and Vessel Injury

full (through-and-through) pins, although half pins could, of course, be safely inserted from either direction. A pin is labeled *caution* if a major nerve or vascular structure is located on the opposite side of the bone at a distance equal to or greater than the diameter of the bone itself. In this respect, the designation refers only to half-pin placement. A full pin may be labeled *caution* if the direction or angle of pin insertion is critical to avoid neurovascular injury.

A pin is labeled *danger* if a major neurovascular structure is between one-half and one bone diameter away from the bone on its opposite side. It is wise to insert such pins under radiographic or fluoroscopic control. A pin is also considered a *danger* pin if it must be inserted adjacent to a neurovascular structure on the near side of the bone. Generally this requires open pin insertion—a longitudinal incision to identify the location of the structure prior to pin insertion.

As illustrated in Figure 61, pin placement is measured in degrees, rotating around the bone from anterior to posterior, with the center of the bone always presumed to be the center of pin placement. Thus, the direct anterior position is considered to be 0°, and the direct posterior position is considered to be 180°. Pin placement from directly lateral to directly medial is considered to be *90° lateral* and a pin placed from directly medial to directly lateral is considered to be *90° medial*. In the forearm where there are two bones available for pin placement, the pin position for each is noted separately. The limb must be in the anatomic position during pin insertion if the atlas is to be used correctly. The humerus should be in neutral rotation, and the forearm supinated to correlate with the location of the anatomic structures indicated.

Fluoroscopy

I recommend that image intensification fluoroscopy be used during pin insertion. The correct assessment of the position and depth of the pin can best be determined if the pin is seen in its true lateral projection. (In the true lateral projection of the pin, the central beam of the x-ray tube must be perpendicular to the pin itself.) At times, there is a tendency by surgeons to judge pin position through use of an oblique projection because a true lateral projection of the pins is difficult to obtain when the patient is supine on a large operating table. The surgeon may have to use considerable ingenuity to position a limb for fluoroscopy with a C-arm image intensifier. It may be necessary, for example, to rotate the limb 45° or more, while rotating the C-arm in the opposite direction in order to obtain a true lateral projection. In order to determine the exact location of a pin within a bone, it is necessary to direct the central beam of the x-ray tube along the pin itself. A perfect axial projection of the pin will result in a small circular image equal to the diameter of the pin. In this manner, the position of the pin relative to the cortices can be determined. (If fluoroscopy is utilized during application of the fixator, it is essential for the surgeon to wear a lead apron for protection.)

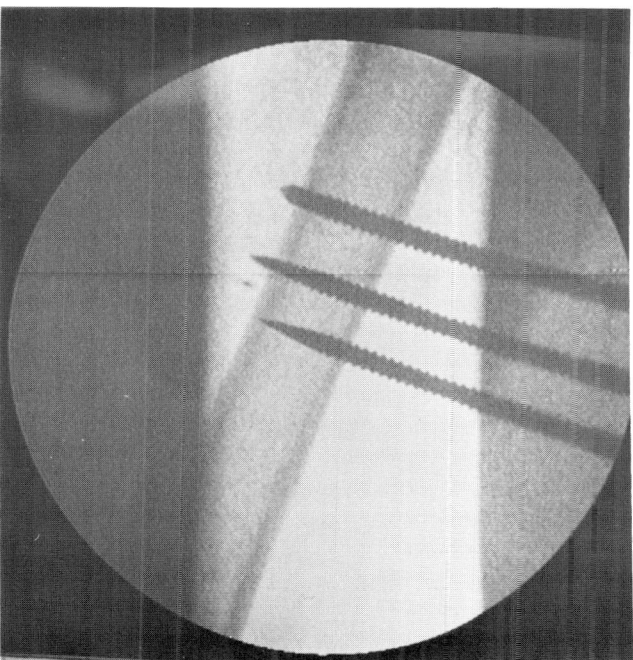

Figure 62: When determining pin depth with fluoroscopy or roentgenograms, the central x-ray beam should be perpendicular to the pins.

Figure 61: Pin placement notation.

If roentgenograms, rather than fluoroscopy are used, the initial evaluation can be obtained after the first pin is inserted to the presumed proper depth. Before the roentgenogram is taken, it is safer to be too shallow than too deep. If a pin is inserted too deeply, there is the obvious danger to neurovascular structures. Also, "backing out" a pin reduces its fixation in bone. When the depth of the first pin is satisfactory, additional pins of the same length can be inserted to the same depth. This strategy for pin insertion can also be employed to reduce x-ray exposure to the operating room personnel when image intensification fluoroscopy is utilized. Only a brief exposure is necessary to determine the position and depth of the first pin. Thereafter, pins can be inserted to the same depth without checking the progress of each pin individually.

THE ATLAS

38 *Complications of External Skeletal Fixation*

Using the Atlas

The cross-section atlas in this book was specifically created to aid the surgeon in the operating room.* Proper orientation of the cross-section diagrams to a patient on the operating table depends on the location of easily palpable landmarks. Each limb section in the atlas is treated in an identical manner. Each anatomic area is divided into four equal zones. Palpable bony landmarks identify the upper and lower limits of each anatomical area under consideration.

- In the *thigh*, the proximal bony landmark is the lateral prominence of the greater trochanter of the femur; the distal landmark is the lateral prominence of the lateral epicondyle of the femur.
- In the *lower leg* section, the proximal landmark is the medial tibial joint line; the distal landmark is the medial prominence of the medial malleolus.
- In the *upper arm*, the proximal landmark is the lateral prominence of the greater tuberosity of the humerus, which is one thumb's width below the lateral tip of the acromion process. Distally, the landmark is the lateral epicondyle of the humerus.
- In the *forearm*, the proximal landmark is the lateral prominence of the radial head, which is one thumb's width distal to the lateral epicondyle of the humerus. The distal landmark is the lateral prominence of the radial styloid process.

Technique of Identifying Landmarks

Stretch a surgical towel between the landmarks described above and indicate the position of the landmarks on the towel with a surgical pen (Fig. 65). Fold the towel so that the marks touch each other and mark the midpoint of the fold. Lay the towel against the limb again and mark the midpoint on the limb using the towel as a guide. In this manner, the limb section will be divided in half. Repeat the procedure and find the

---

*Several sources were consulted in making these plates. [92, 109, 110, 177, 178, 279] The Rancho anatomy laboratory provided a final check of all pin placement positions.

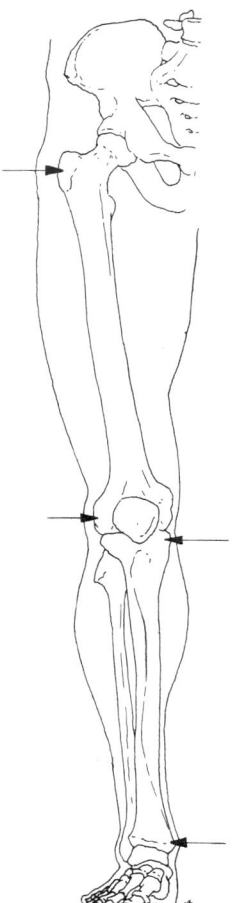

Figure 63: Lower extremity landmarks.

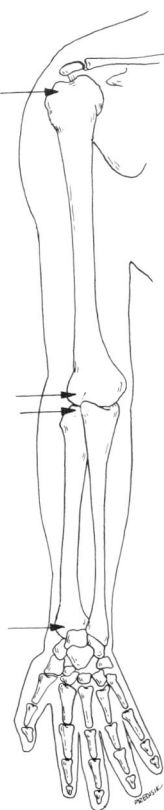

Figure 64: Upper extremity landmarks.

midpoint of each half, thus dividing the limb segment into four equal zones. The zones are labeled A, B, C, and D, with A proximal and D distal. The zones approximate, but are not exactly, the quarters of each limb segment. The atlas illustrates cross-section anatomy in the top, middle, and bottom of each zone. Key diagrams on each plate orient the reader to the zones illustrated.

Plates

For purposes of clarity, bones, nerves, arteries and veins have been emphasized in relief. Muscle planes are indicated, but the muscle masses themselves are not labeled. Small cutaneous nerves, veins, and muscular branches of arteries have been omitted. Major arteries are shown with one vein even if they are usually accompanied by two. In the forearm, deep veins have been omitted completely. Some neurovascular structures have been further emphasized by making them slightly larger than natural size. Many structures are labeled only once on each page, rather than on each slice. Mental reconstruction of the zone will fill in labels on the unlabeled slices.

Unfortunately, some anatomic features are not easily presented in cross-section views. These are the transverse vessels and nerves that wind around the bone at one level. Furthermore, the atlas plates do not take into account variations in anatomy that can occur at any level. For these reasons, the atlas illustrations must be considered schematic, rather than representational.

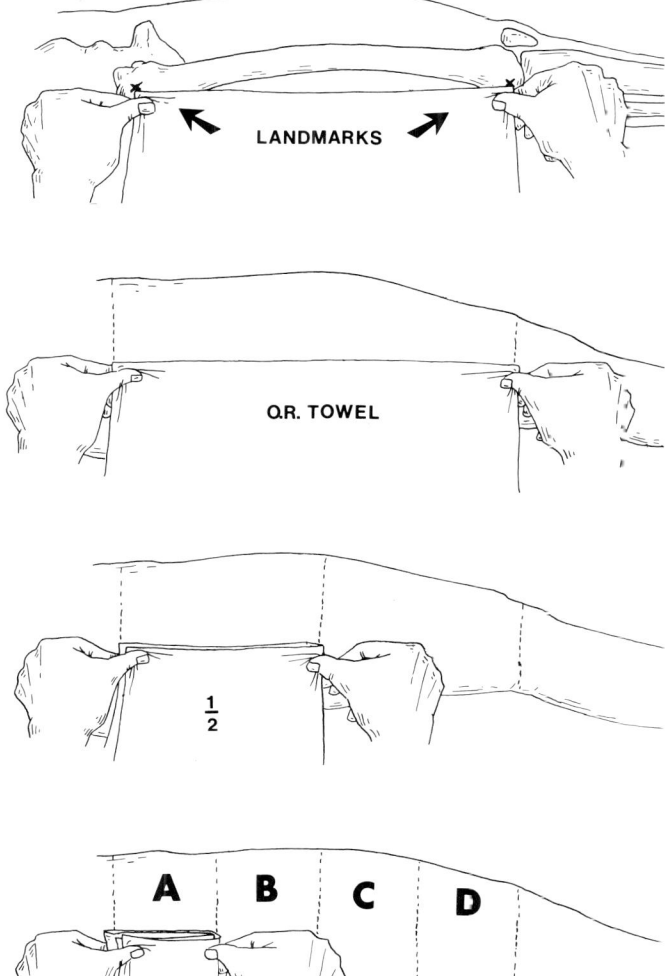

Figure 65A: Marking zones.

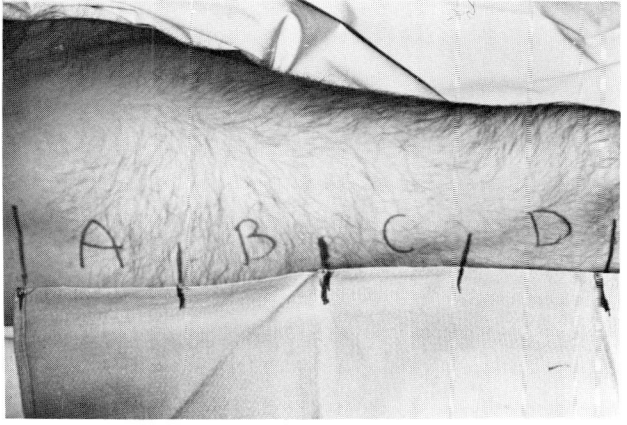

Figure 65B: Marking zones on thigh with a surgical towel.

Thigh—Zone A

Anatomic Considerations

1. The femoral shaft is quite lateral in the proximal thigh.
2. The sciatic nerve remains medial to the femur throughout Zone A, always separated from the bone by a portion of the adductor muscle group equal in distance to the diameter of the femur.
3. The superficial and deep femoral vessels enter Zone A anterior and medial to the femur. The deep femoral artery comes to lie medial to the femur in the lower end of Zone A, separated from it by the origin of the vastus medialis muscle, but only one-half bone width away.
4. The lateral femoral cutaneous nerve is in line with the lateral cortex of the femur.
5. The anterior femoral circumflex artery winds around the lateral cortex of the femur at the base of the greater trochanter.

Pin Placement

1. Pin placement from 0° anterior can penetrate the bone and, in fact, pass out the posterior aspect of the thigh without danger.
2. Half-pin insertion from the 90° lateral position can be done with caution in the upper two-thirds of Zone A, and with extreme caution in lower Zone A.
3. With care, and image intensification control, additional pin insertion can be obtained throughout a wide range in the upper portion of this zone.

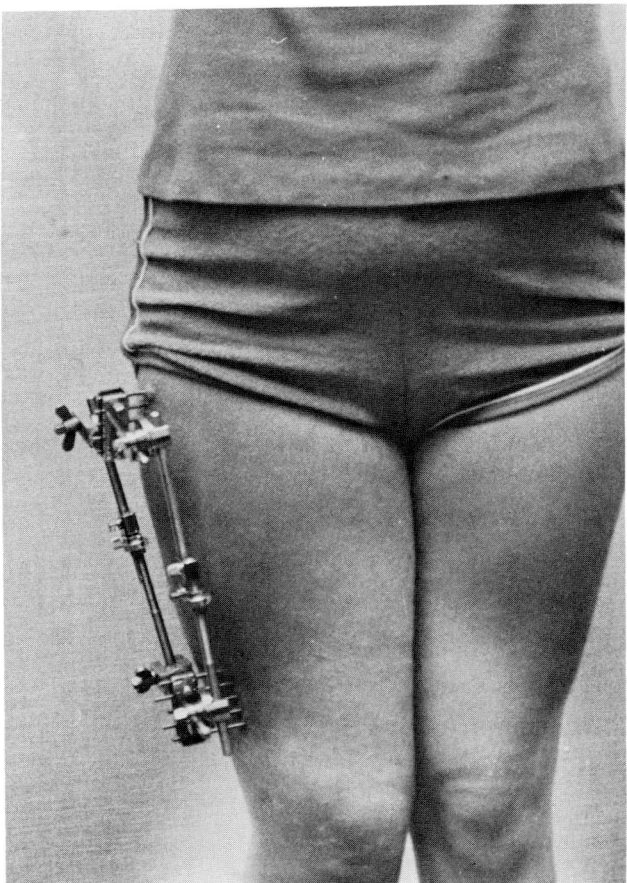

Figure 66: Half pin placement: Zone A (90° lateral); Zone C (120° lateral).

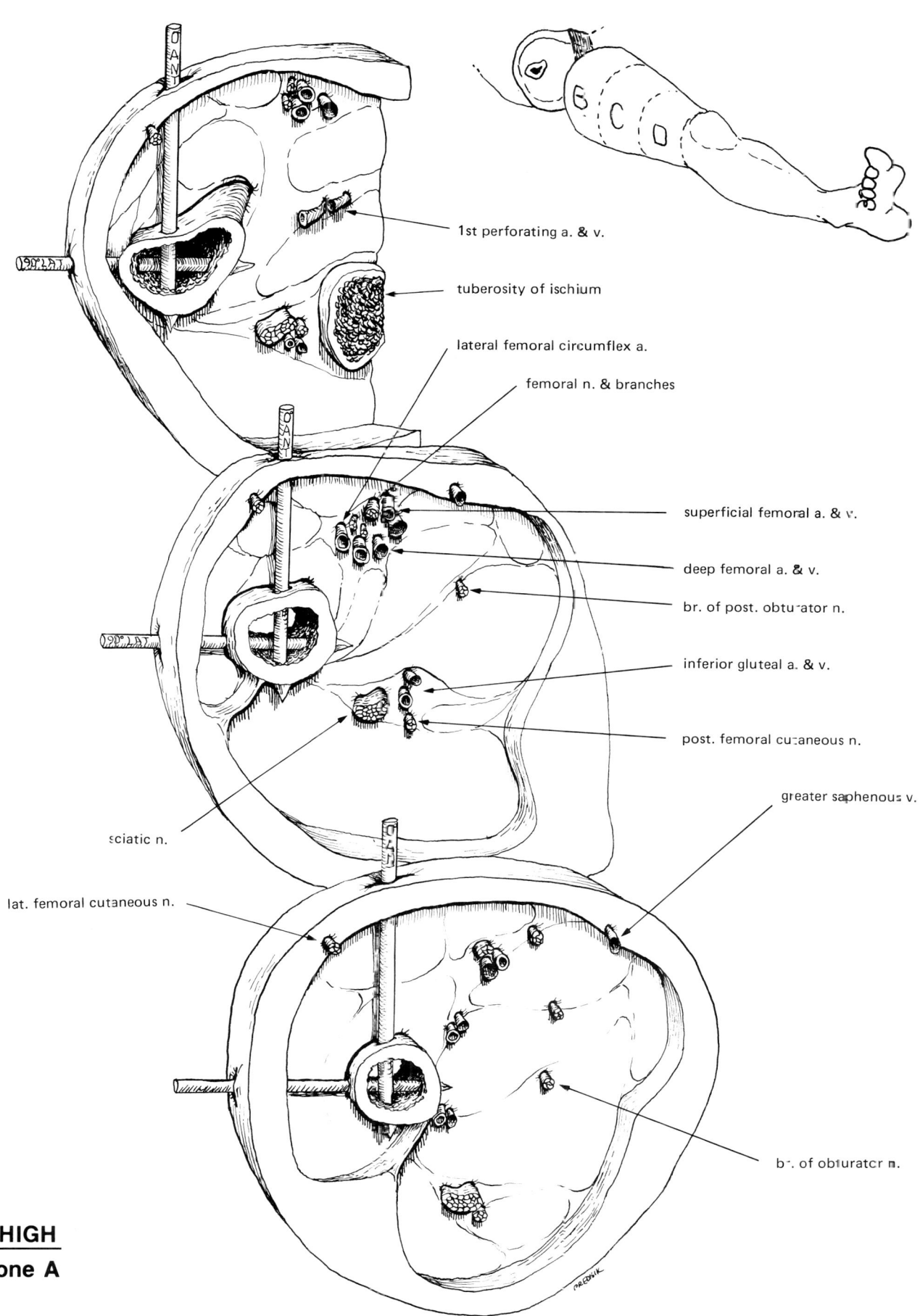

THIGH Zone A

Thigh—Zone B

Anatomic Considerations

1. The femur is laterally placed in the cross-section throughout Zone B.
2. The sciatic nerve is posterior and medial to the femur, separated from it by a portion of the adductor magnus muscle equal in size to the diameter of the bone.
3. The superficial femoral artery passes from antero-medial to directly medial (with respect to the femur) in this zone. It crosses the coronal plane of the femur into the posterior portion of the thigh at the junction between Zone B and Zone C.
4. The deep femoral artery and vein are medial to the femur in proximal Zone B, and posterior to the femur in distal Zone B.
5. The lateral femoral cutaneous nerve is anterior to the femur.

Pin Placement

1. Full-pins (and half-pins) may be inserted with caution from the 30° medial position or the 150° lateral position. Extreme caution is necessary in proximal Zone B with this placement, because the superficial and deep femoral vessels are in a straight line and can both be injured with a pin placed too far medially, as noted in the cross-section diagram indicating proximal Zone B. Bear in mind that the femur is lateral in the cross-section of the thigh at this level, requiring rather lateral pin insertion into the limb.
2. Additional half-pins may be placed in other directions, but the intimate association of the deep femoral vessels to the shaft of the femur should be kept in mind.

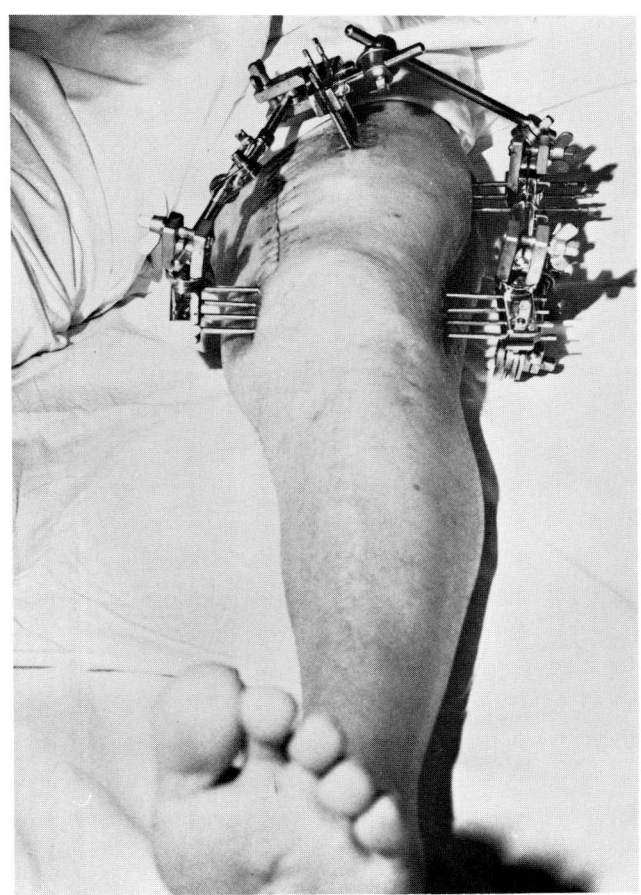

Figure 67: Half pin placement: Zone B (30° medial); Zone A (90° lateral). Full pin placement: Zone D (90° medial →90° lateral).

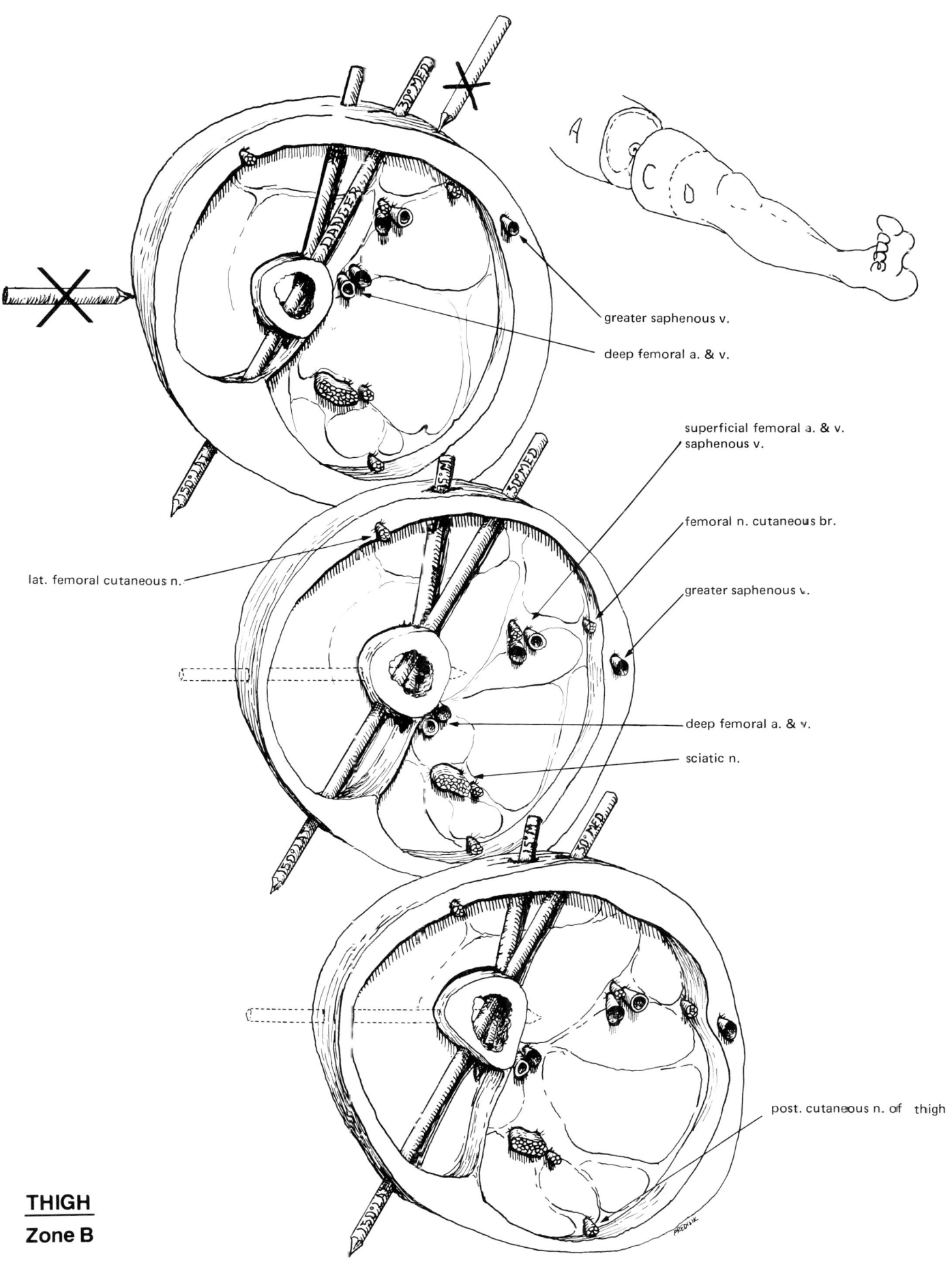

THIGH
Zone B

Thigh—Zone C

Anatomic Considerations

1. The femur is more centrally placed on cross section, although it remains anteriorly situated.
2. The sciatic nerve passes from medial to lateral behind the femur, approximately one bone width away.
3. The superficial femoral artery passes the coronal plane of the femur in Zone C and is posterior to the bone at the lower end of this zone.
4. The deep femoral artery and vein are adjacent to the posterior surface of the femur, but terminate at the lower end of Zone C.

Pin Placement

1. Full or half-pins can be inserted from the 60° medial or 120° lateral position.
2. Half-pins can be cautiously inserted from the 0° anterior position in distal Zone C because the deep femoral artery and vein are no longer present (not shown).
3. Full or half-pins can also be inserted 90° medial or 90° lateral in distal Zone C.

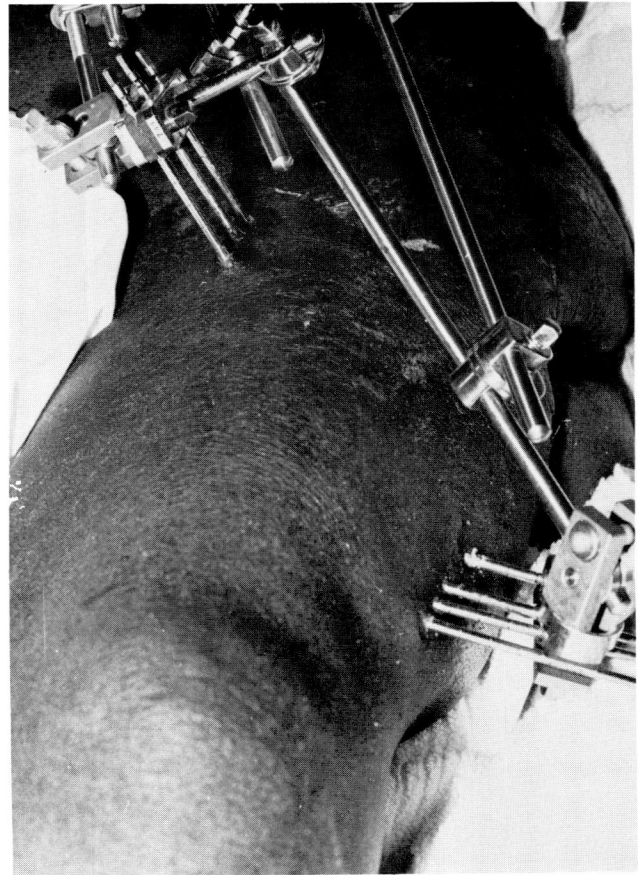

Figure 68: Half pin placement: Zone C (120° lateral): Zone B (30° medial).

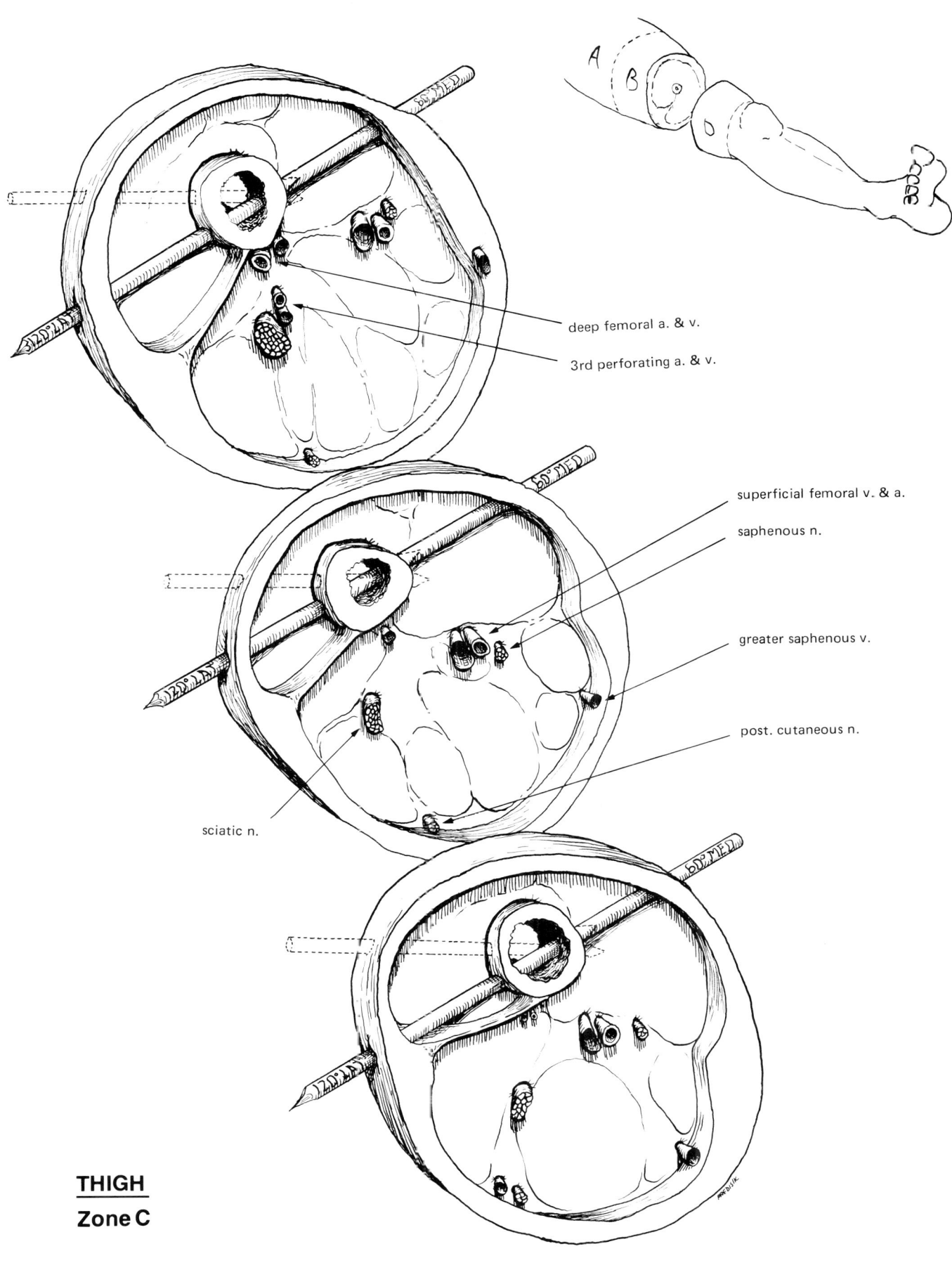

THIGH
Zone C

Thigh—Zone D

Anatomic Considerations

1. The femur is an anterior structure until the flair of the condyles.
2. The sciatic nerve is posterior to the femur in proximal Zone D crossing to the lateral side while dividing into the tibial and peroneal divisions.
3. The femoral artery becomes the popliteal artery and, with the popliteal vein, is immediately posterior to the femur in Zone D.
4. The synovial cavity of the knee joint enlarges to encompass the anterior half of the femur immediately above the joint line.

Pin Placement

1. Full-pins or half-pins from 90° medial or 90° lateral are safe.
2. Half-pins from the 90° lateral position have the additional advantage of not transfixing the vastus medialis muscle.
3. At the level of the epicondyles, the synovial cavity is present anteriorly and posteriorly, leaving only one inch of bone, which is extrasynovial. Three or four pins may be placed close to each other in a transverse plane through the bone at this level, although if four pins are placed, the most posterior pin may pass through the synovial cavity.

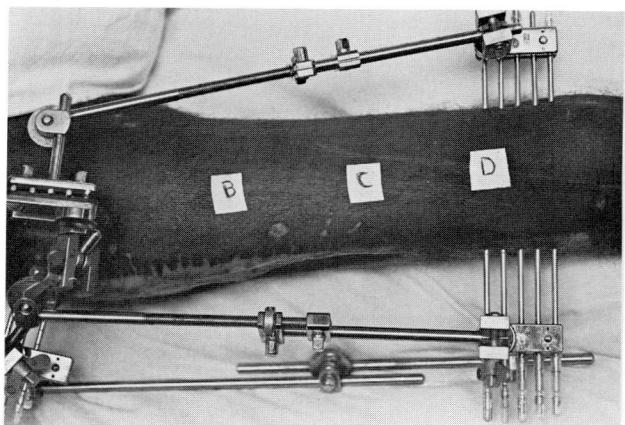

Figure 69: Zone D placement: full pins (90° medial⟶90° lateral).

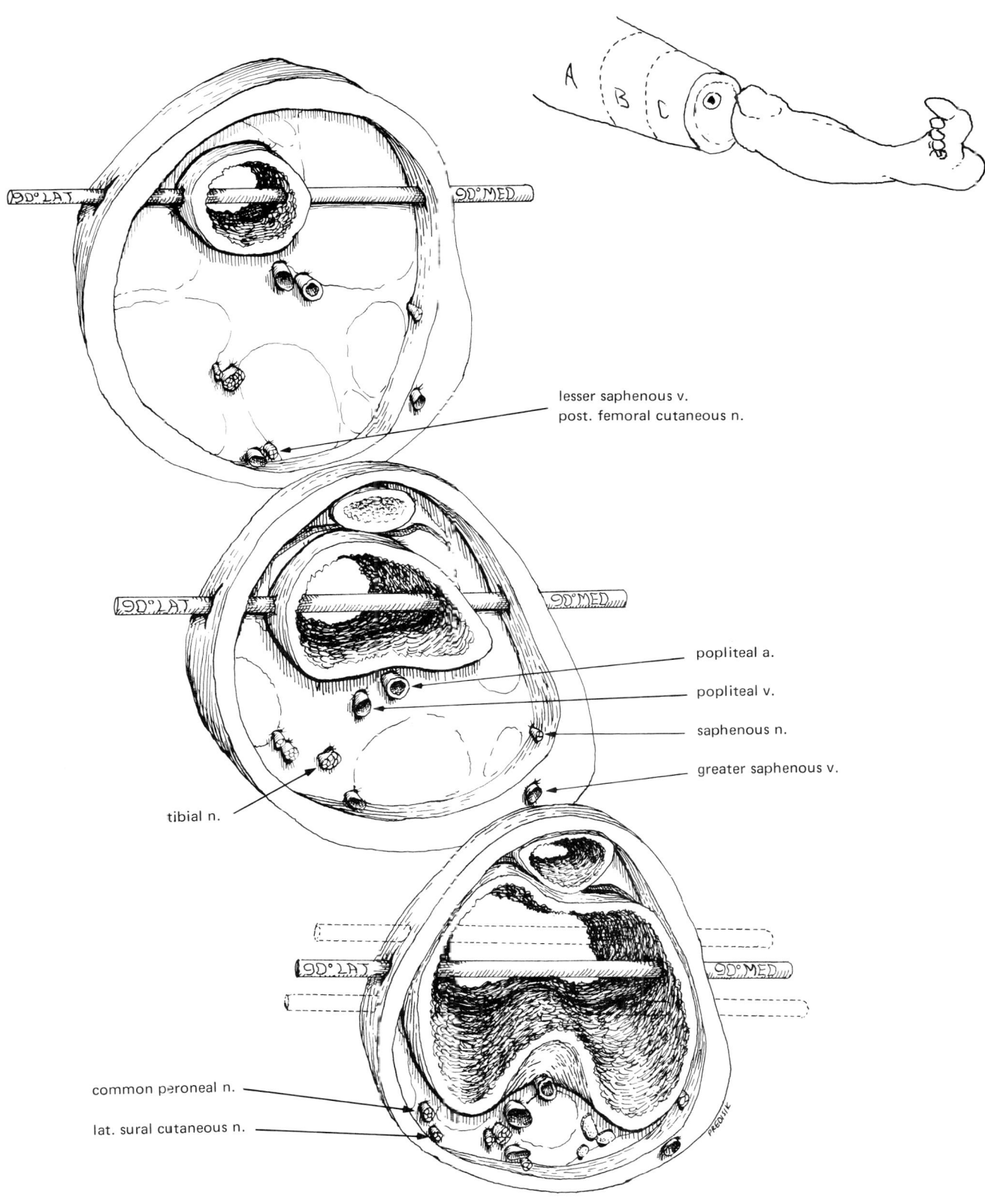

THIGH
Zone D

Complications of External Skeletal Fixation
Leg—Zone A

Anatomic Considerations

1. The shape of the tibia changes rapidly through this zone.
2. The popliteal artery is posterior to the tibia where it divides into its terminal branches.
3. The superficial and deep peroneal nerves are lateral to the fibula as they wind around the fibular neck.
4. The saphenous nerve and greater saphenous vein are posterior to the tibia on the medial side of the limb.
5. In distal Zone A, the anterior tibial artery is on the anterior surface of the interosseous membrane and the peroneal and posterior tibial arteries are posterior to the tibia, accompanied by their associated veins.

Pin Placement

1. Full-pins (or half-pins) can be placed in the 90° medial—→90° lateral direction throughout Zone A.
2. Pins can be placed parallel to the joint line (and to each other) through the condyles of the tibia in proximal Zone A.
3. Great caution is necessary when placing 60° medial half-pins into med Zone A, because the anterior tibial artery crosses between the tibia and fibula at the upper edge of the interosseous membrane.

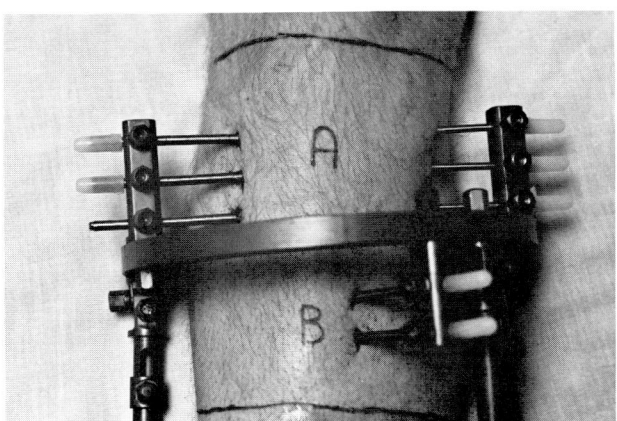

Figure 70: Biplanar Zone A placement: full pins (90° lateral —→90° medial.) Zone B placement: half-pins (30° medial).

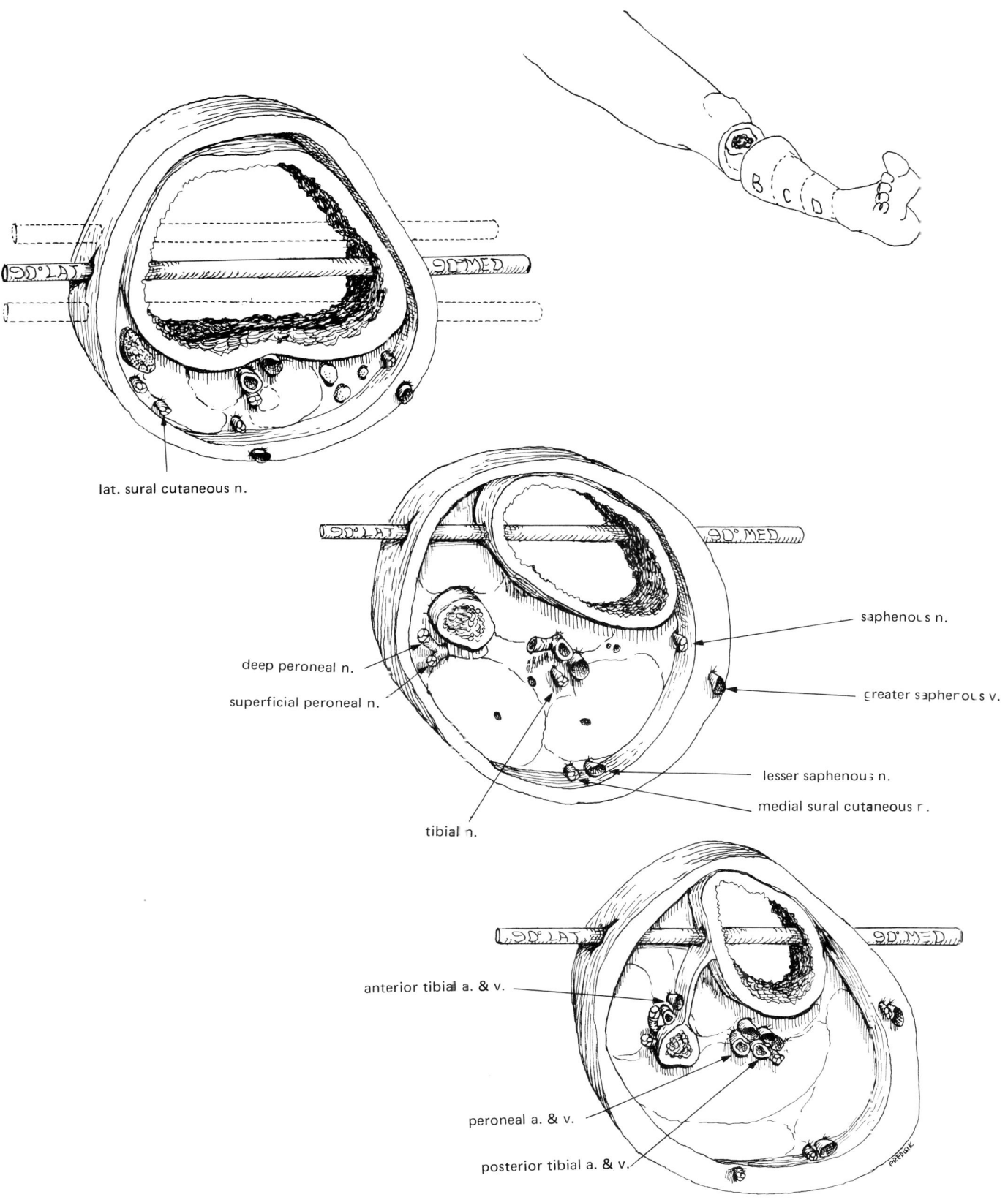

LEG
Zone A

Leg—Zone B

Anatomic Considerations

1. The tibia has a triangular cross section throughout Zone B, with the lateral surface relatively vertical, and the medial surface oblique.
2. The posterior tibial vessels, the tibial nerve, and the peroneal vessels maintain a constant relationship throughout Zone B with respect to the posterior surface of the tibia and the medial surface of the fibula.
3. The anterior tibial artery and vein, and the deep peroneal nerve, lie on the anterior surface of the interosseous membrane in Zone B, traversing from the anterior ridge of the fibula towards the lateral ridge of the tibia.

Pin Placement

1. Full-pins (or half-pins) can be inserted from 90° lateral or 90° medial.
2. Half-pins can be inserted with caution from the 30° medial (or 45° medial) position perpendicular to the oblique medial surface of the tibia. The tip of the pin will penetrate the tibialis posterior muscle. Bear in mind the relationship of the peroneal artery and vein, adjacent to the medial corner of the fibula.

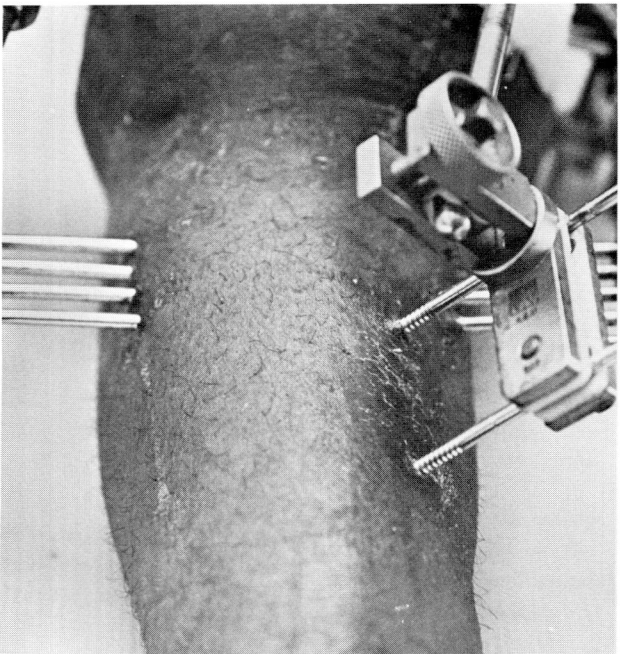

Figure 71: Biplanar half-pin placement for knee arthrodesis: Zone B (30° medial).

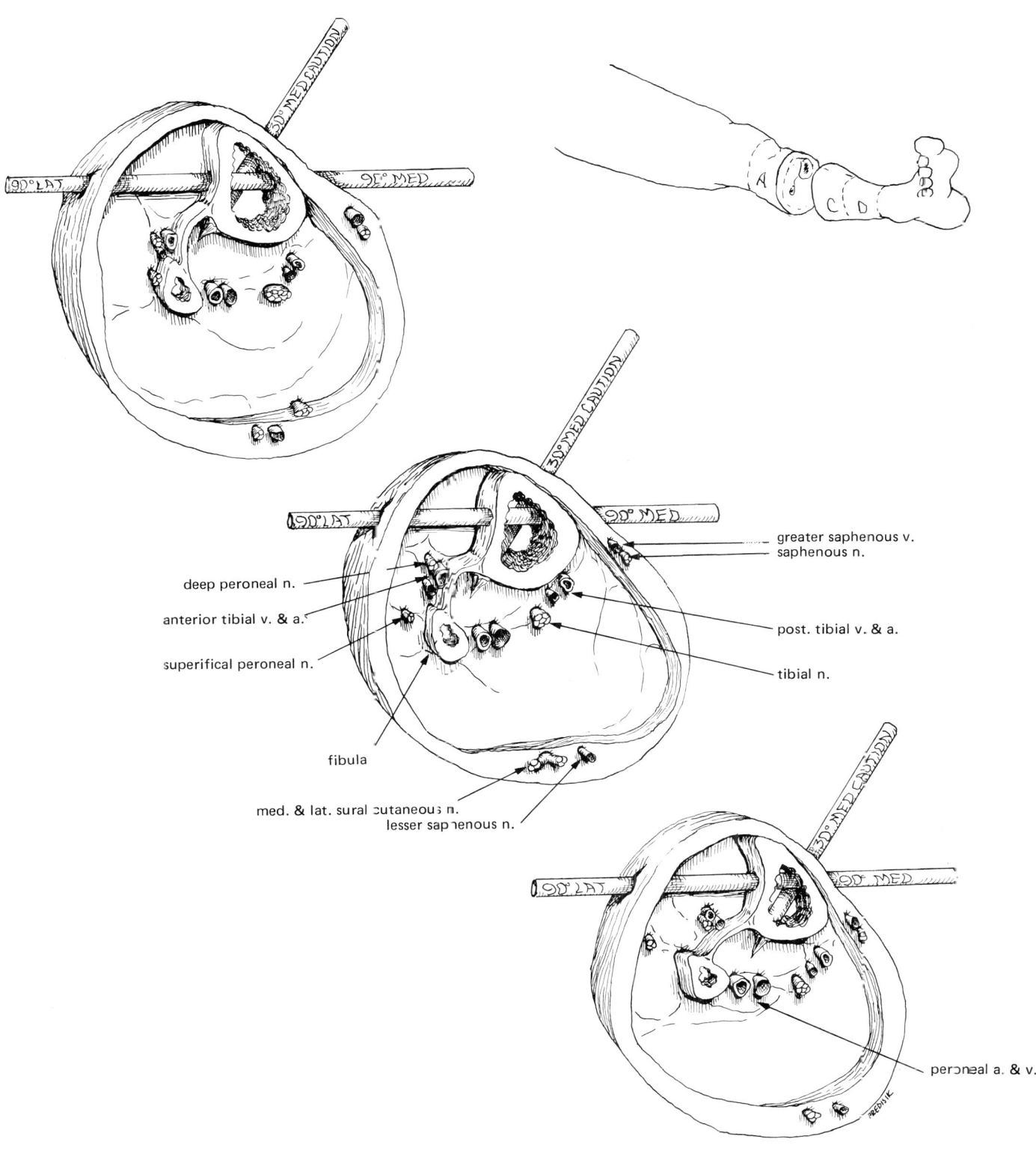

**LEG
Zone B**

Leg—Zone C

Anatomic Considerations

1. The tibia retains its distinctive triangular cross section.
2. The posterior tibial artery and vein and the tibial nerve remain posterior to the tibia and the peroneal vessels remain slightly medial to the fibula.
3. The anterior tibial artery and vein and the deep peroneal nerve have completed their traverse of interosseous membrane and are adjacent to the posterolateral corner of the tibia throughout Zone C. These structures begin to traverse the lateral surface of the tibia in distal Zone C (Fig. 60).
4. The saphenous nerve and greater saphenous vein are located at the posteromedial corner of the tibia in the subcutaneous tissue.

Pin Placement

1. In the upper part of Zone C, full or half-pins can be safely placed from the 90° medial or 90° lateral direction.
2. Half-pins into the oblique medial surface of the tibia are difficult to place in Zone C, because of the intimate relationship of the anterior tibial vessels to the bone. A 0° half-pin would be safe in distal Zone C, but it is technically difficult to place because of the obliquity and thickness of the bone.
3. In distal Zone C, pin placement from the 90° lateral or 90° medial position can endanger the anterior tibial artery and deep peroneal nerve.

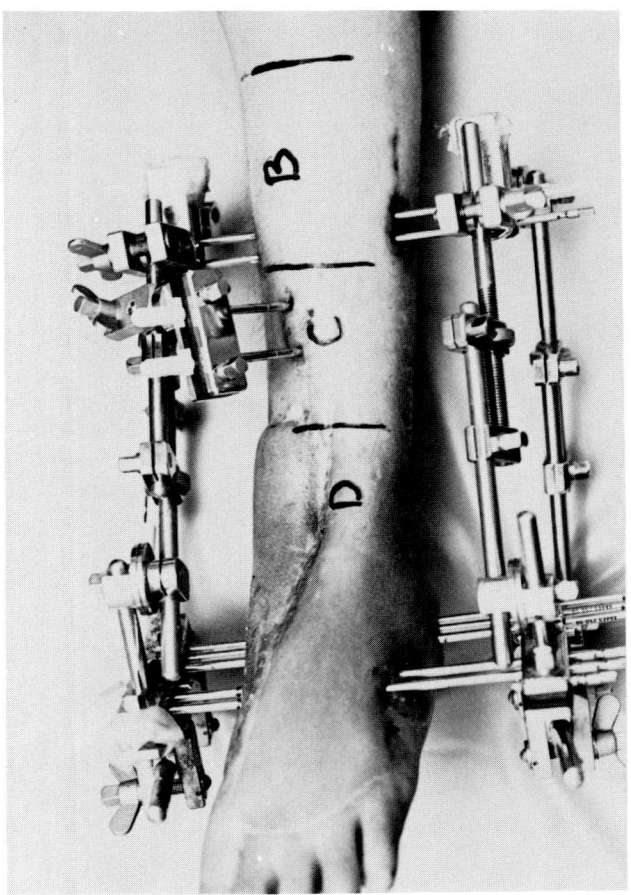

Figure 72: Pin placement for bone graft arthrodesis of the ankle: Zone C (15° anterior—not illustrated in diagrams).

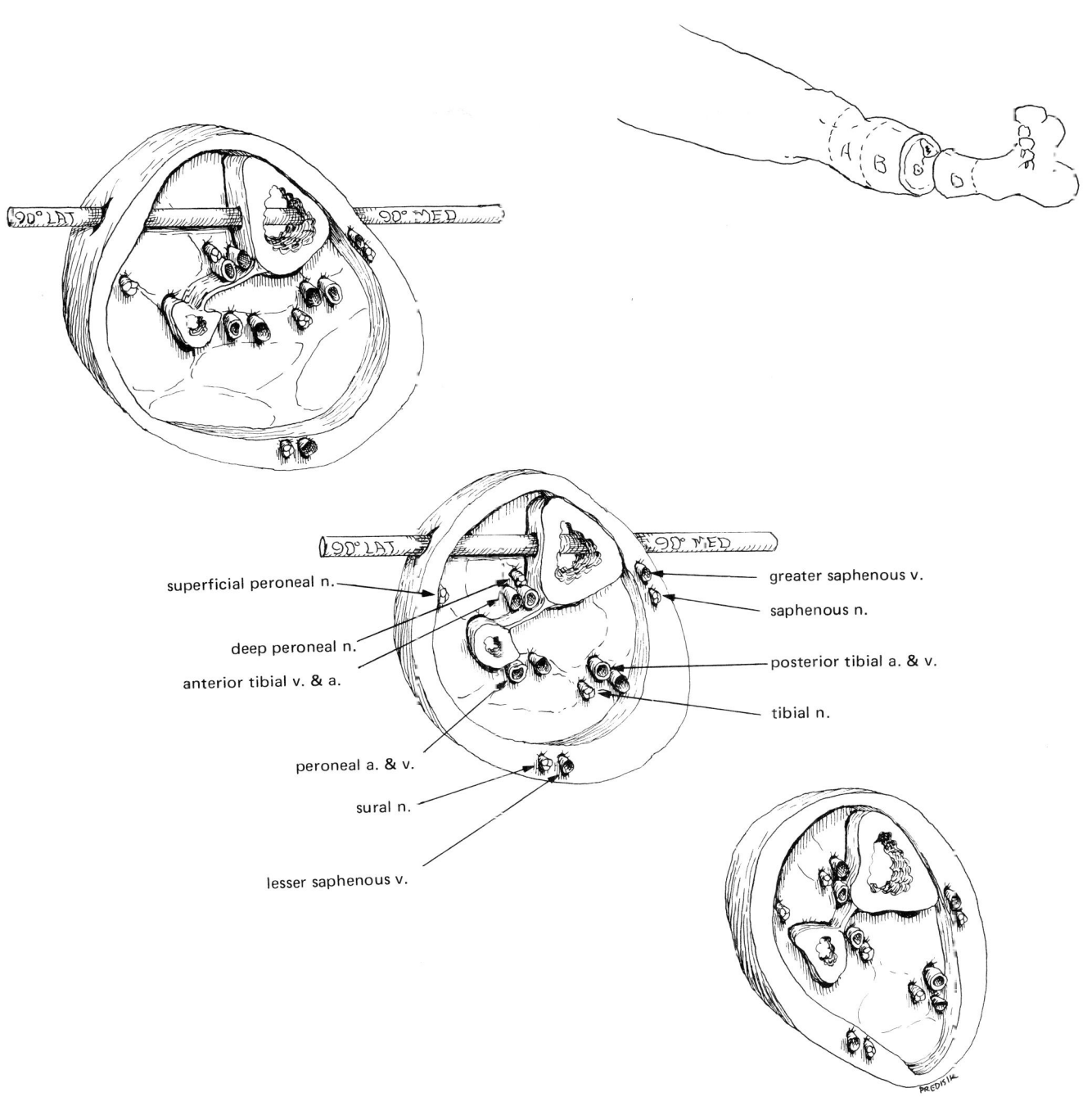

LEG
Zone C

Leg—Zone D

Anatomic Considerations

1. The posterior tibial artery and vein and the tibial nerve remain posterior to the tibia, traversing medially as they approach the ankle joint.
2. The anterior tibial artery and vein, and the deep peroneal nerve, are on the lateral surface of the tibia in proximal Zone D. They lie on the anterior surface of the tibia in distal Zone D.
3. The saphenous nerve and greater saphenous vein are on the medial side of the tibia throughout Zone D.
4. The superficial peroneal nerve has divided into its terminal branches in this zone.

Pin Placement

1. Half-pins can be placed from the 30° medial position into the subcutaneous portion of the tibia.
2. Full-pin placement from the 90° medial and 90° lateral directions can be accomplished in the distal two-thirds of Zone D.
3. Full or half pin placement from 90° medial or 90° lateral can endanger the anterior tibial artery and deep peroneal nerve in the proximal one-third of of Zone D.

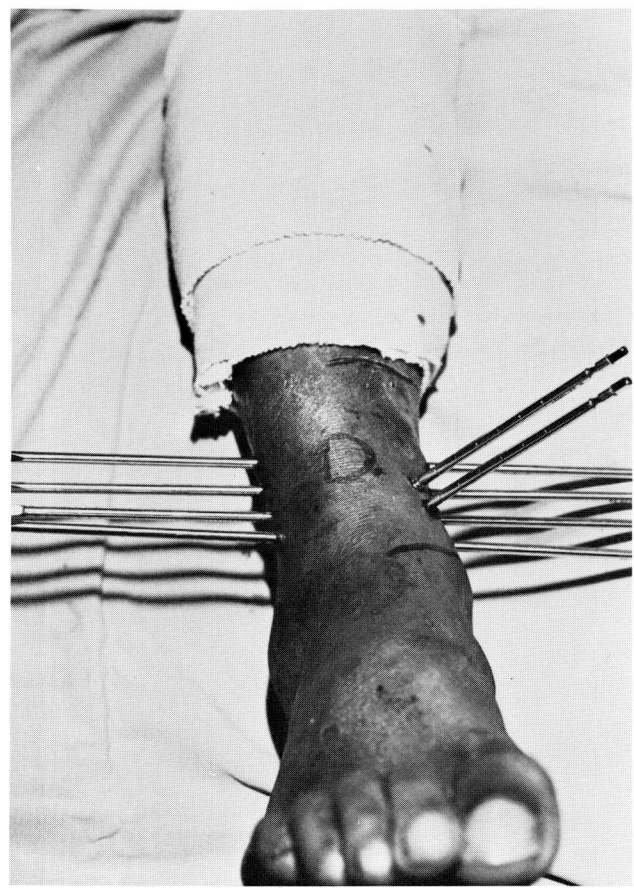

Figure 73: Bi-planar fixation, Zone D: half pins (30° medial); full pins (90° medial⟶90° lateral). Note that no pins are placed in the proximal third of Zone D.

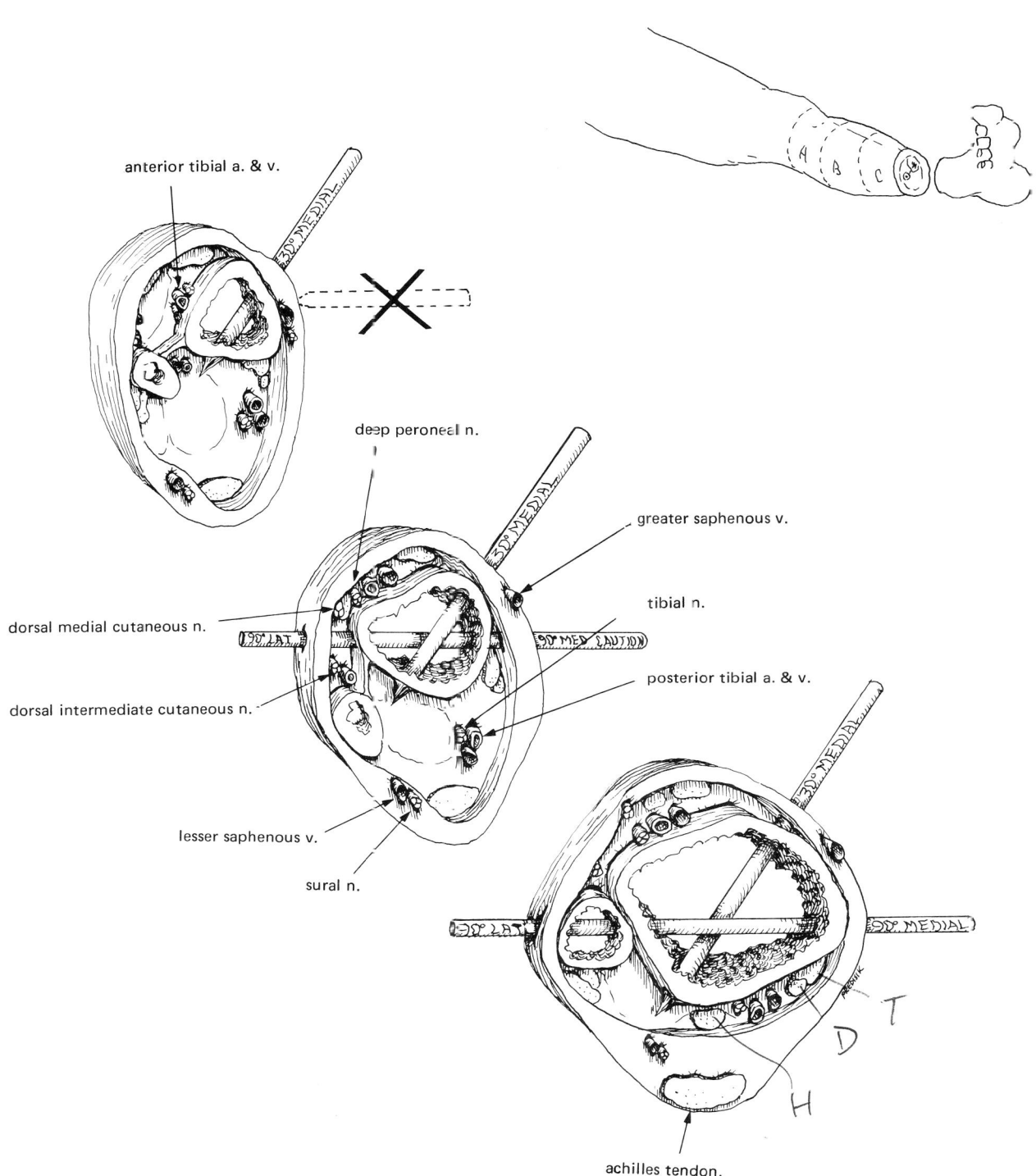

LEG
Zone D

Arm—Zone A

Anatomic Considerations

1. The humeral head is largely intrasynovial, being surrounded by a joint cavity medially and posteriorly and by the subacromial bursa anteriorly.
2. The main neurovascular bundle containing the brachial plexus is medial to the humerus, separated from it by a distance equal to the width of the bone.
3. The anterior and posterior humeral circumflex vessels surround the upper humerus slightly below the surgical neck, accompanied by the axillary nerve.

Pin Placement

1. Half-pins may be cautiously placed in the 90° lateral position.
2. Half-pins can be inserted into the humeral head from 0° anterior around laterally to the 90° lateral position, if the tip of the pin does not penetrate the opposite cortex of the humeral head.
3. Below the level of the surgical neck of the humerus, pin placement may endanger the humeral circumflex vessels and the axillary nerve.
4. Below the neck of the humerus, half-pins can be placed from the 90° lateral position (not shown).

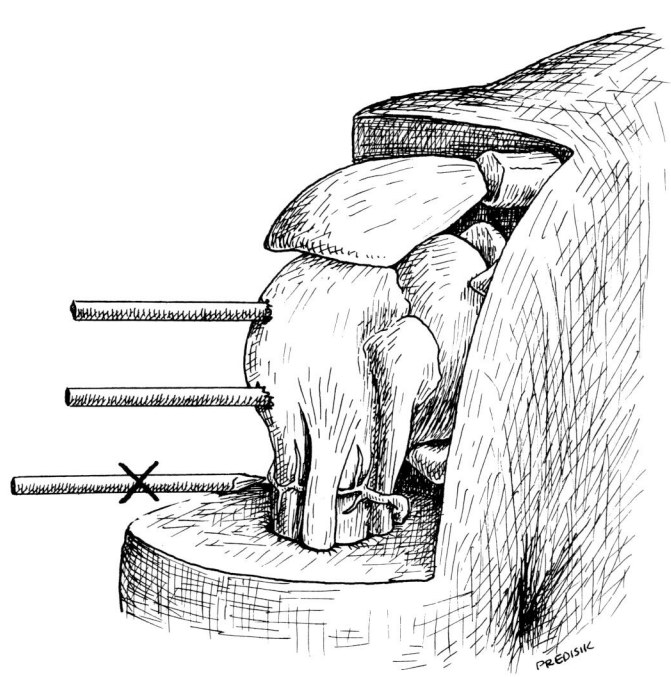

Figure 74: Zone A: Level of anterior humeral circumflex artery and axillary nerve.

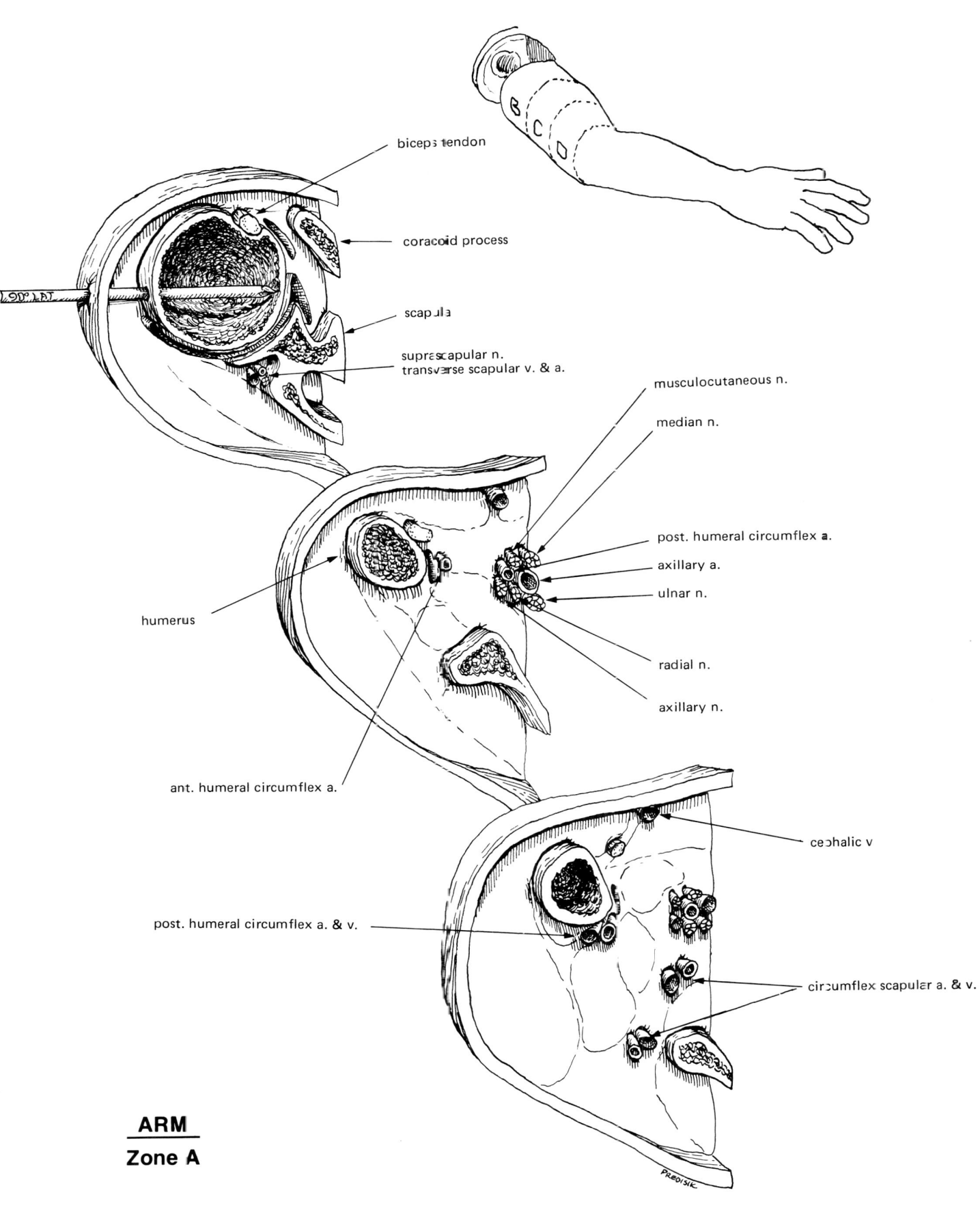

**ARM
Zone A**

Arm — Zone B

Anatomic Considerations

1. The brachial artery and veins and the brachial plexus remain medial to the humeral shaft in this zone.
2. The radial nerve separates from the main neurovascular bundle and passes to the posterior side of the humerus in Zone B, separated medially from the bone by the medial head of the triceps.
3. The musculocutaneous nerve and cephalic vein are anterior to the humerus in Zone B.

Pin Placement

1. Half-pin placement from 90° lateral can be accomplished with great caution in mid-Zone B because of the position of the radial nerve on the medial side of the humerus.

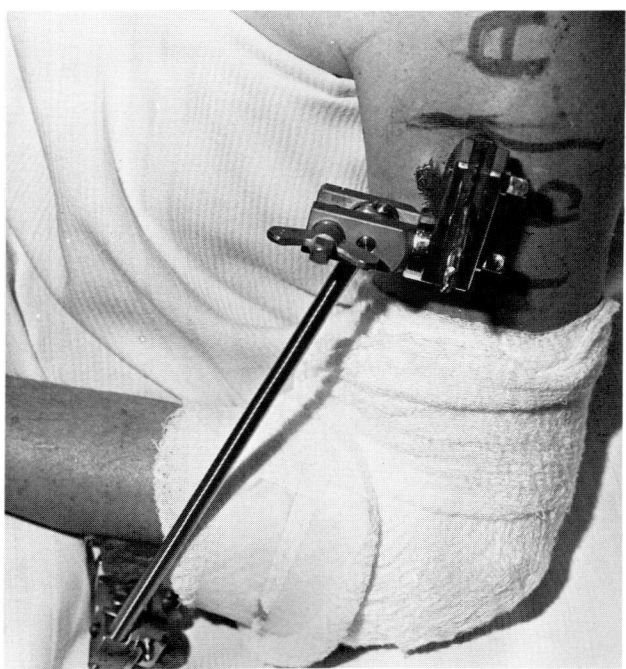

Figure 75: Zone B placement for temporary stabilization of a dislocating total elbow prosthesis. Half pins (90° lateral) inserted with fluoroscopy control because of proximity of radial nerve to medial side of humerus.

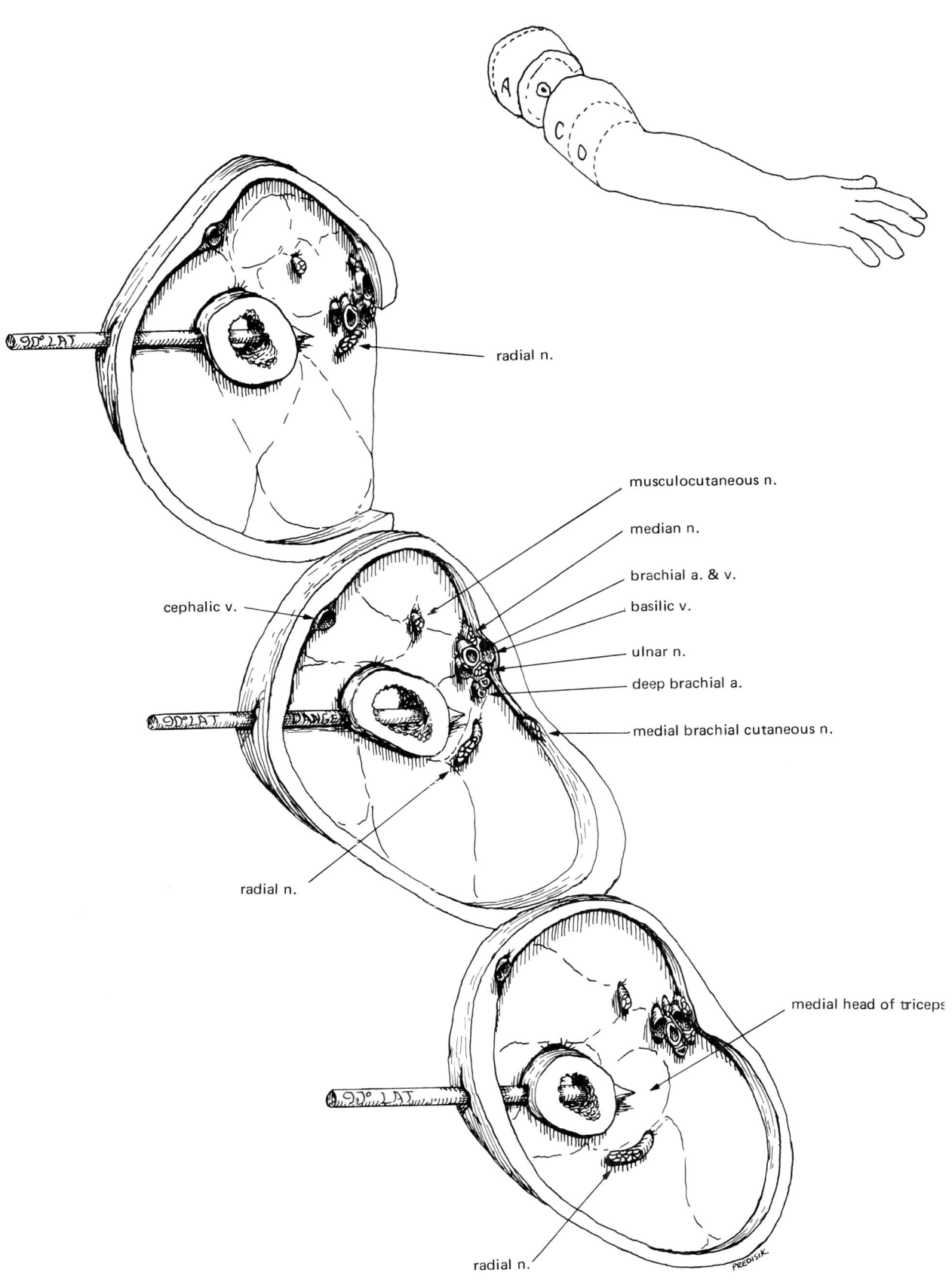

**ARM
Zone B**

Arm — Zone C

Anatomic Considerations

1. The radial nerve winds around the lateral side of the shaft of the humerus in intimate contact with the bone.
2. The brachial artery, veins, and branches of the brachial plexus remain medial to the humeral shaft. The ulnar nerve separates from the main neurovascular bundle in this zone.
3. The musculocutaneous nerve becomes the lateral cutaneous nerve of the forearm and remains anterior to the humerus.

Pin Placement

1. Half-pins should be placed in the 90° lateral position in the humerus with direct observation of the radial nerve through a surgical exposure.

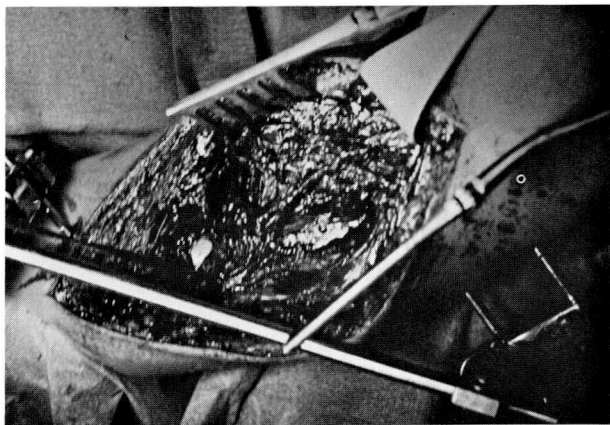

Figure 76A: Surgical exploration of lateral side of arm. Note absence of Radial Nerve—twelve centimeters were "wound-up" on a pin inserted percutaneously (without direct observation). Courtesy of Dr. Franz Burny.

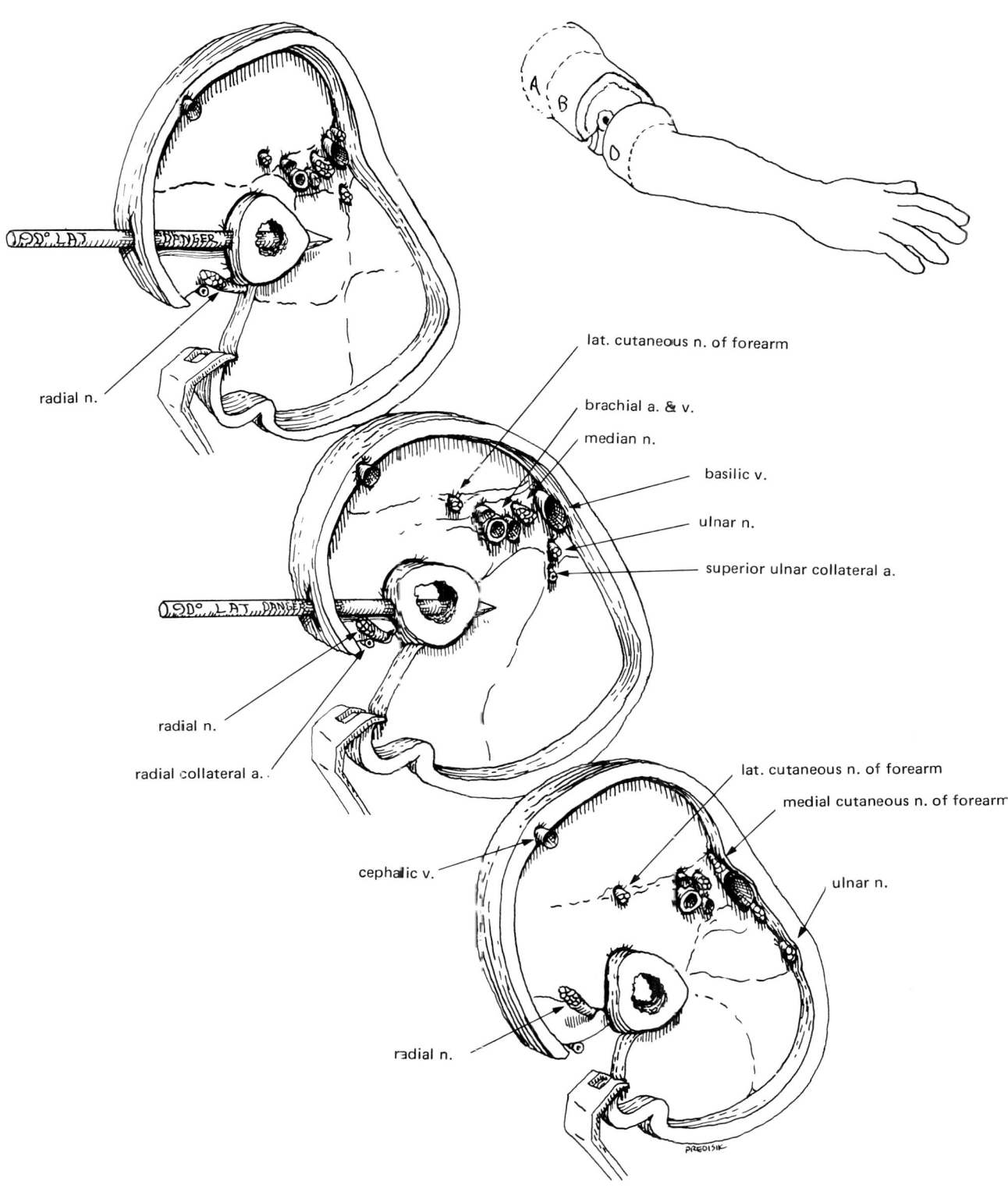

**ARM
Zone C**

Arm — Zone D

Anatomic Considerations

1. The distal humerus assumes a flattened contour but is rotated so that the lateral epicondyle is posterior to the medial epicondyle by an angle of 30°.
2. The radial nerve lies on the lateral side of the radius in proximal Zone D but lies anterior to it in the distal portion of the zone.
3. The median nerve remains anterior and medial to the bone throughout this zone.
4. The ulnar nerve passes posterior to the plane of the distal humerus and lies in contact with the posteromedial corner of the bone immediately above the elbow joint.

Pin Placement

1. Half-pins can be placed with caution from the 180° posterior position. The median nerve and brachial artery are separated from the shaft of the humerus by the thickness of the brachialis muscle in Zone D. Likewise, half-pins can be placed from the 150° medial position.
2. Half or full-pins can be placed from the lateral epicondyle into the medial epicondyle. Unfortunately, the proximity of the ulnar nerve to the medial epicondyle of the humerus makes pin placement in this position somewhat dangerous. It is recommended that the ulnar nerve be exposed for transepicondylar pin placement.

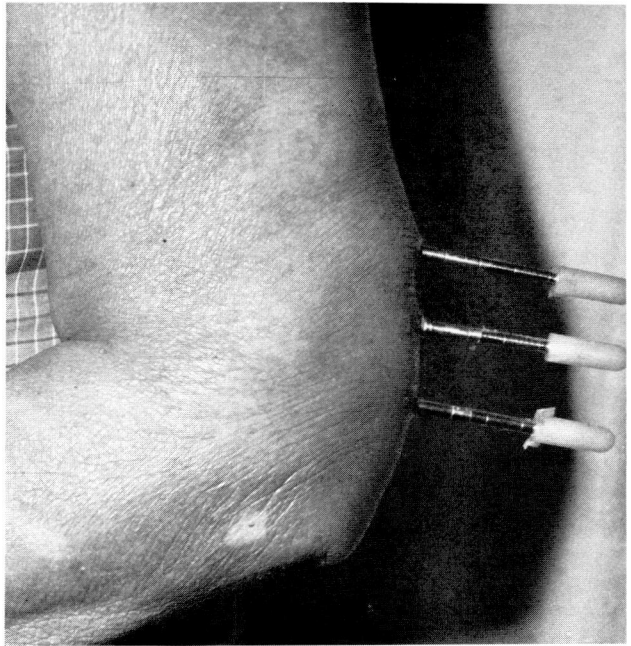

Figure 76B: Zone D placement for humeral fracture: half-pins (180° posterior).

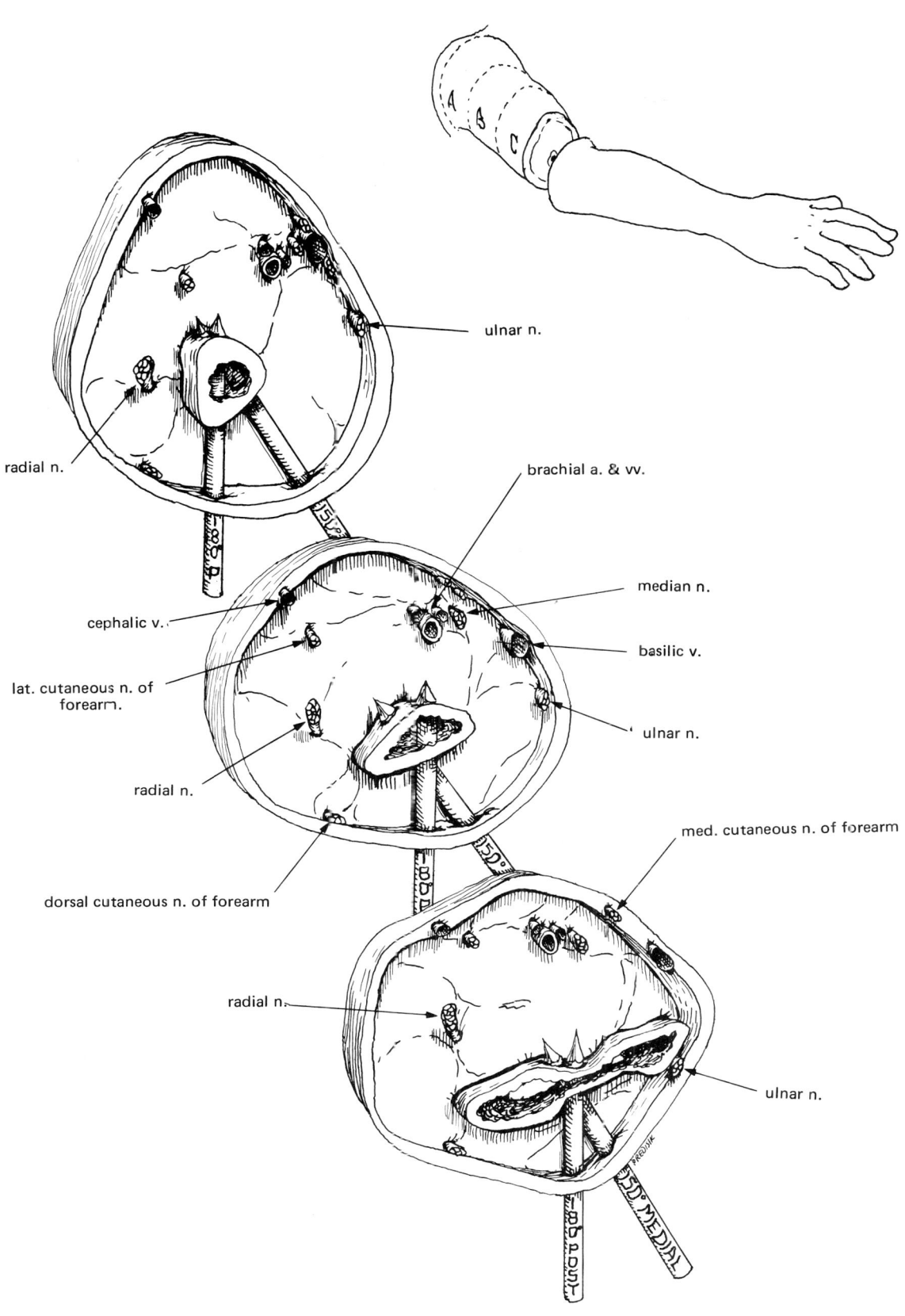

ARM
Zone D

Forearm — Zone A

Anatomic Considerations

1. The deep branch of the radial nerve winds around the lateral side of the humerus within the substance of the supinator muscle.
2. The brachial artery divides into its terminal branches in Zone A, the common interosseous artery and the ulnar artery, with associated veins, are anterior to the proximal ulna in distal Zone A.

Pin Placement

1. Half-pins can be inserted into the proximal ulna from the 150° medial direction. Image intensification fluoroscopy is recommended.
2. Pin placement into the proximal radius is dangerous because of the location of the deep branch of the radial nerve. If it is necessary to stabilize the proximal radius with external fixation, it is wise to identify this structure surgically before pin insertion.
3. In distal Zone A, pins may be placed into the ulna from the 150° lateral position (not shown).

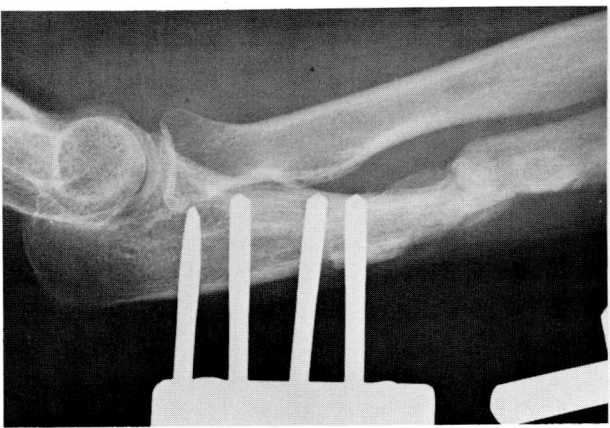

Figure 76C: Pin placement into ulna, Zone A. (150° medial)

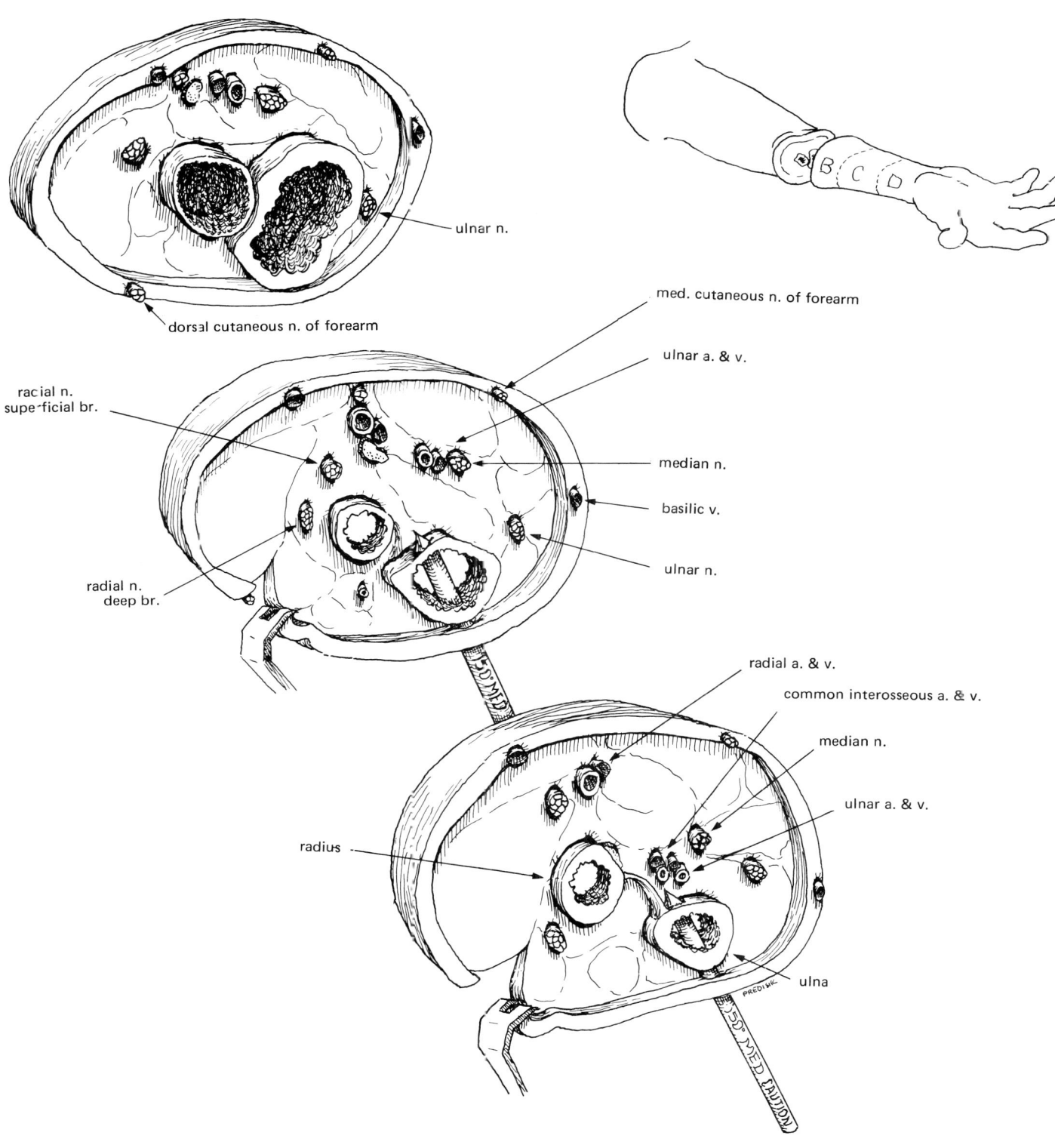

FOREARM
Zone A

Forearm — Zone B

Anatomic Considerations

1. The radial, ulnar, and median nerves remain in relatively constant position throughout Zone B.
2. The anterior interosseous artery and nerve lie on the anterior surface of the interosseous membrane.
3. The deep branch of the radial nerve lies adjacent to the posterior interosseous artery, posterior to the interosseous membrane and separated from it by muscle.

Pin Placement

1. Half-pins can be inserted into the ulna from the 150° medial position. Depth can be assessed with fluoroscopy.
2. Half-pins can be inserted (employing considerable caution) into the radius via the 60° lateral position. As with half-pin insertion into the ulna, fluoroscopy control is recommended.

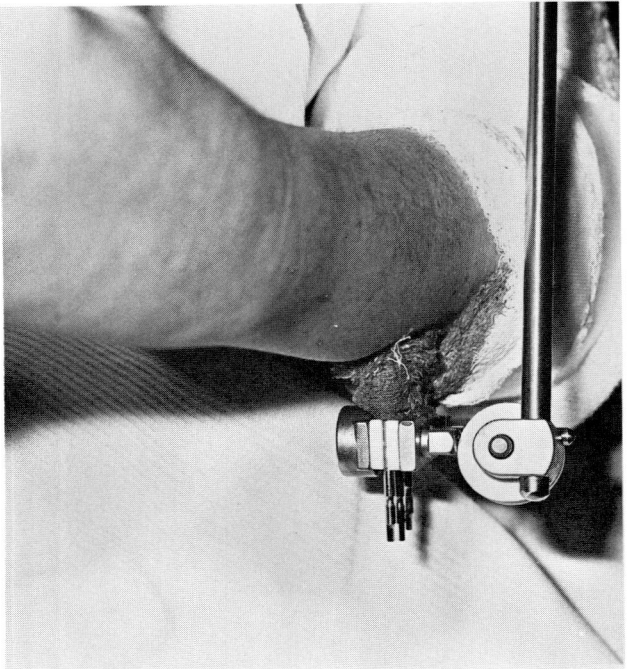

Figure 77: Pin placement into ulna, Zone C: half-pins (150° medial). The pins were inserted with the forearm in the anatomic position (supinated).

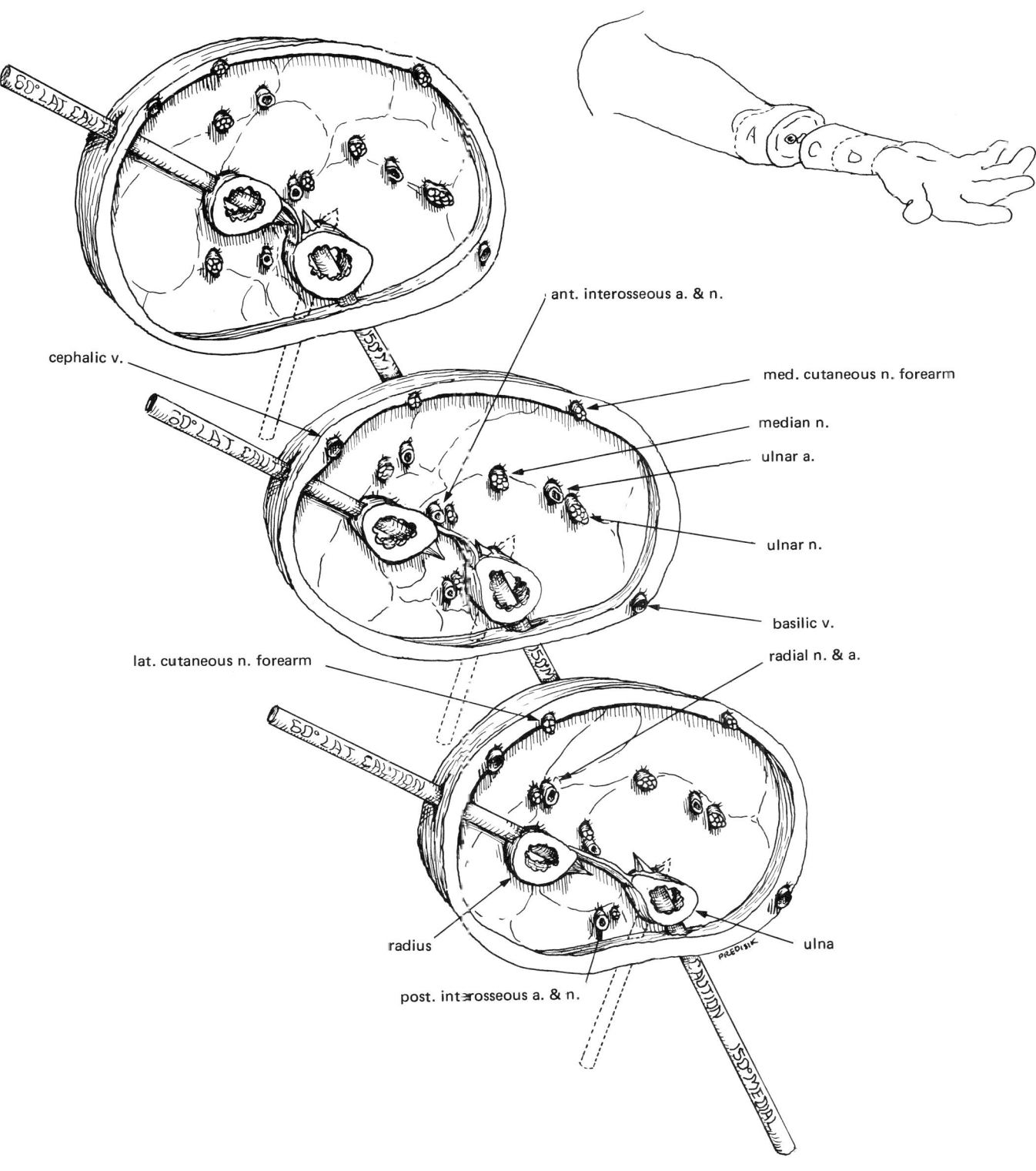

FOREARM

Zone B

Forearm — Zone C

Anatomic Considerations

1. The superficial branch of the radial nerve and radial artery are anterior to the radius in Zone C, becoming more lateral and superficial in the distal part of this zone.
2. The median nerve maintains its position in the middle of the forearm, surrounded by muscle.
3. The ulnar nerve and ulnar artery remain anterior and medial to the ulna throughout Zone C.

Pin Placement

1. Half-pins may be placed into the ulna with caution from the 150° medial direction. In fact, half-pins may be inserted into the ulna from the 180° posterior position and the 150° lateral position as well, being mindful of the position of the extensor tendon as illustrated in distal Zone C.
2. Half-pins may be placed into the radius from the 150° lateral position. Pins may also be placed into the radius from the 180° posterior position if care is taken to avoid impalement of extensor tendons.

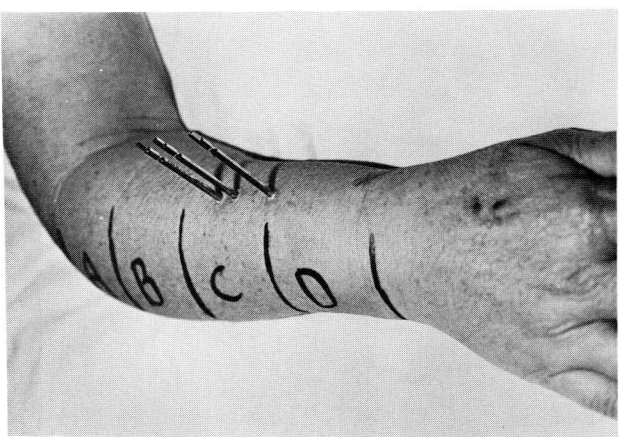

Figure 78A: Pin placement into radius for unstable Colles fracture fixation, Zone C: half-pins (150° medial). The forearm was supinated for pin placement.

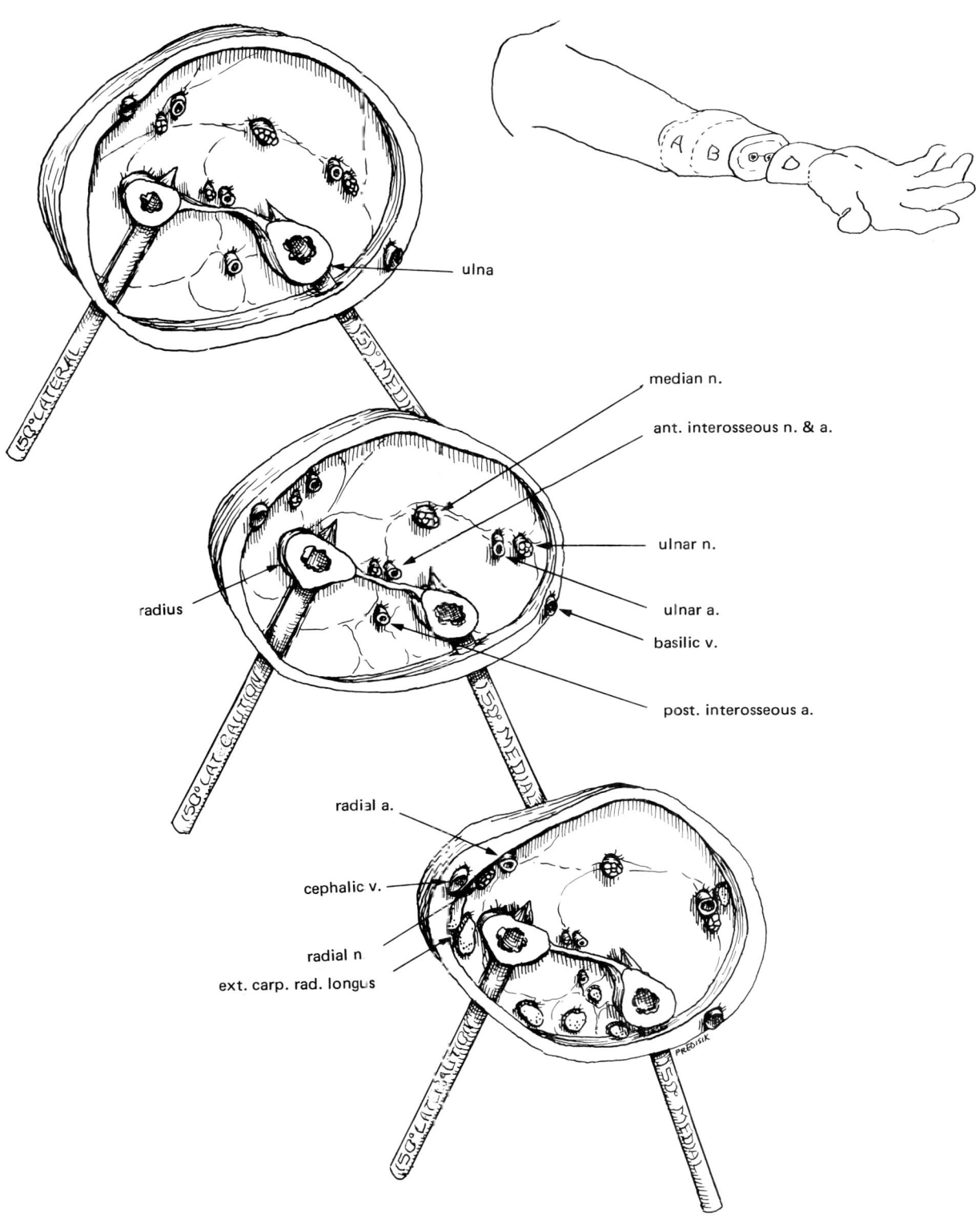

FOREARM
Zone C

70 Complications of External Skeletal Fixation

Forearm — Zone D

Anatomic Considerations

1. The radius and ulna are posteriorly located in the cross section of the forearm.
2. The radial nerve is lateral to the shaft of the radius, dividing into dorsal and volar branches in Zone D.
3. The median nerve remains within the volar muscle mass.
4. The ulnar nerve divides into dorsal and volar branches, the dorsal branch passing to the posterior aspect of the distal forearm.
5. The extensor and flexor muscles become tendinous in Zone D.

Pin Placement

1. Half-pins may be inserted with caution from the 150° medial direction into the ulna.
2. Half-pins may be placed into the distal radius from the 150° lateral direction. Note the relative position of the extensor tendons so they are not impaled by a pin.

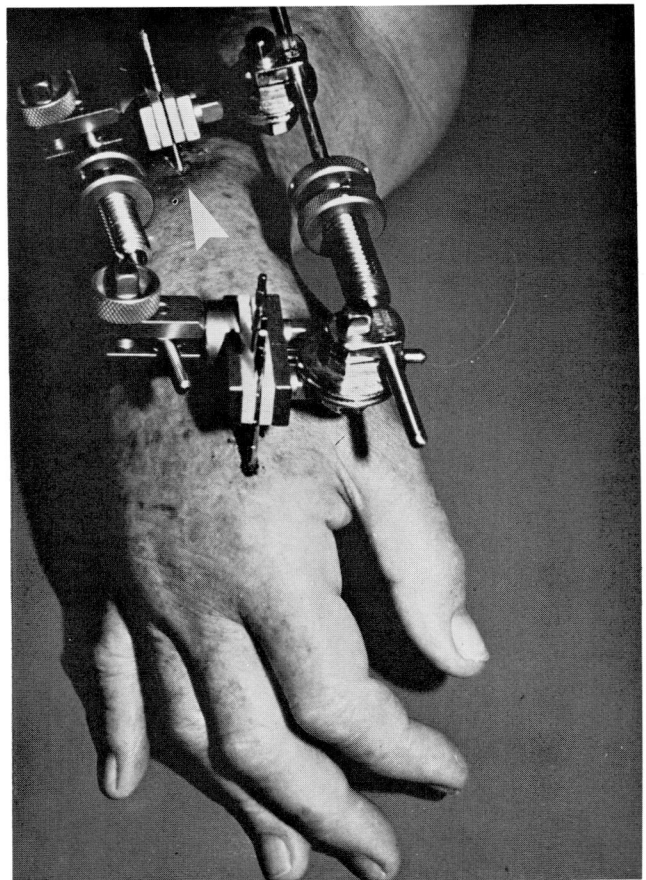

Figure 78B: Pin placement into radius, Zone D (white arrow): half-pins (150° lateral) were inserted with forearm in the anatomic position.

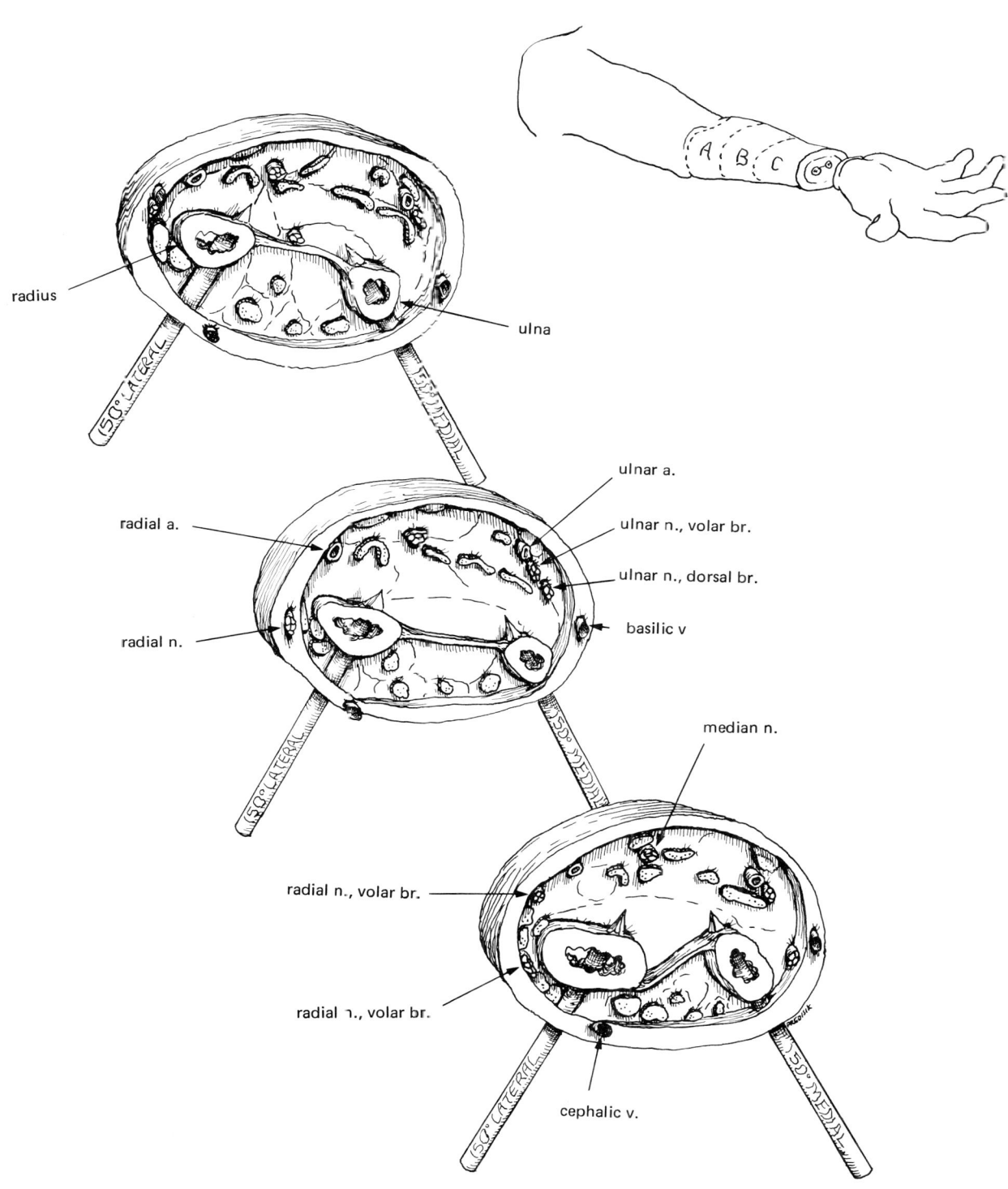

FOREARM
Zone D

Hand

Anatomic Considerations

1. Cross section through the metacarpal shafts demonstrates the close relationship of the radialis indicis artery to the volar surface of the second metacarpal.
2. The palmar metacarpal artery to the second web space is adjacent to the radial volar surface of the third metacarpal shaft.
3. The ulnar artery and deep branch of the ulnar nerve are volar to the fourth metacarpal shaft, separated from it by muscle, a distance equal to the width of the bone.

Pin Placement

1. Full or half-pin placement from the 90° lateral position can be safely passed through the shafts of the second, third, and fourth metacarpals. Extensor tendon impalement may occur as the pin passes through the skin on the medial side of the dorsum of the hand. The oblique lateral surface of the second metacarpal makes pin insertion difficult because the tip of the pin tends to slide on the bone.
2. Half-pin insertion into the second metacarpal from the 150° lateral position can be safely accomplished if done carefully.
3. Half-pin placement into the fifth metacarpal shaft from the 120° medial position can be done with caution, although the curved surface of the bone makes pin insertion difficult.

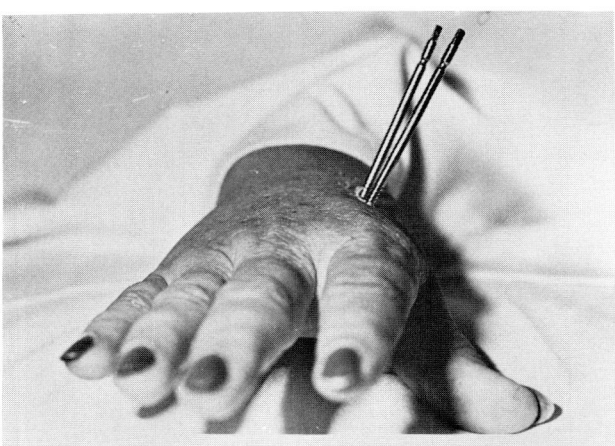

Figure 79: Half-pin placement into the hand.

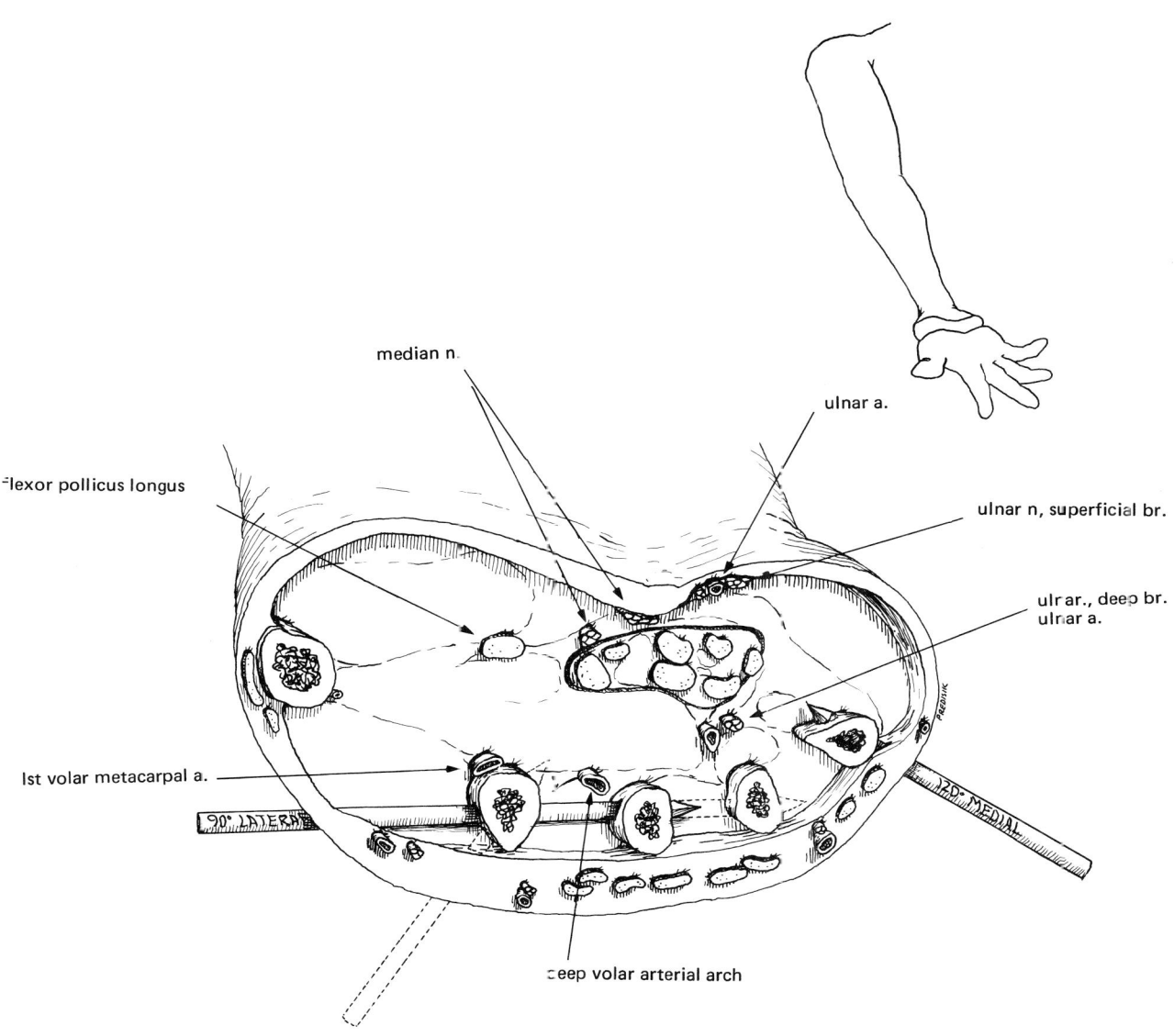

HAND

Foot

Anatomic Considerations

1. Cross section through the metatarsals demonstrates the curvature of the transverse metatarsal arch.
2. The dorsalis pedis artery is between the first and second metatarsal shafts.
3. The plantar arterial arch is crossing beneath the third metatarsal shaft at the level illustrated.
4. The flexor hallucis brevis tendon is adjacent to the lateral inferior surface of the first metatarsal shaft.

Pin Placement

1. A full or half-pin can be inserted from the 90° medial position into the first metatarsal shaft. It will penetrate one or perhaps two other metatarsal shafts but cannot transfix all of them.
2. A 45° medial half-pin can be inserted into the first metatarsal.
3. Other half-pin positions can be safely used into the metatarsal bones, including a 90° lateral pin into the fifth metatarsal shaft (not shown).

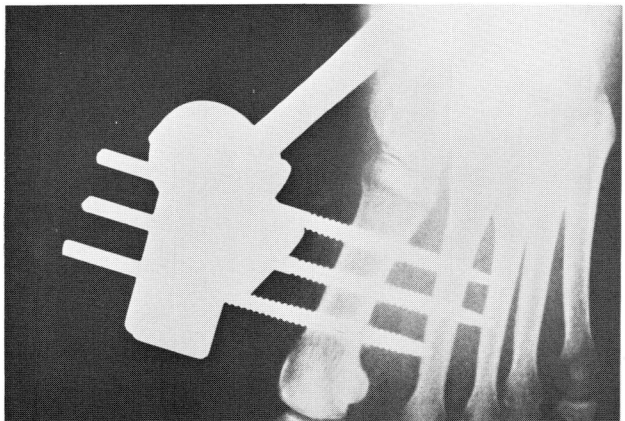

Figure 80: Half-pin placement to stabilize an open talar neck fracture in polytrauma victim: 90° medial.

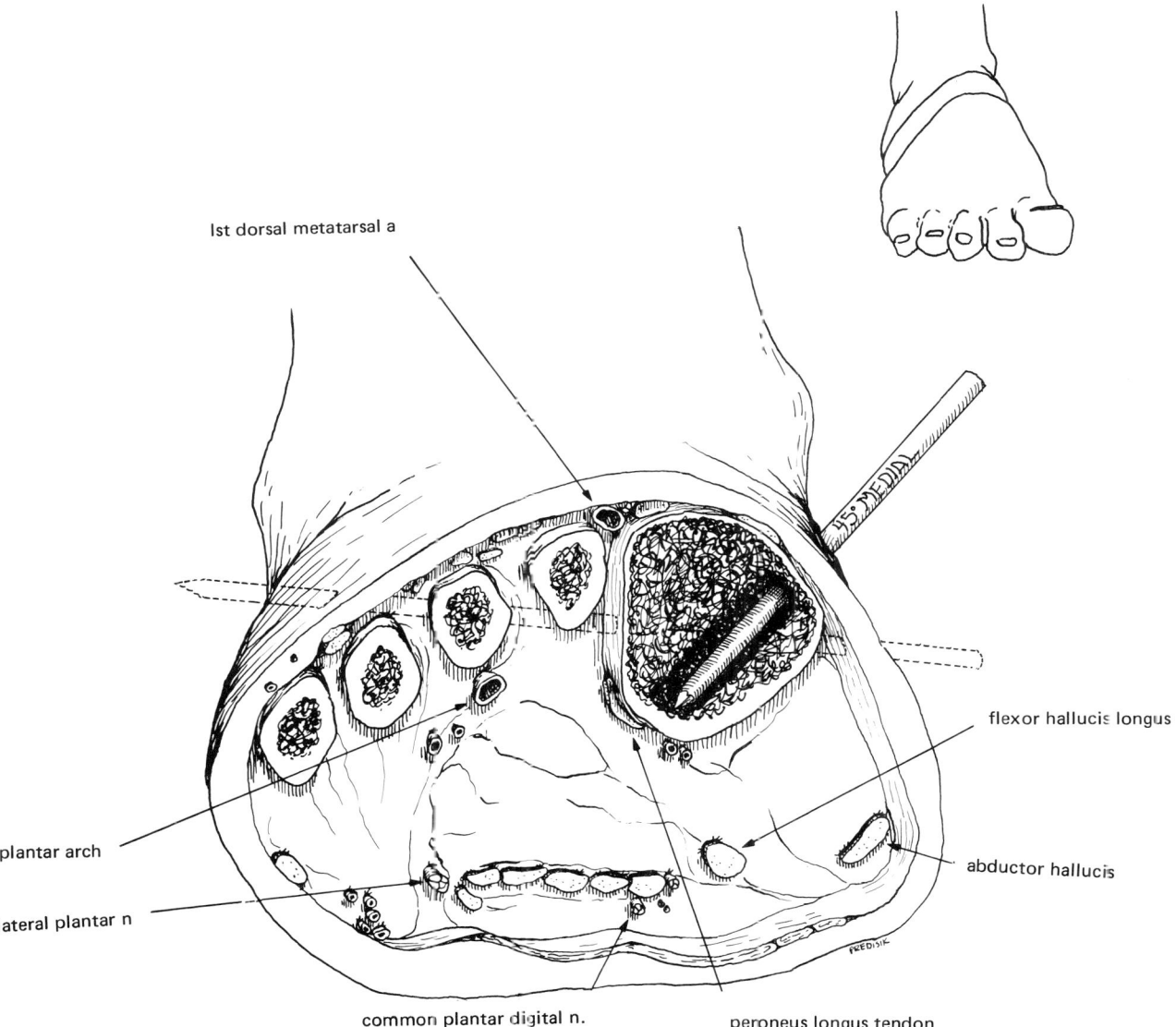

FOOT

Pelvis

Anatomic Considerations

1. The inner table of the ilium is separated from the abdominal contents by the iliacus muscle.
2. The iliac wings are concaved medially.

Pin Placement

1. Half-pins can be inserted along the iliac crest aiming at either the sciatic notch or sacroiliac joint from the 20° lateral position.
2. Full pin placement from the anterior inferior iliac spine to the posterior inferior iliac spine (Mears) requires a special alignment guide.
3. It is important to reduce iliac displacement before pin insertion when treating pelvic fractures.
4. Pin placement will be safer if the tip of the pin penetrates the outer table of the ilium rather than the inner table.

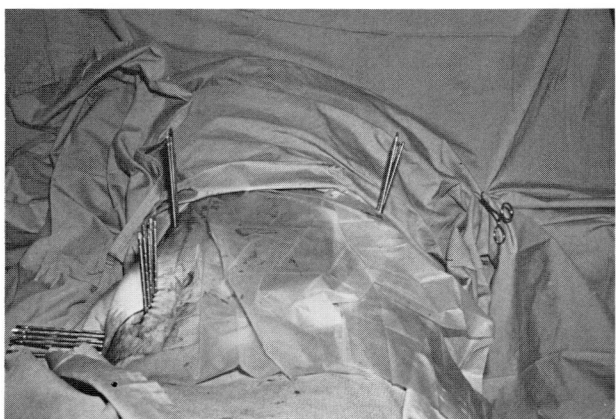

Figure 81: Pin placement into pelvis and femur for hip joint stabilization. The right iliac crest pin group is too vertical, but acceptable because pins will penetrate center table.

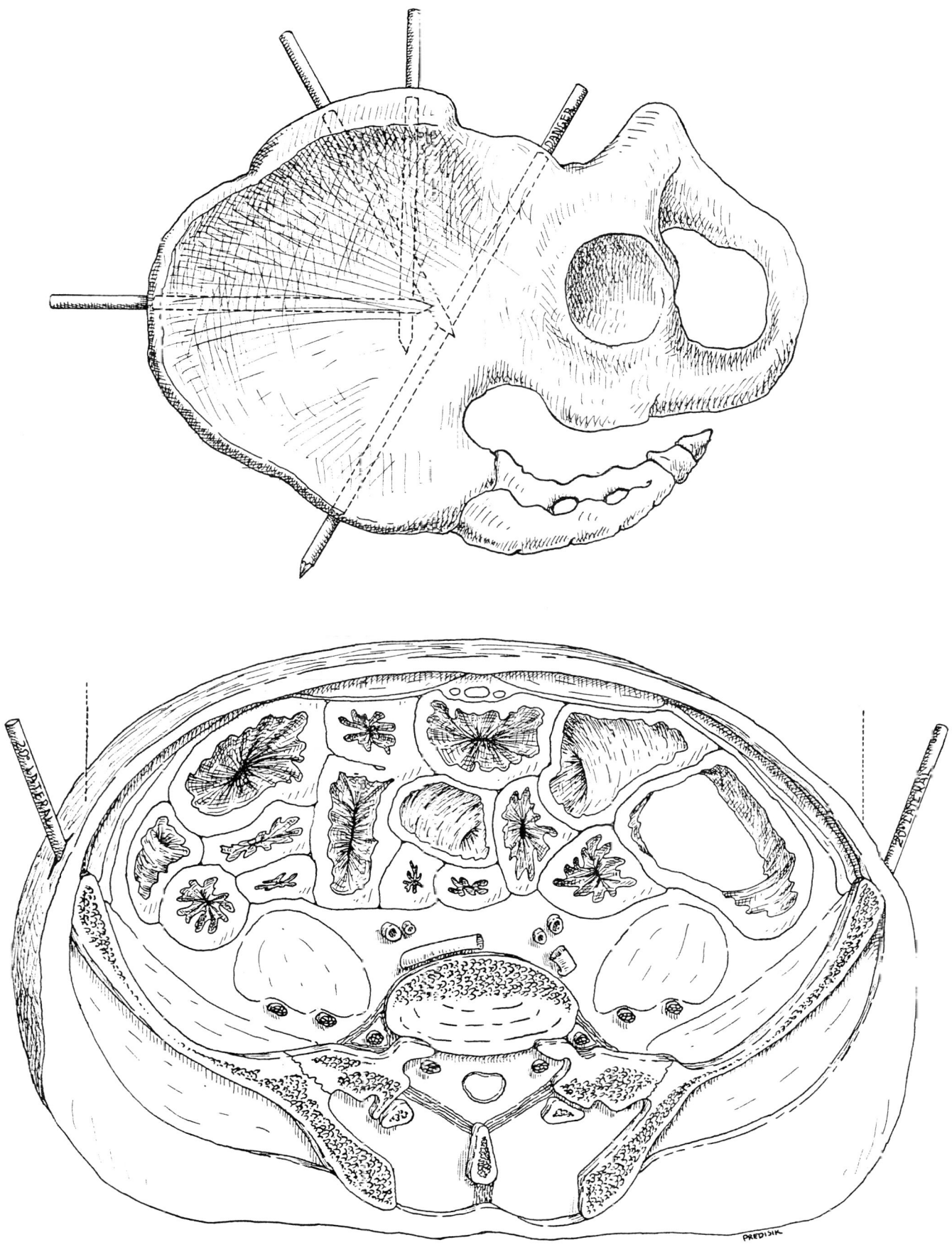

PELVIS

Chapter 4

INTERFERENCE WITH LIMB FUNCTION

Introduction

External fixation is appealing since it provides immediate stability of a fracture while leaving the adjacent joints free to move. Nevertheless, external fixation can, by impaling muscles and tendons, restrict joint mobility. Pin transfixion of synovial membranes can also reduce joint function, as can external fixation spanning a joint. Shortening or angulation of bone, while uncommon, can occur if an unstable fracture is compressed while in an external fixator. Pain, whether from pin sepsis or another cause, will reduce the effectiveness of external fixation in achieving early mobilization of the patient, as will excessive swelling of the limb.

MUSCLE AND TENDON IMPALEMENT

Discussion

Muscle impalement is, to some extent, an unavoidable disadvantage of the use of percutaneous pins. In most situations, it is tolerable, provided measures are taken to insure a satisfactory range of motion for adjacent joints. In some cases, the problem can be avoided altogether by inserting pins into areas where bone is not covered by overlying muscles.

Thigh

A majority of the femoral shaft is completely clothed by muscle, making impalement difficult to avoid. The sartorius and rectus femoris are likely to be transfixed by pins inserted in the anteroposterior direction. The more distal the transfixion of the rectus femoris, the greater will be the restriction of knee joint motion. Full pins tranfixing the vastus lateralis and vastus medialis are more likely to restrict knee motion than are half-pins inserted into the vastus lateralis alone. Permanent restriction of knee motion after fixator removal may be caused by quadriceps adhesion to the fracture site, rather than by arthrofibrosis. In this situation, a quadricepsplasty is worthwhile.[220]

A standard practice is to apply the fixator to the thigh with the knee extended, while the patient is in the supine position, producing impalement of the quadriceps mechanism, which leads to restriction of knee flexion. We tried to avoid this situation in one patient by applying a femoral fixator with the knee flexed. Unfortunately, the patient was unable to fully extend his knee when he awoke from anesthesia. Presently, I apply the fixator with the knee extended and then flex the knee as much as possible to enlarge the pin holes through the muscle.

Leg

In the lower leg, full pins passing though the anterior compartment musculature will interfere with active dorsiflexion of the ankle. The problem can be avoided by making certain that the foot is dorsiflexed at the time the pins are inserted. The position can be maintained in the operating room by an assistant, while the surgeon inserts the pins. Furthermore, the foot must be maintained in the dorsiflexed position for the first few postoperative days until the patient can actively exercise his ankle without undue pain. A padded foot support attached to the fixator should be used. When postoperative pain diminishes, the foot support can be changed from a static cradle to a dynamic splint, by utilizing elastic bands to connect it to the fixator. The support should be in position while the patient is sleeping, because the foot has a tendency to fall to the plantar flexed position, especially if the limb is suspended from an overhead frame.

The problem of impalement of anterior compartment muscles can be avoided by utilizing half-pins inserted into the medial subcutaneous surface of the bone. Such a unilateral mounting may be sufficient for most applications, but it is not rigid enough to deal

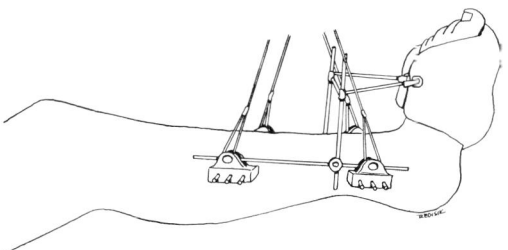

Figure 82: Soft padded foot support attached to outriggers on fixators. A hard foot support should not be used in unconscious or hypesthetic patients.

with those problems requiring considerable stability, such as fracture comminution, the presence of a contaminated wound, or septic pseudarthrosis requiring bone grafting.

Reports of tendon transfixion in the lower limb are extremely unusual. Tendons are generally pushed to the side during pin insertion. Krempen[164] describes one case of transfixion of the extensor hallucis longus through the musculotendinous region.

Humerus

Muscle impalement can be avoided when inserting pins into the distal humerus by placing them into the lateral epicondyle and lateral supracondylar region. Unfortunately, there is danger of injury to the ulnar nerve, which is adjacent to the medial epicondyle. Posterior insertion through the triceps tendon is safer. To prevent significant restriction of elbow motion while the fixator is in place, flex the elbow to 90° before pin insertion, followed by manipulation of the joint through a full range of motion. Passive joint extension will occur with the aid of gravity. A sling should be worn to prevent excessive pin-skin motion. Reporting results with 100 humeral applications of external fixation, Burny[41] noted a 9 percent loss of terminal flexion and extension when the patients were evaluated at follow-up. He routinely inserted distal humeral pins through the triceps tendon.

Forearm

Pin insertion into the distal forearm may impale

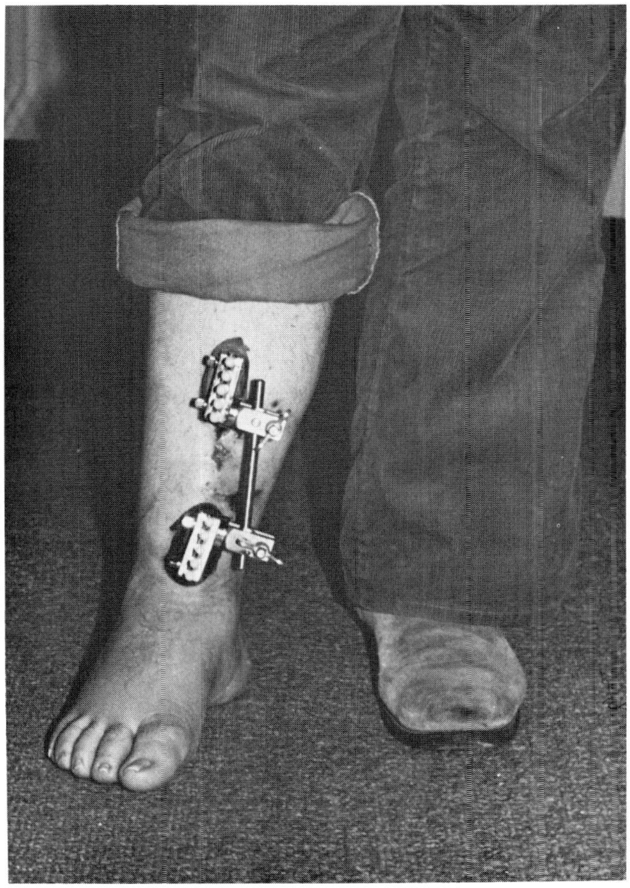

Figure 83: Unilateral half-pin fixator inserted into subcutaneous portion of tibia.

tendons. Vidal,[267] Mears,[191] and others[63] recommend opening the insertion site down to bone prior to pin insertion. Cooney,[67] reporting wrist and forearm applications, has not found this necessary, nor have we. Careful penetration of the soft tissues with the tip of the pin, followed by slight wiggling of the point against the bone, will usually push the tendons to the side.

IMPAIRED JOINT MOBILITY

There are many causes of reduced joint mobility following limb injury, several of which may interact with each other. Some—including injury to muscle and tendon, ischemic contracture, and incongruity of joint surfaces—can occur with any method of treatment. Restricted joint mobility not caused by muscle impalement is likely to be a consequence of the initial trauma. Nicoll,[204] studying the end results of tibial fractures treated conservatively, concluded that residual ankle stiffness was associated with more severe

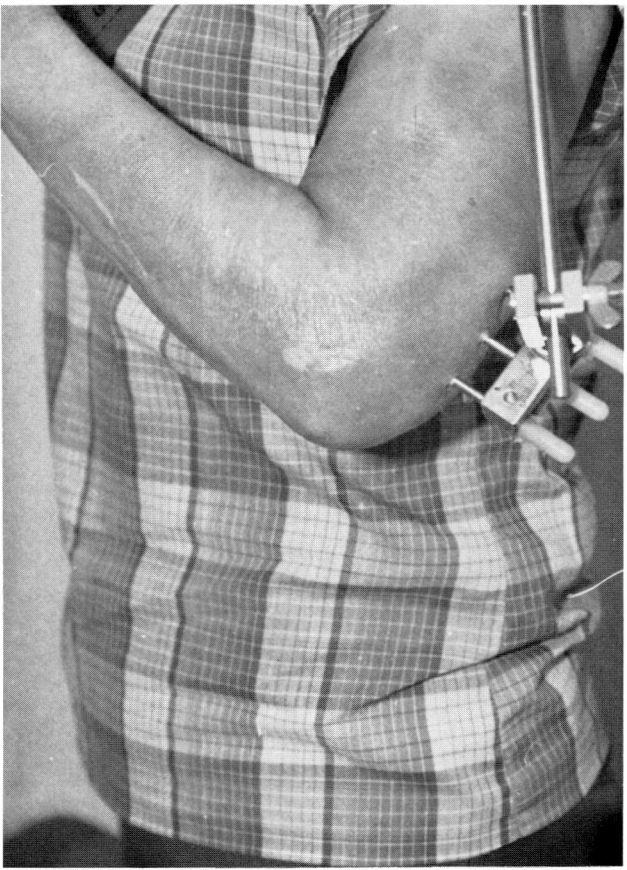

Figure 84A: Range of elbow flexion with transtriceps tendon insertion.

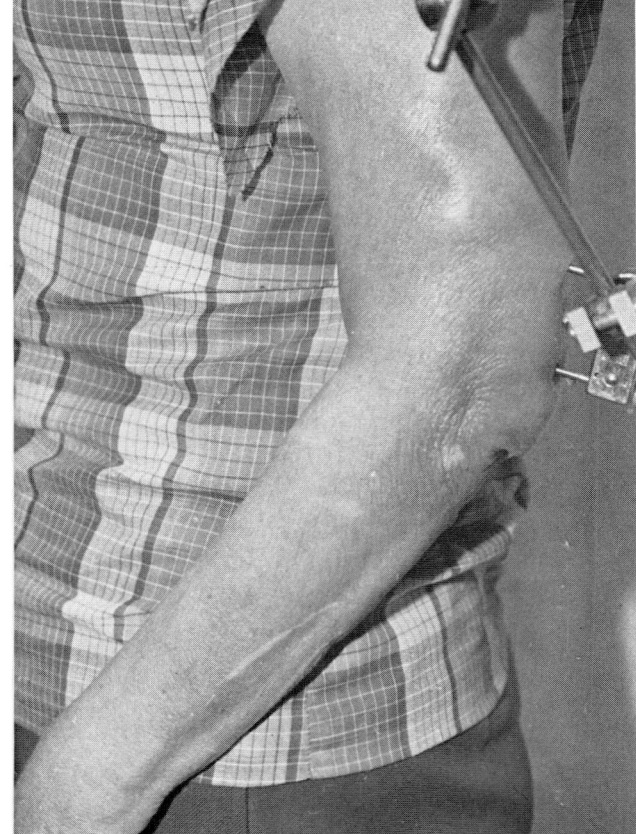

Figure 84B: Range of elbow extension with transtriceps tendon insertion.

fractures requiring prolonged immobilization, but not with less severe fractures requiring the same period of immobilization. He noted significant ankle and subtalar stiffness (more than 25% loss of flexion or extension) occurring in 25 percent of patients with mild injuries, with the incidence rising to 60 percent in the presence of serious trauma and 70 percent in cases where union was delayed. Burny,[41] reviewing 1,421 tibial fractures (673 open) treated with external fixation, noted loss of ankle motion in 18.3 percent. He emphasizes the importance of early physiotherapy to the ankle joint to prevent an equinus contracture from occurring.

In certain situations, it is necessary for an external fixator to span a joint. Some intraarticular and juxtarticular fractures can be reduced and maintained with external skeletal fixation. The capsule and periarticular ligaments aid fracture reduction when the fixator frame is used in a distraction mode. This technique, call ligamentotaxis,[267] has been employed to treat injuries of the hip, knee, and wrist. The short-term results are gratifying thus far. Cooney[68] has reviewed eighty-five comminuted unstable Colles fractures treated with external pin fixation (60 in a Roger Anderson component frame and 25 in a Hoffmann component frame); 95 percent of the patients were asymptomatic at follow-up. The median loss of motion (when compared to the uninjured side) was 7 percent dorsiflexion, 8 percent plantar flexion, 4 percent radial deviation, 10 percent ulnar deviation, and 5 percent pronation and supination. Thus, it appears that six to eight weeks of external fixator immobilization of the wrist joint does not interfere with recovery of wrist motion and has the beneficial effect associated with restoration of normal joint anatomy.

Complex intraarticular-metaphyseal fractures may require a frame configuration across the joint, especially if the bone on the opposite side of the joint is also fractured. The prolonged immobilization necessary to deal with such injuries may be harmful to the joint crossed by the fixator. In this situation, it is reasonable to immobilize the joint temporarily with fixator components that span the joint and that can subsequently be removed, leaving two separate frames on the limb. If there is concern about joint instability (due to ligamentous injury), polycentric hinges can be clamped to the fixators to allow constrained motion in one plane.[88] Needless to say, a bridged joint should never be compressed. Inhibition of articular cartilage nutrition may produce degenerative arthrosis at a later date.[94]

PAIN AND SWELLING

Pain sufficient to preclude physiotherapy may occur after a fixator is applied, but it usually diminishes in a few days. If the pain persists, an assessment of the symptoms is required. Detailed consideration of pain occurring while in an external fixator is presented in Chapter 8.

Limb swelling is not often mentioned as a problem associated with external fixation, but I believe it is more common than is reported. Krempen[165] noted persistent limb swelling in 50 percent of patients with septic pseudarthrosis treated with external skeletal fixation. He recommends managing the problem with a Jobst stocking fabricated to produce 40 mm of pressure.

SHORTENING AND ANGULATION

One of the benefits of external fixation is that the fixator frame prevents shortening and angulation following comminuted limb fractures. Naden,[202] in reviewing 950 applications of the Roger Anderson apparatus, noted only five cases with greater than one-quarter inch shortening (all had loss of bone substance). Two patients demonstrated significant lateral bowing; two patients demonstrated posterior bowing; one patient demonstrated rotational malalignment. In Burny's[37] large series of tibial fractures, 91.8 percent had perfect alignment, while 5.4 percent had less than 10° of angulation and 2.8 percent had more than 10° of angulation.[37] Krempen's[164] series of acute tibial fractures treated with external fixation averaged 1.0 cm shortening, and his series of infected or previously infected tibial fractures treated with external fixation averaged 1.3 cm shortening. Three patients had up to 4.0 cm shortening in the second series due, for the most part, to resection of infected bone. Anderson and Hutchins reported 184 tibial fractures treated with pins-in-plaster technique.[7] Only 2.2 percent of the patients had greater than three-quarters inch shortening, more than 10° varus or valgus deformity, or more than 20° anterior or posterior angulation. Eighty-seven percent of their patients had what was considered good-to-excellent results with less than one-half inch of shortening, less than 5° varus or valgus deformity, and less than 10° anterior or posterior angulation. These results compare favorably to those obtained with conservative weight-bearing techniques.

A potential danger of external fixation is the temptation to compress unstable fractures in the hope of promoting bone union. Weekly compression of an unstable fracture will result in limb shortening but is unlikely to promote union. In the Hoffmann frame, one turn of the compression knob shortens the frame 0.127 cm. Compressing the frame one turn per week will result in 1.27 cm (0.5 inch) shortening in a ten-week period.

Axial malrotation can also occur with external fixation. It happens at the time of frame application because the axial rotation of a fractured limb is sometimes difficult to assess when the patient is covered with surgical drapes. If a fixed-plane fixator frame is applied with axial malrotation of the limb, one pin group must be removed and replaced in a different position after rotational alignment is corrected. This problem is likely to occur if limb lengthening fixators are used for fracture treatment. Limb lengthening fixators, such as the W. V. Anderson and Wagner apparatuses, are ideally suited for their primary purpose. In limb lengthening surgery, the bone is osteotomized between the upper and lower pin groups, while the fixator maintains the original axial alignment of the bone. This feature, so important for limb lengthening, becomes a handicap when applied to fracture problems, because malrotation may first be discovered after the fixator is applied. The variable frame fixators, on the other hand, allow axial adjustments to be made without changing any pins.

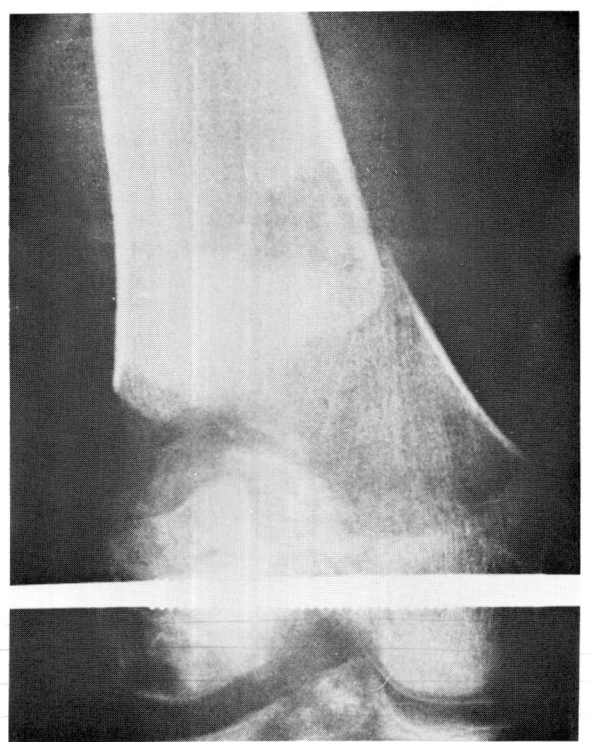

Figure 85A: Unstable distal femur fracture with absent medial cortex.

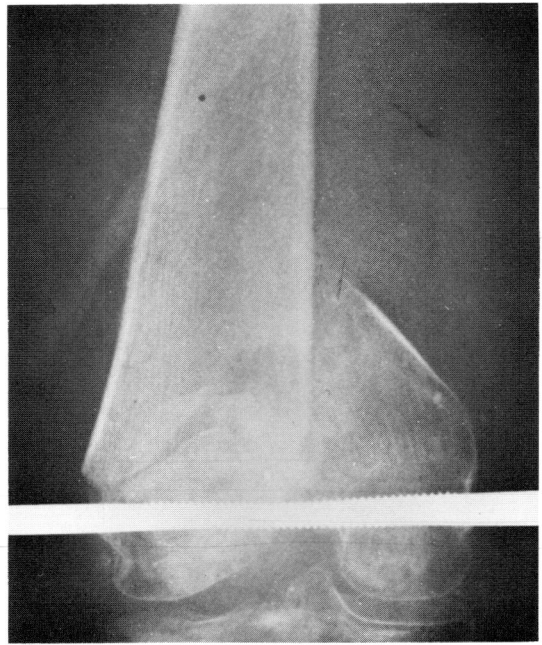

Figure 85B: Weekly compression resulted in one and one-half inch shortening but no union.

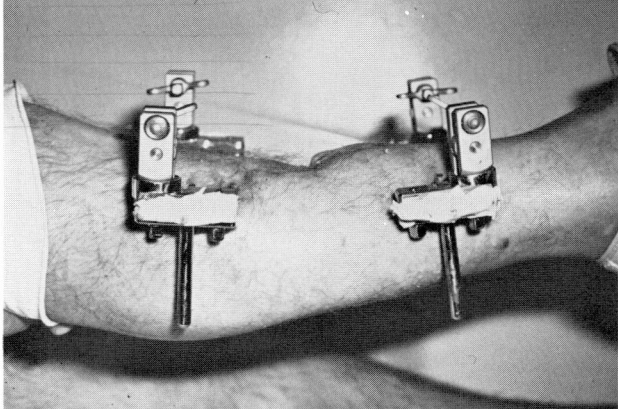

Figure 86: Angulation of tibia produced by unsymmetrical compression. The anterior compressor bar was closer to the tibia than the posterior compressor bar.

Interference with Limb Function

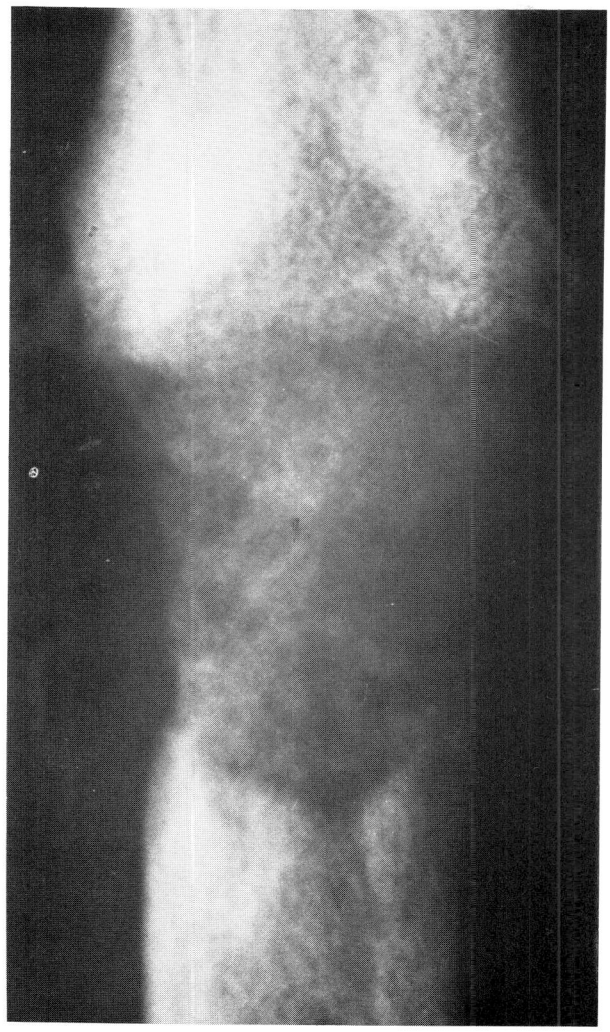

Figure 87A: Maturing cancellous bone graft.

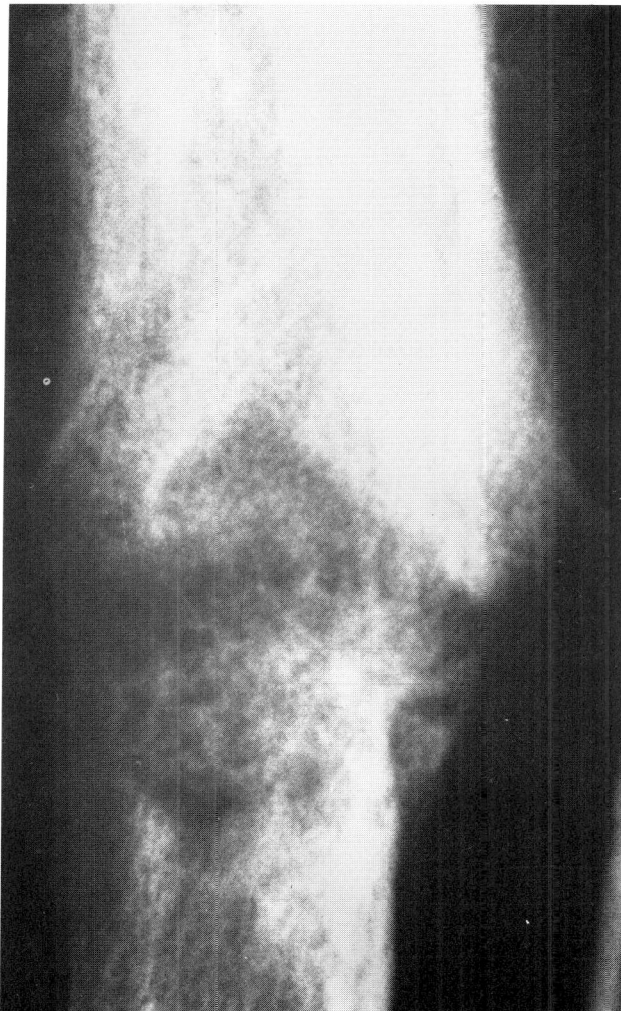

Figure 87B: Weekly compression of the graft mass resulted in excessive shortening.

Chapter 5

FAILURE TO OBTAIN UNION

Introduction

External fixation has a reputation for causing delayed unions and nonunions. The prospect of nonunion was the third most common factor (after pin tract infections and pin hole osteomyelitis) leading to discontinuing the use of external fixation, according to Johnson and Stovall's survey in 1950.[140] Six years earlier, Siris,[248] in his generally unfavorable report on external skeletal fixation, noted that the length of time required to obtain bone union was prolonged for fractures treated by external fixation. He concluded, "from our experience, we are fully in accord with these [observations] that external pin fixation is a contributing factor in delayed unions by maintaining the fragments in distraction." Yet, when one reviews recent European and American reports on external fixation, delayed union and nonunion do not appear to be an especially troublesome problem.[37, 44, 70, 89, 93, 154, 266] A reason for this is the new awareness of the danger of retarded bone healing by workers in the field of external fixation, and the development of strategies designed to prevent the problem from occurring.

Definitions

There is a difference of opinion as to what constitutes a delayed union and what can be called a nonunion. A widely accepted definition of *delayed union* is *failure to obtain fracture consolidation within six months following injury*. Unfortunately, this definition does not take into account variations in healing time of different bones or, for that matter, different degrees of trauma. I am inclined to agree with Brighton,[24] who indicates that delayed union, like beauty, is in the eyes of the beholder. Union is delayed if healing is considered retarded by the treating physician, based on his experience with similar fractures. Brighton defines a nonunion as the absence of evidence of progression towards union on three successive roentgenograms, taken at monthly intervals. Certain fractures (carpal navicular, and subcapital femur) can, at times, clearly be considered nonunions as soon as three months following injury.

INCIDENCE OF RETARDED BONE HEALING

No controlled prospective investigations have been made comparing the union rates of fractures treated with external fixation with those treated by other methods. In order to design such a study, investigators would first have to select patients who fit clearly defined criteria for exernal fixation, and then randomly assign the patients to treatment groups with and without external fixation. This protocol would eliminate the bias (inherent in most reported series of external skeletal fixation) due to fixators being used for the most serious musculoskeletal injuries—those cases for which the probability of delays in union would already be quite high.

The use of "historic controls" (comparing results to previously reported series) is an unsatisfactory way of determining if external skeletal fixation contributes to the problem of retarded bone healing.[163] Unfortunately, it is the only evaluation available to us. The question is which historic controls to use. As Burri[42] has noted (in evaluating the incidence of sepsis following fracture treatment):

> Supporters of one or the other method of treatment often publish a small series of cases with a low rate of infection meant to prove the success of a method or of the author's special procedure, while in other publications, the great number of complications is pointed out with the purpose of warning the reader of this procedure.

Because there are several philosophies of fracture management, the results of each will be evaluated separately for comparison with external skeletal fixation. For the sake of simplicity, only tibial fractures will be considered.

Closed Treatment

Pierre Delbet,[82] a Frenchman who published his experience in 1914, is responsible for the concept of early ambulatory management of fractures of the lower limb. Delbet employed slab-splints molded around tibial fractures, which left the ankle and knee free to move. Based on Delbet's results, Dehne[79] popularized the ambulatory treatment of tibial fractures utilizing a long-leg cast. He initially reported 349 cases, with an average time to union of 4.3 months, and no nonunions in his series.[80]

In 1966, Weissman, Herold, and Engleberg[287] reported a well-documented study of fractures of the tibial shaft treated with long-leg casts. They reviewed 140 consecutive fractures (91 closed, 49 opened) involving the middle two-thirds of the tibia, treated with an initial non-weight-bearing long-leg cast that was converted to a snug-fitting ambulatory plaster cast in 6-8 weeks, followed by a short-leg gaiter. The mean time to union (for adults) was 5.08 months. Twenty-four fractures took longer than six months to heal. There were no nonunions.

Sarmiento,[236] employing techniques derived from the fabrication of prosthetic limbs, perfected a closed method of fracture treatment that resembles Delbet's original concept. His remarkable series of 847 tibial fractures (99 open) resulted in only one nonunion, although the most complex injuries were excluded from his series.[237] Transverse fractures healed in 3.75 months, oblique and comminuted fractures healed in 3.5 months, and segmental fractures healed in 4.25 months.

It would appear that fractures treated with plaster of Paris healed quickly with a very low incidence of nonunion. There are, however, reported series where the incidence of retarded bone healing is significant with conservative methods of treatment. Rosenthal et al.[231] reported a retrospective analysis of 104 open fractures of the tibial shaft treated at a civilian teaching center. Their series contained many patients with complex injuries who could have been candidates for external skeletal fixation had this modality been used. They reviewed 104 open fractures, 60 of which were treated with closed technique and early ambulation. Of these, ten (16.6%) resulted in nonunion. They noted that "all ununited tibias were associated with fractures of the fibula and extensive avulsions of skin and dirty wounds." Carpenter and associates[47] analyzed a series of displaced open comminuted tibial fractures and found a nonunion rate of 75 percent.

Sakellarides and coworkers[235] reviewed one-hundred cases of delayed union and nonunion of tibial shaft fractures treated at Massachusetts General Hospital. Twenty of the patients were from the fracture service of their institution, having derived from 1,460 patients treated for tibial fractures. Eighty other patients were referred from other facilities. Forty-five fractures (12 closed, 33 open) had been treated by closed methods, without internal fixation. Thus, almost half of the nonunions and delayed unions that came to treatment at Massachusetts General Hospital were treated with methods that have been reported to result in a low incidence of nonunion. (Unfortunately, it is not known how large a population of patients is represented by these forty-five delayed unions and nonunions.)

Internal Fixation

The incidence of delayed union and nonunion following internal fixation is difficult to assess. Internal fixation with plate and screws requires rigid stabilization and interfragmentary compression, which may result in no external callus formation.[75, 198, 218] (See "Primary Bone Healing" in this chapter.) With the intramedullary nail, slightly better assessment of delayed union may be possible because of the presence of some fracture callus, but a nonunion may first become identified when the patient develops a bent or broken intramedullary device.

Lottes,[180] in reviewing his experience with intramedullary fixation of tibial fractures, reported 13 (2.3%) out of 573 cases that failed to unite. His series included 204 open single fractures, of which five (2.4%) developed nonunion. Of forty-seven open segmental fractures, three (6.3%) developed nonunion. However, reviewing roentgenograms in his article, one is left with the distinct impression that Lottes did not manage the most severely comminuted open fractures with an intramedullary nail. It is this type of injury that might be placed in an external fixator by today's standards. Plato's[221] series of intramedullary nailing of tibial fractures revealed a pseudarthrosis rate of 6.4 percent in surgically treated closed fractures, and 11.8 percent in surgically treated open fractures. (Interestingly, these figures are one-half the pseudarthrosis rate obtained with fractures treated by closed methods.)

Compression plate fixation of tibial fractures may also lead to nonunion. Batten[18] reported a 5 percent nonunion rate following compression plating of tibial fractures, and Burwell[43] reported 4.4 percent nonunions in a series of plated tibial fractures. Karlstrom and Olerud[153] noted delayed healing in 4.4 percent (6 of 135 tibial fractures) following internal fixation. Solheim,[249] on the other hand, reports a nonunion rate of only 2 percent of tibial fractures treated with the ASIF compression plate method. About half of the fractures in his series were caused by road accidents.

In the review from Massachusetts General Hospital cited above, Sakellarides and associates[235] noted that fifty-six of the one-hundred cases of retarded bone healing they evaluated occurred following open reduction and internal fixation. They felt that many of the cases of delayed healing and nonunion showed evidence of suboptimal plating, inadequate fixation, and distraction of the fragment ends.

Rosenthal,[231] in the series previously described, noted twelve (37.5%) nonunions out of thirty-two treated with open reduction and internal fixation. The series involved a significant number of patients with extensive skin and soft tissue loss. Rosenthal noted that the risk of nonunion was twice as great with internal fixation compared to plaster casts. I believe Rosenthal's results are extremely important because they represent the experience of patients treated by a number of different staff physicians, working with residents, at two different institutions in the same city. In many ways, the series reflects the patient population and level of fracture care available in the United States today.

Chapman and Mahoney[50] summarized the English language literature regarding internal fixation of open fractures from 1945 to 1978. A total of 936 fractures were evaluated, including 578 tibia-fibula fractures. The overall nonunion rate was 11.3 percent, with a range of 0 to 45 percent. They noted that most of the nonunions occurred in serious open fractures with either comminution or extensive soft tissue damage. In most series they cited, nonunion was a consequence of wound sepsis.

In a recent report, Van der Linden and Larsson[260] compared plate fixation to conservative treatment of tibial shaft fractures. They randomly assigned one-hundred patients to an AO-plate fixation group or a conservative (cast) treatment group. In each group there were six open and forty-four closed fractures. The median time to union was three months in the group treated with an AO-plate, and 4.25 months in the group treated conservatively. There were four nonunions in the AO-plate group, and three nonunions in the conservatively treated group.

External Fixation

The problem of delayed union with external skeletal fixation has been recognized for a long time. Roger Anderson,[8] reporting his pins-in-plaster technique in 1934, noted that "too much traction [distraction] must be avoided, as nonunion can frequently be attributed to this factor." Siris[248] analyzed the union rates in twenty-one compound tibial fractures treated with external fixation. He noted that seven united in three to six months, nine united in seven to twelve months, and two united in fifteen to nineteen months. Three were ununited fractures at the end of two years, two of which had been grafted. (Siris also noted that union in femoral fractures was delayed, requiring three to four months for union, compared with two to three months in Russell's traction.)

Naden,[202] from Vancouver, in reporting his extensive experience with 950 applications of the Roger Anderson apparatus, treated 156 *closed* tibial fractures. He noted two (1.2%) nonunions. Among thirty-five open fractures of the tibial shaft, there were four (12.5%) nonunions. The average length of time required for union in the cases that did unite was 4.2 months for closed tibial shaft fractures and four months for open tibial shaft fractures. Union was delayed in eight (5.1%) closed tibial shaft fractures, ranging from six months to 9.5 months. Five (13%) of the compound fractures that eventually united demonstrated delayed healing ranging from 5.5 months to 8 months.

Anderson and Hutchins[7] reported their experience with 107 tibial fractures (59 closed, 48 open) utilizing the pins-in-plaster technique. They noted that 95.3% of their fractures went on to union, and 2.8 percent became nonunions. The average time in plaster was 5.7 months for the 102 fractures that united, and 9.6 months (before bone graft) for the fractures that demonstrated delayed union and nonunion. The authors removed the transfixion pins at five to six weeks after fracture.

Schaar and Kreuz[244] noted only one tibial nonunion (and one femoral nonunion) in 110 fractures in a group of American Navy personnel and Navy shipyard workers treated with the Stader apparatus. They did note, however, that all transverse fractures showed delayed healing. The explanation of this observation was that transverse fractures usually occur as the result of a direct blow to the bone and were associated with soft tissue injury. Oblique fractures, on the other hand, were usually the result of torsion and are not generally associated with significant soft tissue injury.

Fellander[93] applied the Vidal-Adrey quadrilateral frame to forty-nine fractures in forty-seven patients. However, thirty-four fractures in the series were established nonunions before application of the fixator. Of the fifteen fresh fractures, two (13.3%) failed to unite.

Karlstrom and Olerud,[154] in 1977, reported their experience with forty-eight patients with open tibial fractures treated with external skeletal fixation. Thirty-two patients had extensive skin and soft tissue damage (Grade III). The patients were managed in a quadrilateral frame fabricated from Hoffmann components. The mean time in the fixator was 4.5 months, and the mean healing time to full weight-bearing (without external support) was 7.8 months. Three (6%) of their fractures were still not united at the end of twelve months. Many of the patients had bone grafting procedures while in the fixator frame. Fourteen of the fractures were bone-grafted once, five were bone-grafted twice, and two were bone-grafted three times. The cases that were not united at the end of treatment were managed with either internal fixation or reapplication of a fixator.

Burny's[40] experience with Hoffmann fixation of 450 "simple" fractures of the tibia (including 25% compound fractures with minimal soft tissue damage) reported consolidation in 359 (79.8%), temporary nonunion in 29 (6.4%), and definitive pseudarthrosis in one (0.2%). It should be noted that approximately one-half of the cases in his series were nondisplaced or minimally displaced. In comparing simple to complex fractures —based on the degree of comminution and displacement—Burny noted healing of simple fractures in three months and of complex fractures in six months. In a much larger series of 1,421 fractures, Burny[37] noted that 62 percent went to union in the external fixator, 25 percent required application of a plaster cast after the fixator was removed; 1.2 percent were treated with a plate following external fixation; 1.7 percent were treated with an intramedullary nail following external fixation.

Vizkelety,[277] of Hungary, recently reported 292 cases treated with several different external fixators, including the Russian ring fixators in 10 percent of his applications. There were two (0.6%) nonunions and eight refractures in his series. Edwards[87] has treated forty-four complex open tibial fractures with external skeletal fixation. Of these serious injuries, 73 percent had bone loss or major comminution, and 55 percent had loss of soft tissue. Of his series, 57 percent went to primary union with good skin coverage, 39 percent required bone-grafting, 30 percent required muscle flaps, and 48 percent required skin-grafting. The median time to union was 7.5 months.

As indicated in the beginning of this section, comparing the incidence of delayed union and nonunion resulting from external fixation to that obtained with either internal fixation or plaster treatment is virtually impossible. Recent reports of series utilizing external fixation for the most serious compound fractures, summarized above, yield union rates that are equal to those obtained with other methods. With less complex fracture problems, the question has not yet been resolved. Many investigators in the field of external fixation have adopted certain measures designed to prevent retarded bone healing. Most of the recent series reported in this chapter incorporate one or more of these measures as part of the treatment protocol. Thus, a valid comparison between external fixation by itself and other treatment modalities cannot realistically be made.

PHYSIOLOGY OF FRACTURE HEALING

To better comprehend the mechanism by which fractures heal while in rigid external fixation, consider the physiology of fracture healing. The classic concept of physiologic fracture healing assumes the formation of a *callus* that progressively stabilizes the fracture fragments as it matures into bone.[121, 72]

Robert Danis,[75] a brilliant Belgian surgeon, first realized that displaced fractures that are resorted to their original anatomic configuration and held in place by compression will behave like nondisplaced (hairline) fractures: they will produce no significant external callus and will heal instead by "internal welding" (soudure autogen). Thus, there are two mechanisms by which fractures heal: external callus formation and internal welding, now called primary bone healing.

A third mechanism, endosteal (or internal) callus formation, occurs when fracture callus fills the medullary canal to line the endosteal surface of the bone It is not a strong bond if it is not supplemented by either external callus or primary bone healing. For this reason, it will not be considered here. Consideration will be given, instead, to the physiology of fracture healing as it pertains to external callus formation and primary bone healing.

Healing by External Callus

A displaced fracture not only breaks the anatomic continuity of the bone itself but also transects numerous haversian blood vessels that traverse the cortex.[75, 121] Blood vessels lining the endosteal and periosteal surfaces of the bone are also damaged, as is the periosteum. Damage to the surrounding soft tissues can also occur in the more serious injuries. Damage to local blood vessels may cause ischemia and necrosis of the cells they have been supplying with oxygen and nutrients.

Shortly after the injury, blood flowing into the wound area forms a coagulated mass (fracture hematoma) around the fracture. The combination of a fibrin meshwork of coagulated blood and the presence of necrotic tissue stimulates an intense local reaction: the *inflammatory phase* of fracture healing.[72,121] The acute-phase inflammatory cells (polymorphonuclear leukocytes) invade the area and begin to phagocytize cellular debris. An invasion of chronic inflammatory cells (lymphocytes and their descendents, the plasma cells)[183] soon follow. Local mast cells release histamine, causing vasodilation of local blood vessels and transudation of intravascular fluid into the wound area. This results in additional swelling and edema. The invading inflammatory cells produce other substances that control the migration and production of yet another class of phagocytic cells, the macrophages, that stimulate further lymphocytic infiltration.

The inflammatory phase is gradually replaced by a *reparative phase* as the hematoma undergoes organization. Capillary buds, arising from the vessels surrounding the area of injury, grow into the hematoma. Pluripotential mesanchymal cells, derived from ingrowing capillaries, give rise to cells that form immature fibrous tissue, immature cartilage, and immature bone.[258] (The exact origin of these pluripotential cells is uncertain, but they may be derived from the vessel wall endothelium,[258] monocytes, or other stem cells.[72]) The stimulation of the pluripotential cells to produce one or another type of tissue is probably related to the local cell environment. Oxygen tension,[14] tissue motion,[88] and vascularity[229] have all been shown to play a role in determining what type of tissue the mesanchymal cells will become.

As the fracture callus matures, the cartilaginous tissue present undergoes enchondral ossification to become bone.[121] With appropriate mechanical and biochemical stimulation, bone can also form directly from the pluripotential mesanchymal cells. Bone is also produced by the osteoblasts lining the inner (cambrium) layer of the disrupted periosteum.

The newly formed bone, called woven or fiber bone, possesses randomly oriented collagen molecules that gradually become calcified. Tiny crystals of calcium hydroxyappatite precipitate into specific locations ("holes") along the collagen molecules.[107] The collagen molecules gradually stiffen as they become laden with the mineral crystals, lending increased rigidity to the fracture callus. Sequential roentgenograms of the fracture will reveal greater radiodensity resulting from this process.

Remodeling, the final phase of fracture healing, begins while the reparative phase is still in progress. The fracture callus, transforming into fiber bone, begins to exhibit characteristics related to Wolff's law[292] (the observation that bone, in its internal architecture and external contour, conforms to the stresses placed upon it by daily life). Mechanical stresses, associated with motion of the fracture callus, are translated into electric impulses by the piezoelectric properties of bone crystal and collagen. Bone-forming cells (osteoblasts) respond to these impulses by producing bone along stress lines. Meanwhile, osteoclasts remove bone where it is no longer needed. Through this process, immature fiber bone is gradually remodeled into mature compact lamellar bone. The entire remodeling sequence may take several years to complete. In the end, a well-formed shaft of dense cortical bone replaces the bulkier mass of immature fiber bone that originally was the fracture callus. As the remodeling phase comes to a close, the medullary canal becomes reestablished as a hollow space. (Fractures through cancellous bone near or into joints will heal more quickly. The remodeling phase is shorter since much less compact cortical bone needs to be created.)

Primary Bone Healing

Following a fracture after which open reduction and internal fixation with compression is performed, fracture healing starts as an advancing cone of osteoclasts that bore holes directly across the fracture site from one fragment into the other.[218] The osteoclastic "drill heads" remove lamellar bone in their path. As they advance across the fracture line, they are followed by vascularized soft tissue containing a conical surface of osteoblasts, which produce new osteons (haversian systems) in which the osteoblasts mature into osteocytes. The osteocytes are connected to each other and their blood supply by a network of canaliculi, as they are in any mature bone.[61]

Any gaps between contact areas that may be present (if they are less than 100 microns wide) will fill with tissue that forms mature lamellar bone, deposited at a right angle to the normal direction of haversian systems within the bone.[218] This bone is eventually replaced by haversian remodeling that is correctly oriented, parallel to the long axis of the bone.

Primary bone healing takes longer than fracture healing by external callus formation because it is, in effect, a slow remodeling of cortical bone in the region of the fracture.

Motion in Fracture Healing

Many factors—chemical, hormonal, mechanical—can influence the bone healing process favorably or unfavorably.[72] For example, corticosteroids, vitamin B, and vitamin A can either promote or retard fracture healing, depending upon the concentration of the substance and other variables.

Failure to Obtain Union

With rigid plate fixation, fractures are expected to heal by "primary bone healing." Primary bone healing has been shown to be inhibited if there is motion at the fracture site in excess of 5 to 10 microns.[218] (A red blood cell is 7.2 microns in diameter.[121]) In the absence of motion at the fracture site, there is frequently no roentgenographic evidence of external callus formation. In fact, clinicians with extensive plate fixation experience regard the appearance of significant external fracture callus as evidence of suboptimal fixation resulting in motion at the fracture site.[192,198,230]

Rigid fixation with an external skeletal fixator also retards fracture callus formation because it reduces tissue motion at the fracture site. Unlike plate fixation, however, continuous compression with less than 10 microns of motion at the fracture site is difficult to achieve with an external fixator. Thus, primary bone healing may also be inhibited. With inhibition of motion by the fixator retarding external callus and suboptimal bone-to-bone contact reducing primary bone healing, it is a wonder that externally stabilized fractures heal at all. The fracture healing that does occur possibly does so because of the slight contribution to fracture healing by the disrupted periosteum, especially in the younger patient. Also, micromotion occurring at the fracture site in spite of fixation may be sufficient to stimulate fracture healing by external callus formation. Finally, the human body has a remarkable capacity to heal itself, outsmarting even the most capable orthopaedists in the process.

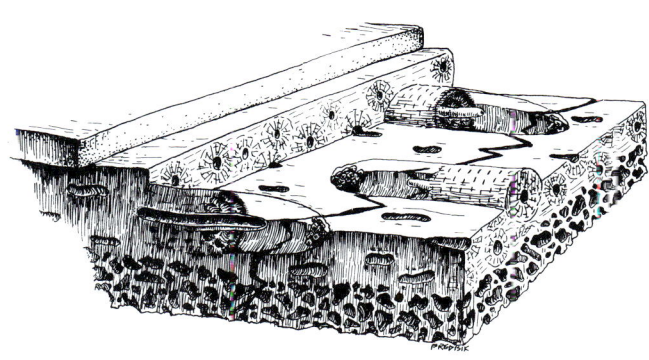

Figure 88: Primary bone healing: cones of osteoclasts drill across the fracture site, followed by vascularized tissue, which forms new osteons (haversian systems).

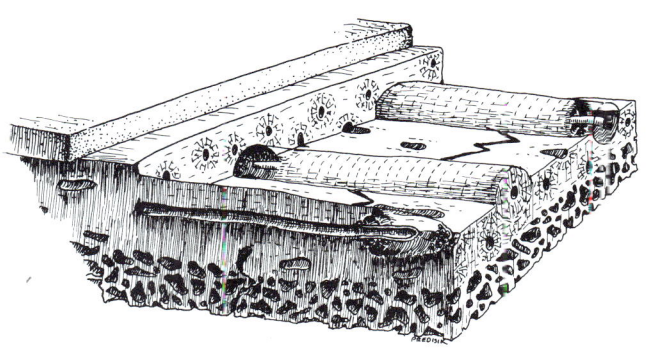

Figure 89: Fracture bridged by newly formed mature osteons.

FACTORS CONTRIBUTING TO RETARDED HEALING

Several factors have been identified that are known to contribute to delayed bone healing and nonunion. In their classic review of 842 patients with nonunion of long bones, Boyd, Lipinski, and Wiley[20] noted that nonunion more frequently occurs in fractures that demonstrate one or more of the following: compounding, infection, segmental fragment or comminution, insecure fixation, immobilization for too short a time, treatment by ill-advised open reduction, or distraction by traction, plates, or screws. Brighton[23] recently analyzed the causes of nonunion in patients referred to his institution for treatment with electrical stimulation. He noted that only 20 percent of the cases could be assigned to one of the well-known categories described above. The remaining 80 percent of the nonunions had to be categorized as cause unknown.

Workers in the field of external fixation have noted the factors associated with retarded bone healing that are likely to be a problem when a fixator is utilized. The factors include distraction at the fracture site, displacement of fracture fragments, comminution of fracture fragments, extensive soft tissue damage, and wound sepsis.

AVOIDING RETARDED UNION

As noted earlier, several strategies are available to reduce the likelihood of developing delayed or inadequate bone healing while in a fixator. These include accurate reduction; the use of compression; early removal of the fixator; bone grafting; sequential removal of the fixator frame; flexible external fixation; biocompression; supplemental electrical stimulation; and internal fixation after external fixation.

Accurate Reduction

Lawyer[171] studied the average healing time in closed fractures treated with external skeletal fixation that had been accurately reduced and compared the results to fractures that had not been accurately reduced. The anatomically reduced fractures healed in 4.4 months, while the fractures that were not anatomically reduced healed in 7.4 months — a difference of three months. His observation substantiates the view that accurate reduction is important when an external skeletal fixator is applied.

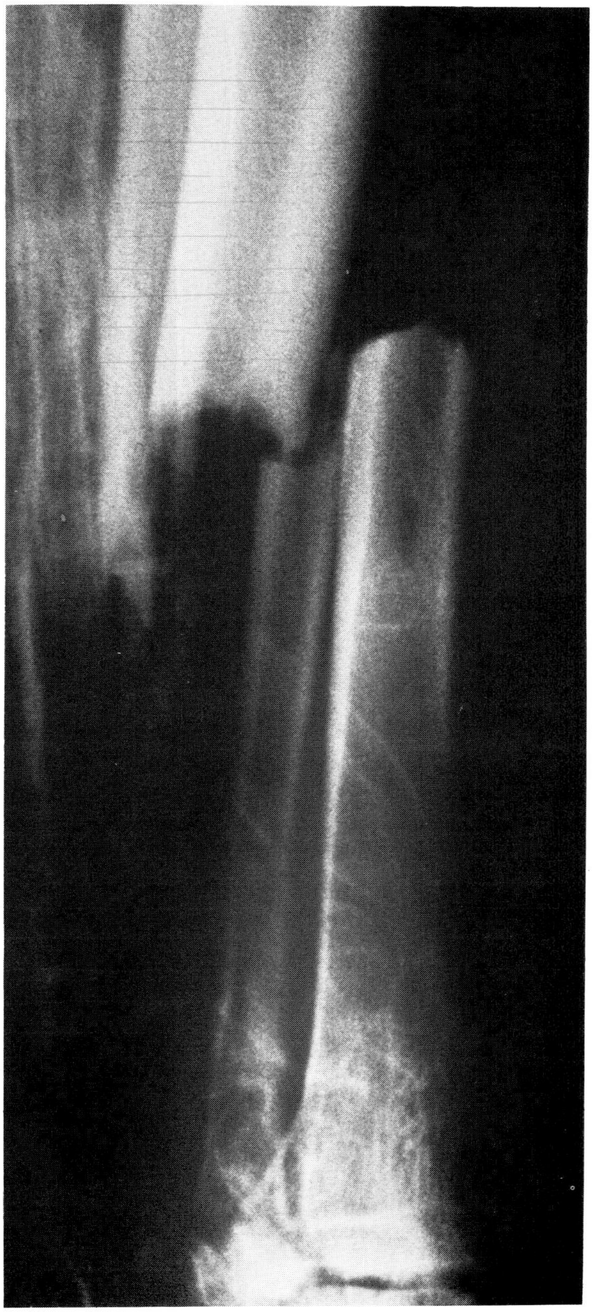

If the fracture is open, obtaining an anatomical reduction is easy. The fracture is reduced after the fixator is applied but before the configuration of the frame is fixed. If the fracture is closed, an open reduction may be necessary to approximate the fracture ends. The reduction requires a minimal incision (3.0 to 4.0 cm in length), after which the surgeon dissects around the fracture with his finger in the extraperiosteal plane. The fracture is manipulated into place by the surgeon as the assistant applies traction. If the fracture is oblique, a screw can be inserted across the fracture for interfragmentary compression, utilizing appropriate "lag screw" techniques. If fracture dis-

Figure 90: Displaced tibial fracture.

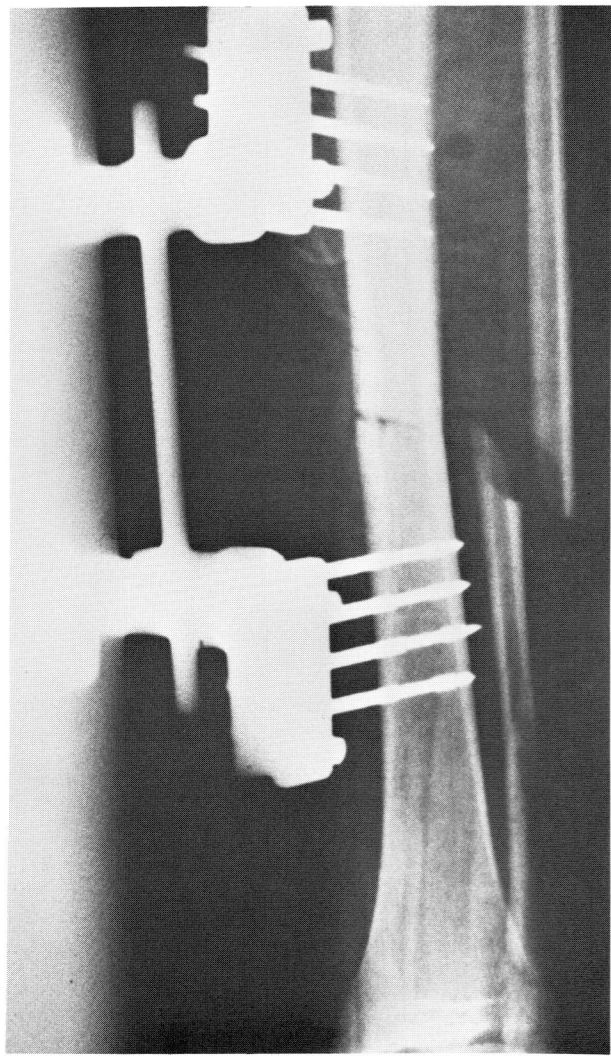

Figure 91: Anatomic reduction, unilateral frame. Slight gap at fracture site should have been compressed (union was delayed).

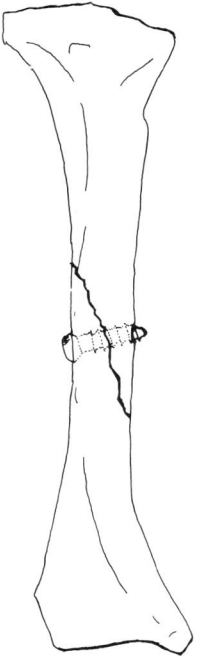

Figure 92: An oblique fracture can be stabilized with a lag screw before fixator application.

placement extends into a joint, internal fixation should be used to restore the anatomy of the articular surface.

Compression

If the fracture configuration permits, compression across the fracture site is worthwhile. It not only stabilizes the fracture, but also enhances primary bone healing across the fracture site by preventing micromotion. Furthermore, compression at the fracture site has been shown[4c] to reduce the stresses on the pins by a factor of 70×. Early fracture consolidation may permit conversion to treatment of another type without loss of position after three or four months in the fixator. If primary bone healing is the goal of the fixator application, the frame configuration must be extraordinarily rigid, and steady compression across the fracture site maintained through the entire application.

In order to achieve such a stable situation, the bone surfaces to be approximated must be perpendicular to the axis of the compression, or nearly so. An oblique fracture line will cause the fragments to slide past each other unless a notch is made in one fragment.

Comminuted unstable fractures cannot be compressed because the limb will shorten while interfragmentary compression cannot be achieved.

When applying external fixation, it is important that the frame not be excessively compressed. A deflec-

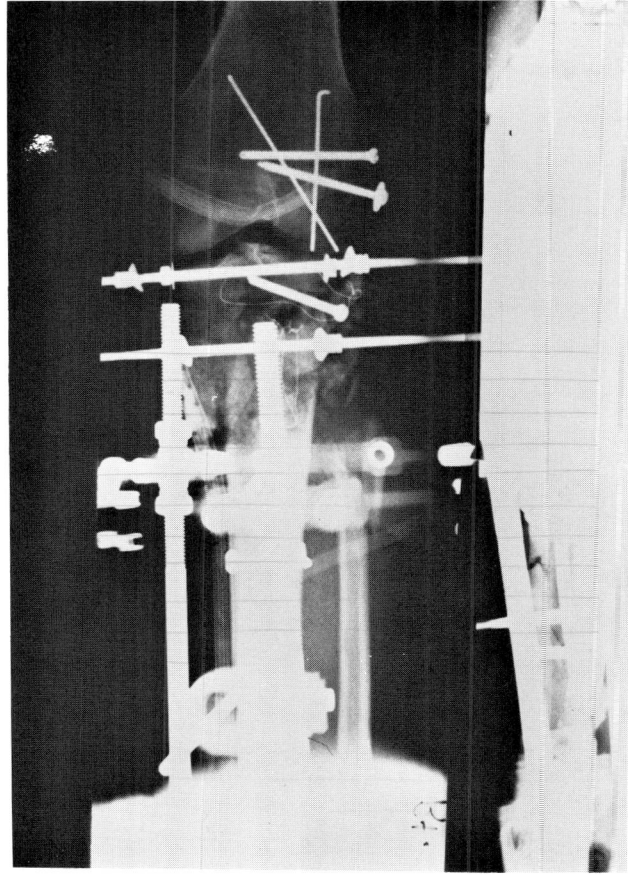

Figure 93: External fixation can be combined with internal fixation, especially for intraarticular and juxtaarticular fractures.

tion of 1° or 2° of the pins is sufficient. Many fixator systems provide wrenches to aid in frame compression; however, it is safer for the surgeon to use his fingers for the compressive effort.

Early Removal of the Fixator

External fixators are frequently applied to deal with acute fractures associated with either extensive skin and soft tissue damage, or gross comminution with instability, or both. It is reasonable in such circumstances to remove the fixator as soon as there is resolution of the problem for which it was applied. If, for example, external fixation is utilized to treat a fracture associated with considerable soft tissue and skin loss, remove the fixator when the soft tissues permit management of the fracture by other methods.

Modern fracture management encourages ambulatory care in molded orthotic devices or plaster casts.[238]

92 *Complications of External Skeletal Fixation*

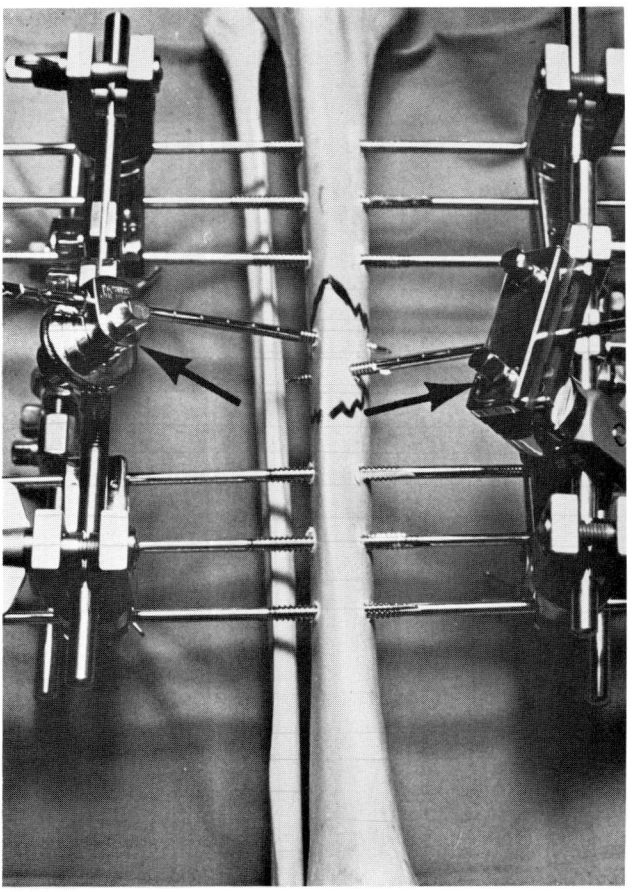

Figure 94: Intermediate fragment control is important in segmental fractures.

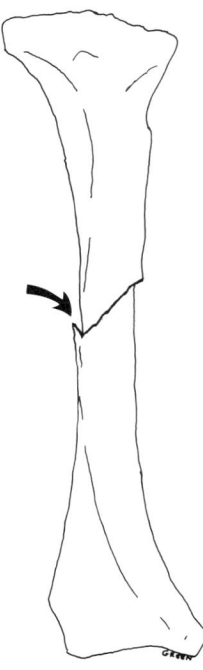

Figure 95: A "back cut" will permit compression of oblique fractures.

Failure to Obtain Union 93

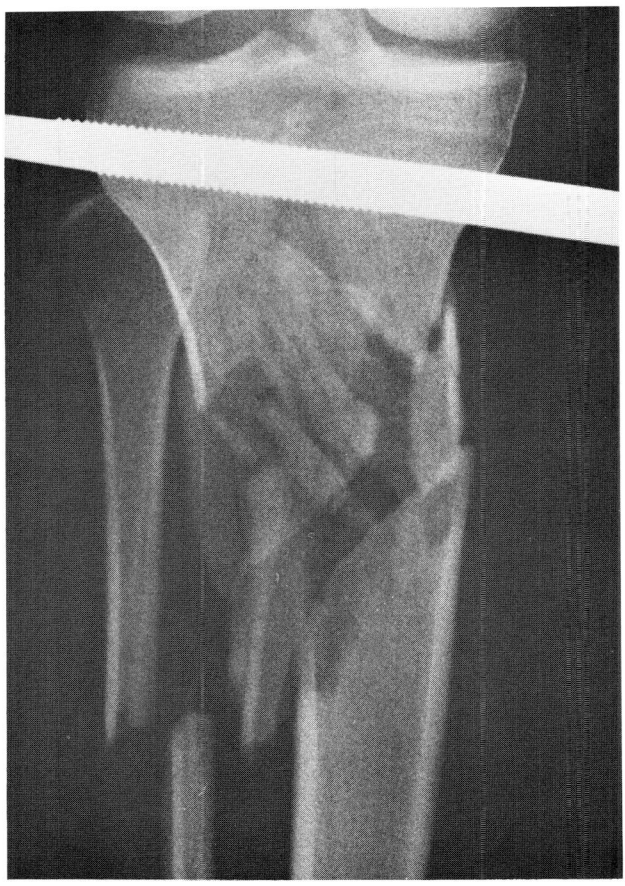

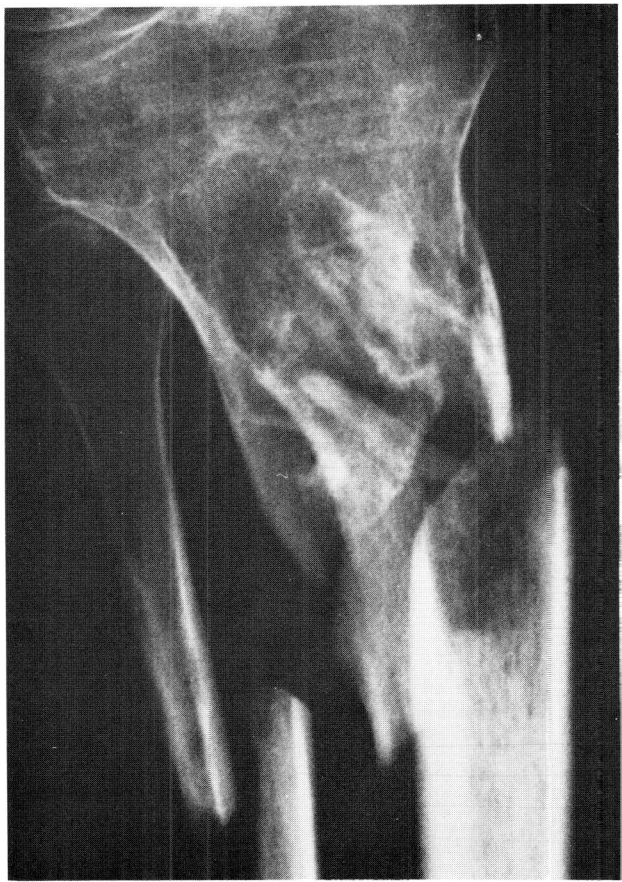

Figure 96: Comminuted fracture placed in external fixation. Note gap between the fragments.

Figure 97: Fracture displaced when fixator was removed six months later. Note absence of fracture callus. A bone graft should have been performed early in the course of treatment, or the fixator removed.

Such appliances can be utilized as soon as soft tissue healing permits, thus staging the treatment plan to include *both* external fixation and orthotic application. If the fixator is to be removed early, aggressive soft tissue management must be included in the treatment protocol. Frequent repeat debridement of devitalized soft tissue, followed by early skin coverage, hastens the process. Split-thickness skin grafts or full-thickness flaps should be performed as soon as tissue conditions permit. (See Soft Tissue Coverage in Chapter 6.) Because fixator placement may interfere with plans for definitive skin coverage, consult a surgeon knowledgeable in soft tissue reconstruction techniques prior to fixator application, if possible.

When dealing with comminuted and highly unstable shaft fractures, it is reasonable to utilize external skeletal fixation to maintain length and alignment of the shaft, while the early phases of fracture callus formation proceed. In this manner, a capsule of tissue will come to surround and stabilize the fracture fragments. In principle, as soon as the fracture has been stabilized in this manner, it may be managed without external skeletal fixation in a cast or functional molded orthosis.

Unfortunately, the optimum time for conversion from an external fixator to an external support has not been clearly established. Siris[248] recommends that pins should be extracted within three to five weeks in tibial fractures and five to seven weeks in femoral fractures if there is no loss of bony substance. King[159] recommends removal of the external fixator at six weeks, and Anderson and Hutchins,[7] as they gained experience with pins-in-plaster, found that six to eight weeks was the appropriate length of time before pin removal.

One danger of this approach deserves mention here:

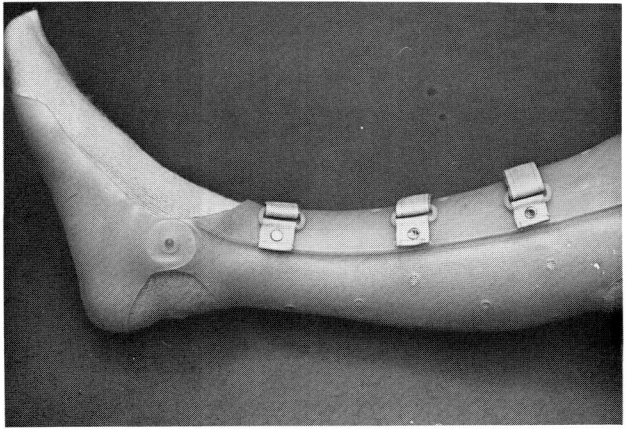

Figure 98: Snug fitting molded polypropylene orthosis is useful when the frame is removed early.

the possibility of permanent inhibition of fracture healing by the presence of mature scar tissue. Such tissue contains mature fibrocytes that have lost the ability to become bone-forming cells.[87] Thus, one would expect no significant fracture healing after the limb is placed in an othosis. Clinical and experimental data are still needed before it will be known when and how tissue around a rigidly stabilized fracture matures irreversibly into fibrous scar tissue.

Bone Grafting

Fresh autogenous bone grafting can supplement the tardy callus formation seen with external skeletal fixation. It is imperative to plan the bone grafting as part of the overall surgical management at the time the fixator is applied. This approach prepares both the surgeon and patient for the eventuality of a bone-grafting procedure. The increasing recognition of the importance of early bone grafting in fractures managed with external skeletal fixation is supported by the observation that 39 percent of the patients in the series reported by Edwards in 1979[89] received bone grafting early in the course of treatment, compared to 0.4 percent of the cases reported by Naden in 1949.[202]

Bone grafting must be a part of the management in cases where there is absolute loss of a segment of bone substance. It is also necessary if there is loss of cortex involving one-third of the diameter of the bone, or more. In the case of absolute bone loss or cortical bone loss, the graft can be applied immediately[268] or delayed one or two weeks.[152] If, after bone grafting, any evidence of osseous defect persists, the area should be regrafted at four to six week intervals. As Mears states, "procrastination will not eliminate the need for supplementary bone graft."[191]

The incidence of pin hole sepsis increases with the passage of time while the patient is in a fixator.[40] The presence of pin hole sepsis makes bone grafting more dangerous because of the possibility of graft infection, since it is difficult, if not impossible, to completely isolate a draining pin hole from the operative field at the time of surgery. For this reason, I recommend that only fresh autogenous *cancellous* bone be grafted if the surgery is to be performed anywhere near the fixator pins. Cancellous bone is less likely to become a sequestrum than will cortical or corticocancellous bone grafts.

Prophylactic parenteral antibiotics are an important adjunct to the bone grafting procedure, even if no infection is present. If any seepage, irritation, or redness is present, a pin hole culture will serve as a guide for the selection of antibiotics. If not, a cephalosporin antibiotic should be administered prior to,

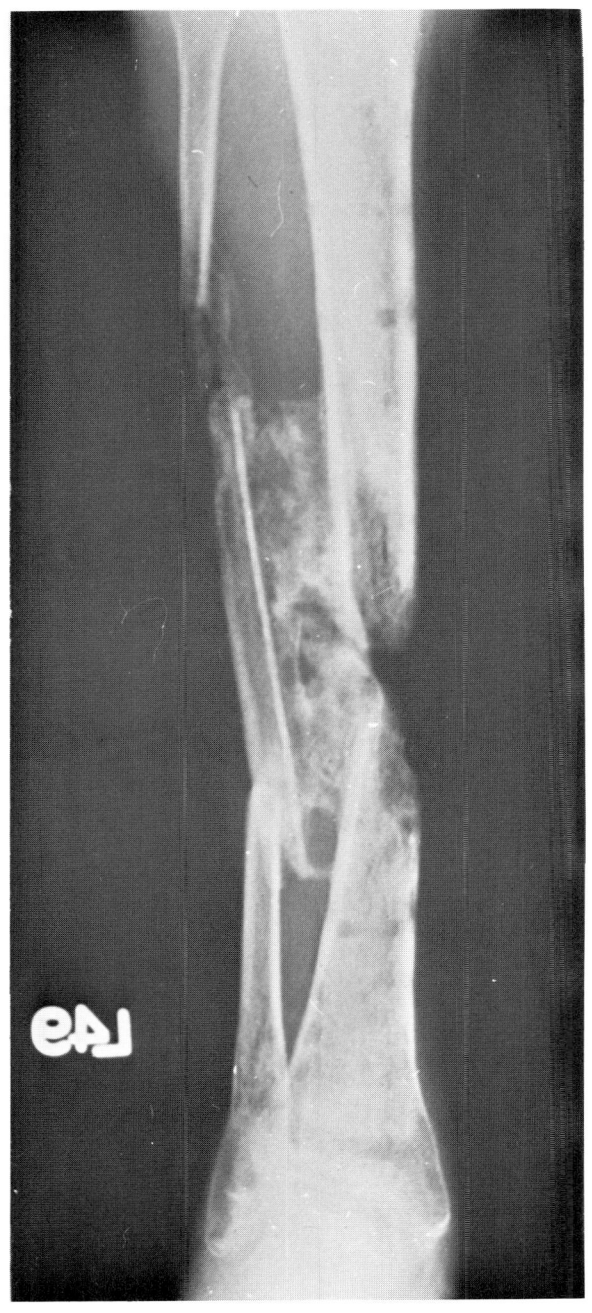

Figure 99: Posterolateral bone graft is indicated when one-third or more of the cortex is deficient.

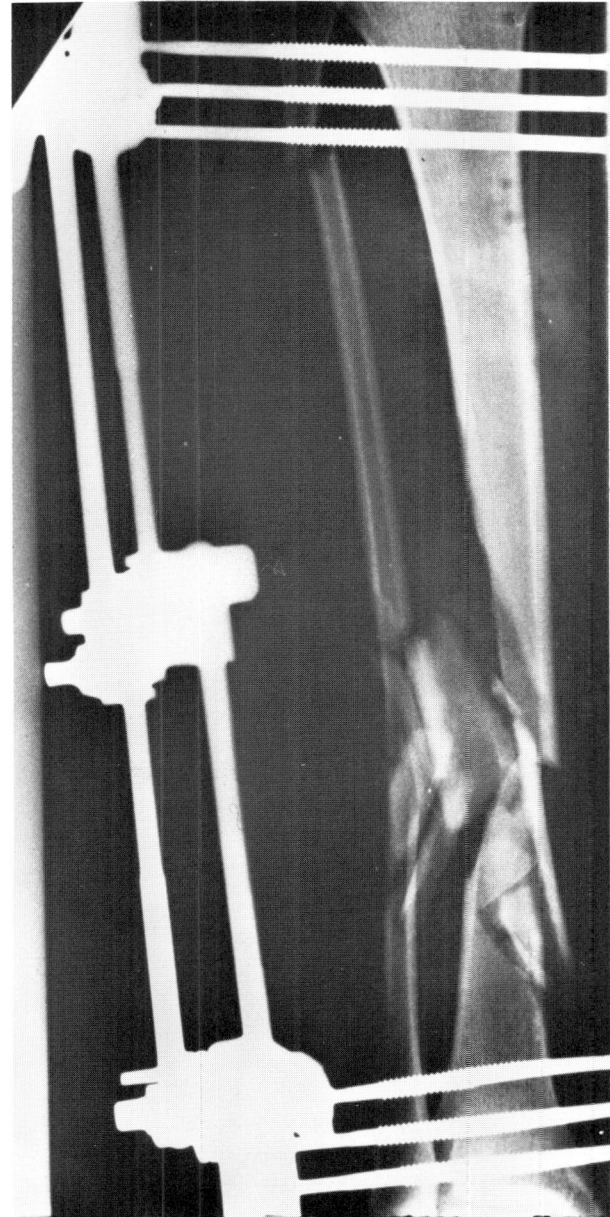

Figure 100: Open comminuted tibial fracture. The surgeon thought the comminuted tibia "looked like a bone graft," so supplementary graft was not done. A nonunion developed.

during, and for three to five days after the surgical procedure.

Technique for Bone Graft

POSITIONING THE PATIENT: It is desirable to place the patient in the position that will permit sufficient bone to be obtained and transplanted into the involved limb without unnecessarily turning the patient during the surgery. Fortunately, there are enough donor sites with good quality cancellous bone to permit this strategy. With the patient in the supine position, bone can be obtained from both greater trochanters, both tibial metaphyses, and both anterior iliac crests. If the patient is placed in the lateral decubitus position, use the uppermost greater trochanter and the anterior and posterior iliac crests. If the aforementioned donor areas have already been used, the patient may have to be turned over on the operating table. This is not harmful to the patient but increases the operating time and, more importantly, increases the length of time the bone graft is outside the body.

DRAPING THE PATIENT: Considerable ingenuity is required when draping for surgery a patient who is wearing an external skeletal fixator. The drapes should be arranged in a manner that will allow the bone graft to be harvested and transferred to the intended limb without contamination of the donor site by microorganisms from the pin hole. Prepare the fixator first. Remove all gauze bandages and adhesive tape from the fixator, and take off all protectors from the ends of the pins. Disassemble as much of the frame as possible, without compromising its stability. Remove any components that block easy access to the operative field.

Next, scrub the fixator, the pins, and the limb with a povidone-iodine soap, rinse them clean with alcohol, and paint them with the povidone-iodine solution, which is allowed to dry. Care should be taken during the cleansing to wash and prep the pin holes themselves thoroughly. The pin holes can be temporarily sealed with an application of a little antiseptic ointment. If any sharp pin points are present, wrap a sterile lap pad over the ends and hold them in place with sterile rubber bands. Place drape towels over the pin groups and components of the fixator, holding them in position with towel clips.

Finally, prepare the donor site in the usual manner. Both operative fields can be draped with skin towels

Figure 101: Fresh autogenous cancellous bone, obtained with a currette. An assortment of straight and curved currettes is very helpful in obtaining the graft.

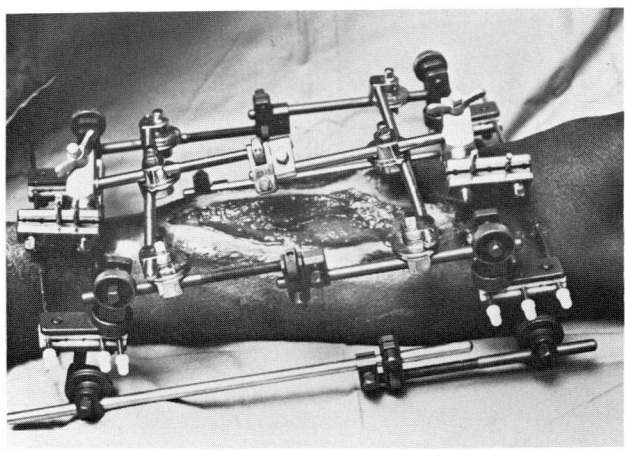

Figure 103: Draping a fixator for surgery.

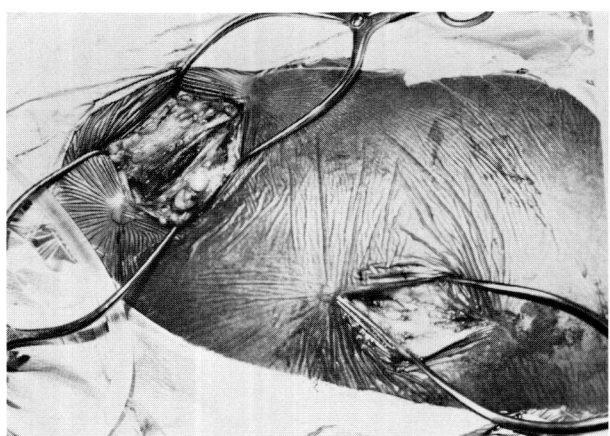

Figure 102: Obtaining the bone graft from the iliac crest and greater trochanter at same time. We now utilize shorter skin incisions.

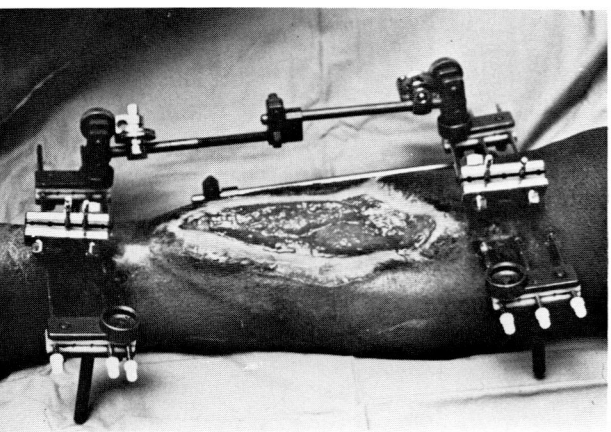

Figure 104: Remove all components blocking access to operative site. Remove pin covers and clean fixator with a povidone-iodine solution.

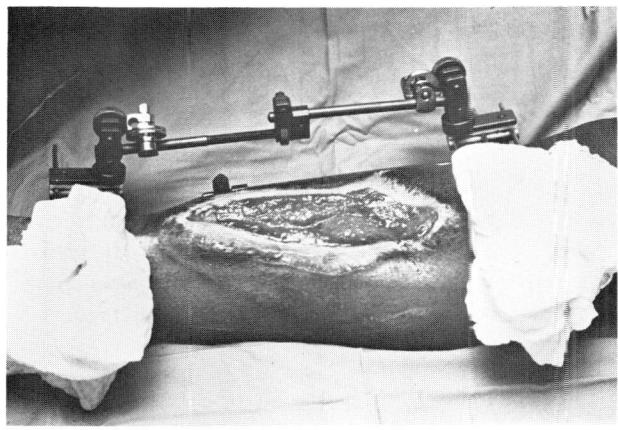

Figure 105: Cover pin grippers with sterile wrap pads.

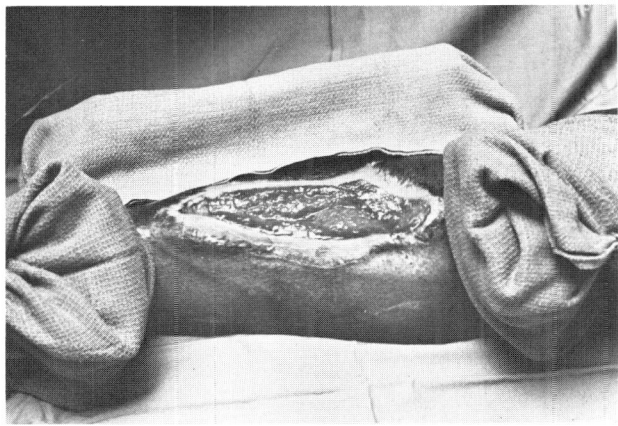

Figure 106: Cover frame with towel and use self-adherent sheets around the wound to hold them in place.

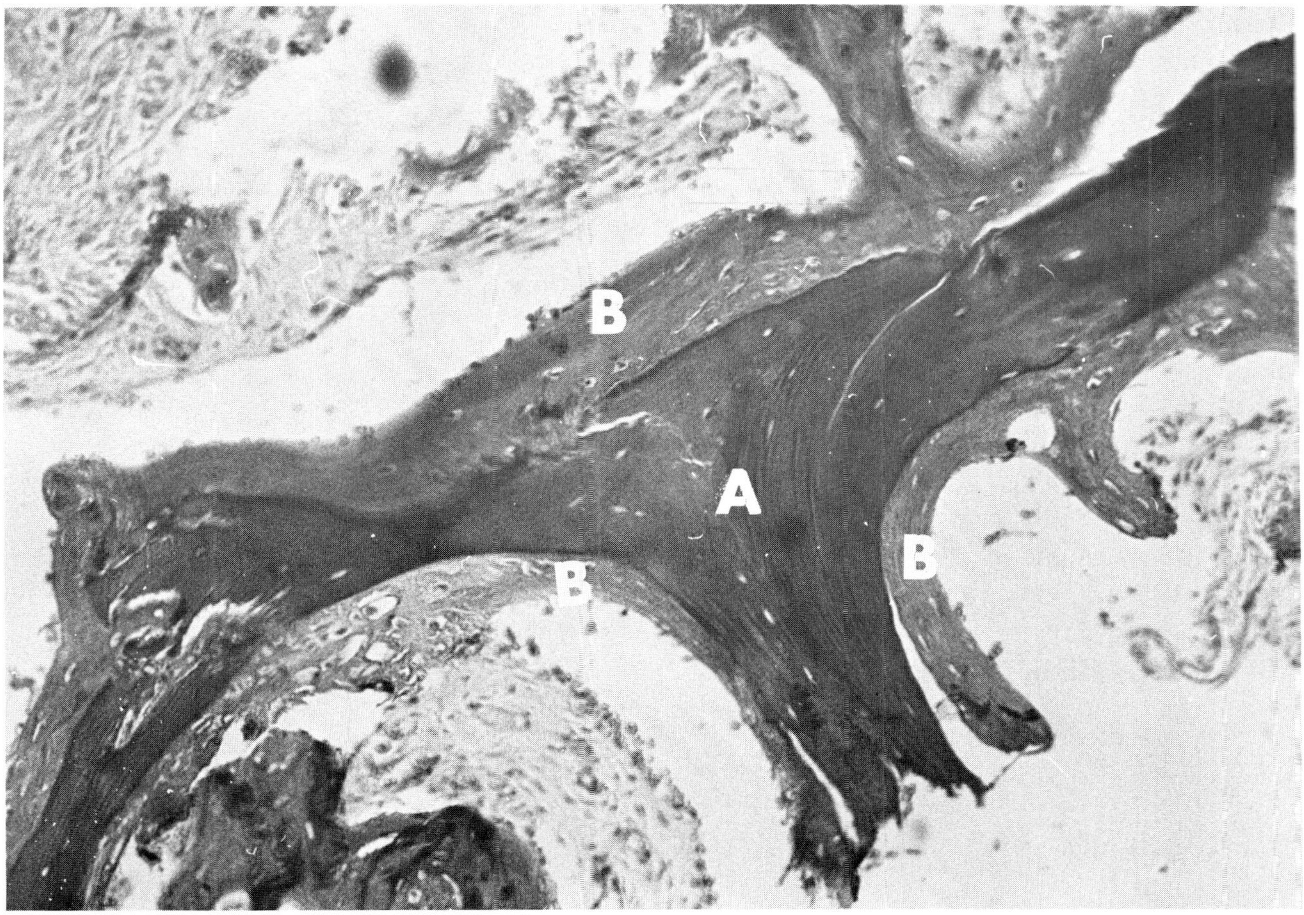

Figure 107: Fate of a fresh autogenous cancellous bone graft. Osteoblasts that survived transplantation form new bone (B) on surface of dead trabeculae (A), which are not absorbed.

and sheets, in such a manner that two separate operative areas are produced. Spray the operative site on the recipient limb with a tacky skin adhesive, and press a clear plastic self-adherent sheet in place over the surgical field. This effectively isolates the operative field from the pins and pin holes. Pass sharp towel clips through the plastic sheet into the skin to anchor the corners, preventing it from lifting off the limb during surgery. If a tourniquet cuff is needed, it may be necessary to use a sterile one, applied over the limb after the prepping and draping are completed. Cover the recipient limb with a sterile towel, and proceed to obtain bone from the donor site.

Obtaining the Graft

A fresh autogenous bone graft will rapidly be revascularized by ingrowth of surrounding tissue and will demonstrate new bone formation within one month of transplantation if some of the cells survive the ordeal of transplantation.[15] For this reason, every effort should be made to preserve as many viable cells as possible after the graft has been harvested from the donor site. Because of limited diffusion of nutrients to the cells of the graft, reduce the width of the graft fragments to 5.0 mm or less.[15]

Obtaining cancellous bone grafts through trap-door cortical windows seems to reduce postoperative pain and morbidity. An assortment of straight and curved curettes are necessary in order to obtain sufficient bone.

GREATER TROCHANTER: Make a two inch incision over the lateral prominence of the greater trochanter. Continue the exposure in line with the skin incision, through the fascia lata, down to the lateral cortex of the trochanter. Do not incise the periosteum, but use it instead as a hinge for a "trap door" in the cortex. Make a rectangle one by one-half inch by drilling four corner holes in the lateral cortex of the greater trochanter. Using an osteotome, create a trap door by making a U-shaped bone flap, based proximally. Crack the fourth (cephalad) side of the rectangle open, leaving the soft tissue hinge in place. Using curettes, remove the cancellous bone from the greater trochanter as far proximally as the apex, and as far distally as possible. Additional bone may be obtained with curved curettes by scraping against the inner surface of the shaft. Be careful not to curette bone from the base of the neck of the femur. The volume of bone obtained in this manner

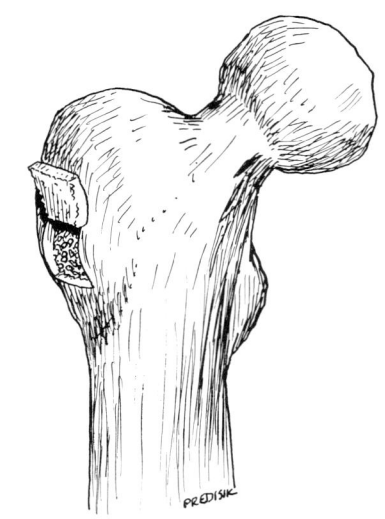

Figure 109: Hinged trap door in lateral cortex of femur. The periosteum must be left intact if the trap door is to be replaced.

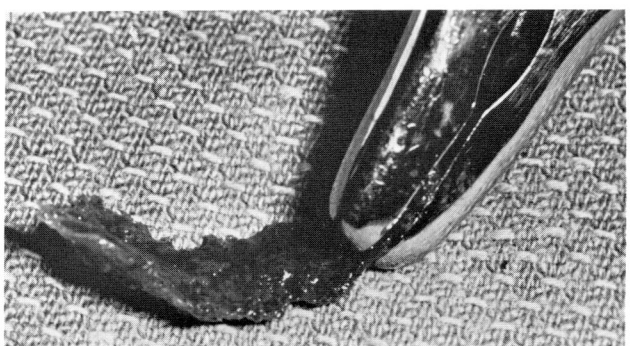

Figure 108: Diffusion of nutrients into graft fragment will not exceed 5 mm; therefore, keep graft fragments thin.

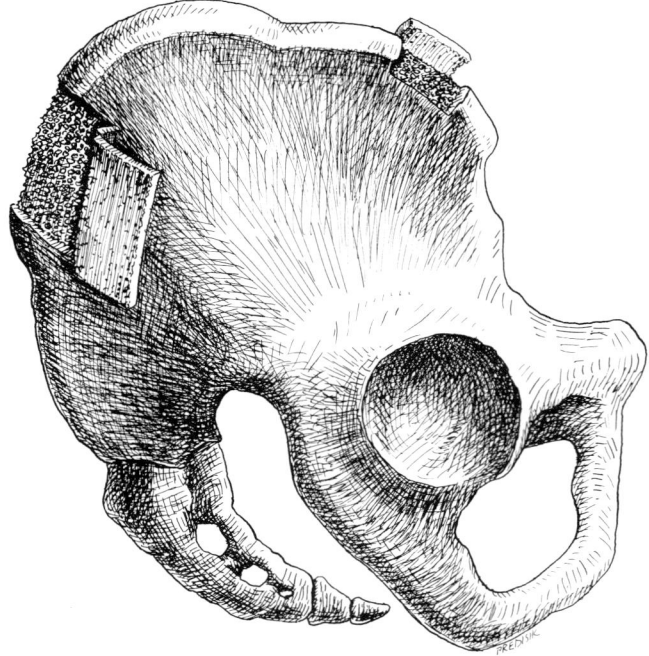

Figure 110: Trap doors for cancellous iliac bone graft. Curved curettes should be used to obtain bone from under the iliac crest.

usually equals the size of a man's thumb, after the graft is compressed.

ILIAC CREST: Pure cancellous bone can be obtained from the anterior iliac crest through a rather small incision by scooping the bone out from between the two tables of the ilium. This technique does not yield as much bone as is obtained when the outer (or inner) surface of the ilium is removed, but the patient experiences only minimal discomfort when the technique to be described is utilized.

Make a two inch incision along the iliac crest, starting one-half inch posterior to the anterior superior iliac spine. Continue the dissection down to the iliac crest, but do not elevate the periosteum. Using a half-inch osteotome, make two cuts perpendicular to the crest approximately one inch apart. Connect the cuts on the lateral surface of the iliac crest with a one-inch osteotome, and pry open the trap door, leaving a periosteal hinge medially. Using curettes, remove bone from between the two tables of the ilium. Curved curettes can be used to obtain additional bone from the area directly under the iliac crest, proximally and distally. Be sure to hollow out the anterior-superior iliac spine completely, as cancellous bone in this area is of good quality. The same procedure can be carried out when obtaining bone from the posterior iliac crest. If it is necessary to obtain a larger volume of cancellous bone, the outer table of the ilium can be removed as a thin sheet, permitting cancellous bone to be scraped from the inner table.

PROXIMAL TIBIA: Make a one inch incision over the proximal-medial tibial metaphysis at the level of the tibial tuberosity. Carry the incision through the skin and subcutaneous tissue down to the periosteum overlying the bone. Make a trap door in the tibia, approximately one-half by one-half inch, hinged with the periosteum proximally, as described above for the greater trochanter. Scoop out bone using straight and curved curettes, being careful not to remove bone from the subchondral region of the knee joint.

When a sufficient volume of bone graft material has been obtained, the donor sites can be closed by replacing the trap door and suturing its overlying periosteum to the adjacent soft tissues. If excessive bleeding is noted, microfibular collagen (Avitene®) can be applied to the exposed surfaces. Alternatively, Gelfoam® soaked with thrombin may be utilized to reduce bleeding. Close the donor site in layers, and apply a sterile postoperative dressing, but do not tape it into place at this time. Cover the donor site field with a sterile towel, and proceed to insert the graft into the recipient limb.

Graft Placement

The graft should be placed on an adequate bed of healthy well-vascularized tissue. Placing the graft in dense scar tissue may inhibit revascularization.[87] Tissue badly damaged by trauma should also be avoided as a graft bed, if possible. This limitation sometimes necessitates the employment of unusual surgical approaches for graft insertion. In the thigh, for example, extensive anterolateral scarring may suggest the posteromedial approach to the femur, as described by Henry.[126] In the lower leg, the usual damage associated with a compound wound is anterior and medial, overlying the tibia. This suggests the posterior-lateral approach described by Harmon,[123] which also permits placement of the graft along the interosseous membrane, thereby creating a tibia-fibula synostosis with the graft mass.

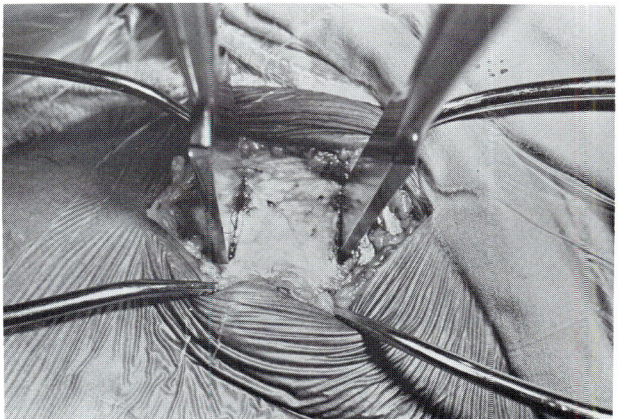

Figure 111: Making a trap door in the iliac crest with two 3/4 inch osteotomes. The inferior edge of the cuts are connected with a 1 inch curved osteotome and the trap door is pried open.

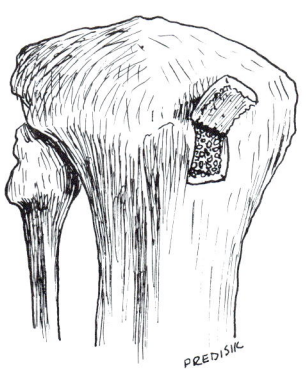

Figure 112: Trap door in proximal medial tibial metaphysis.

Postoperative Management

The orthopaedic surgeon should treat bone grafts with the same care plastic surgeons treat skin grafts. Skin grafts, of course, can be visually inspected during healing while bone grafts, unfortunately, cannot. Open bone grafting (the Papineau[213] procedure) does permit observation of the cancellous graft as it matures. At Rancho, we have been employing the open technique since 1975. Through use of this procedure, we have observed that certain measures promote rapid granulation tissue ingrowth and revascularization of the graft.

Absolute bedrest after the grafting procedure reduces limb swelling and seems to aid "take" of the graft (an observation also made by plastic surgeons). Bedrest during the first five to seven days postoperatively is quite reasonable. Since muscle action is conducive to a healthy graft-bed, physiotherapy should be started during recumbency. Progress of the graft can be followed roentgenographically. Pure cancellous bone is quite radiolucent and is, therefore, sometimes difficult to see on postoperative films. With increase in limb activity, however, the graft appears to come to life, with increasing radiodensity as time passes.

If evidence of wound sepsis develops, the grafted area should be opened immediately to permit free egress of purulent fluid. A cancellous bone graft is surprisingly tolerant of purulence in its midst, provided the wound is allowed to drain freely.

As the graft matures, the surgeon must decide when to remove the fixator. It is not necessary to leave the fixator on the patient until full maturation and trabeculation of the graft occurs. On the contrary, by reducing local mechanical stresses, the fixator may actually retard final remodeling of the graft. Two or three months after grafting, the frame may be removed and the limb supported in a molded orthosis. Alternatively, the fixator frame can be removed in stages.

If the graft has been applied to substitute for a segmental defect, or for aid in controlling a septic process, the fixator should be left on the patient until the graft is quite mature (eight to twelve months is not unreasonable in some situations). A radiolucent line through the center of the graft mass suggests the fixator has not been sufficiently rigid to prevent micromotion (Figure 113). This "pseudarthrosis" of the graft is extremely troublesome and is unlikely to heal with prolonged ambulation in an orthosis. We manage this problem by regrafting the area, and increasing the rigidity of the fixator with supplementary half-pins and additional bars, combined with slight compression of the graft mass. This strategy has resulted in consolidation of the radiolucent line within the graft mass in approximately ten weeks.

Unloading the Frame

Whether or not supplementary bone graft has been utilized, there comes a point in time when the fixator, by virtue of its rigidity and strength, prevents the final stages of bone remodeling from taking place. The healing or grafted fracture may not progress beyond the initial stages of bone remodeling until the mechanical contribution the fixator is making to limb stability has been eliminated. The bone, unfortunately, is frequently

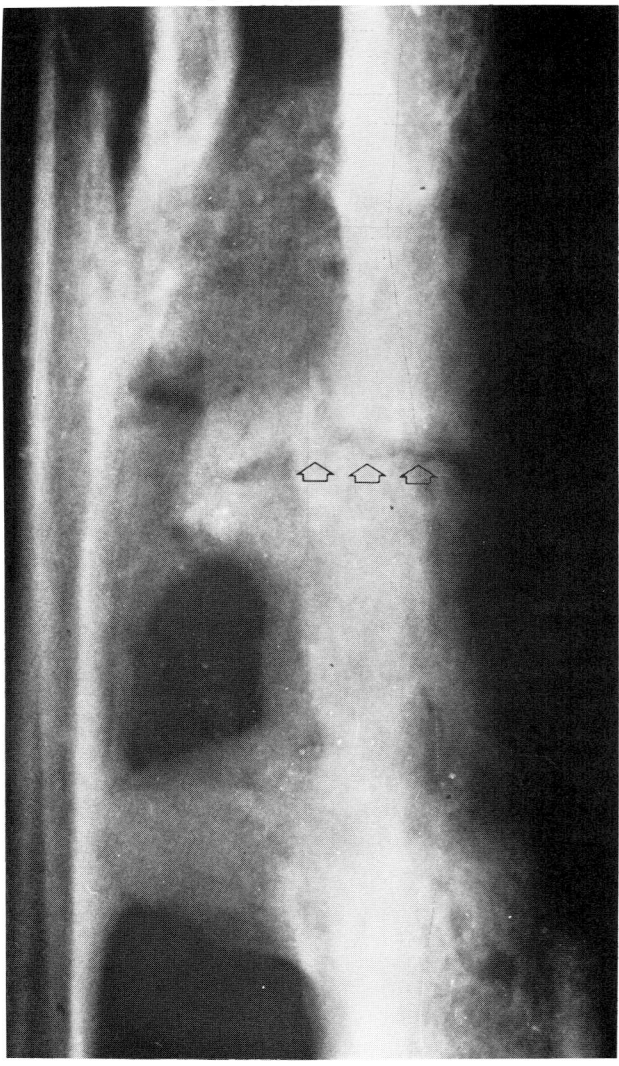

Figure 113: "Pseudarthrosis" of graft mass. Note the stabilizing effect of the synostosis to the fibula.

not strong enough to prevent bending at the level of the immature fracture callus when the fixator is removed. This problem has given rise to a concept called "unloading the frame." With this procedure, an attempt is made to strengthen early tenuous bone union (or early maturation of the bone graft) by sequentially removing connecting bars of the fixator frame.

When utilizing the Hoffmann system in its quadrilateral configuration, uprights are removed one at a time, at two week intervals. (Appropriate modifications for other frames can be made.) In this manner, an increased load is transferred to the bone itself, while stability is maintained in the fixator. No experimental evidence is available that proves this procedure actually results in stronger bone when the fixator is finally removed, but the idea is appealing. The healing bone is presumed to provide some mechanical stability to the bone-fixator system as the frame is unloaded.

The last step of frame unloading is removal of the final bar (Figure 115). Before the transcutaneous pins

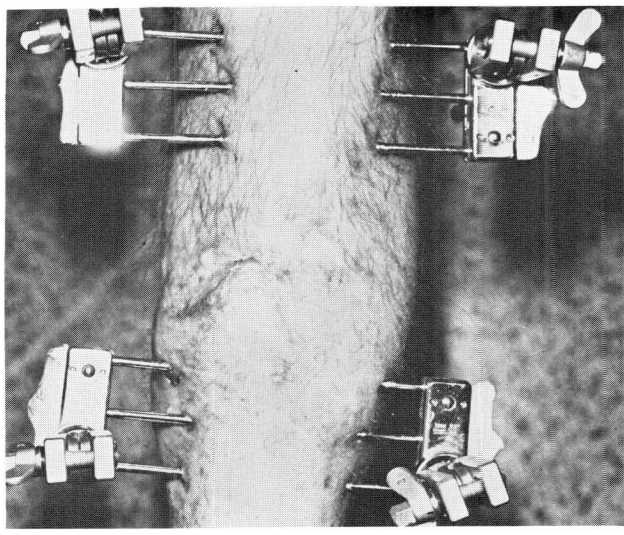

Figure 115: All bars removed—pins and clamps left in place for one week.

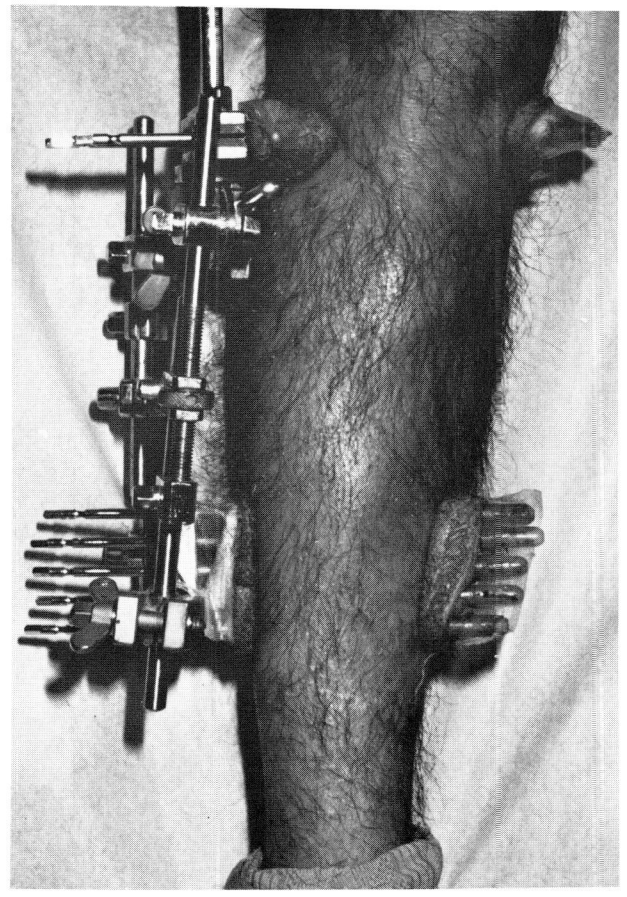

Figure 114: Quadrilateral frame partially unloaded.

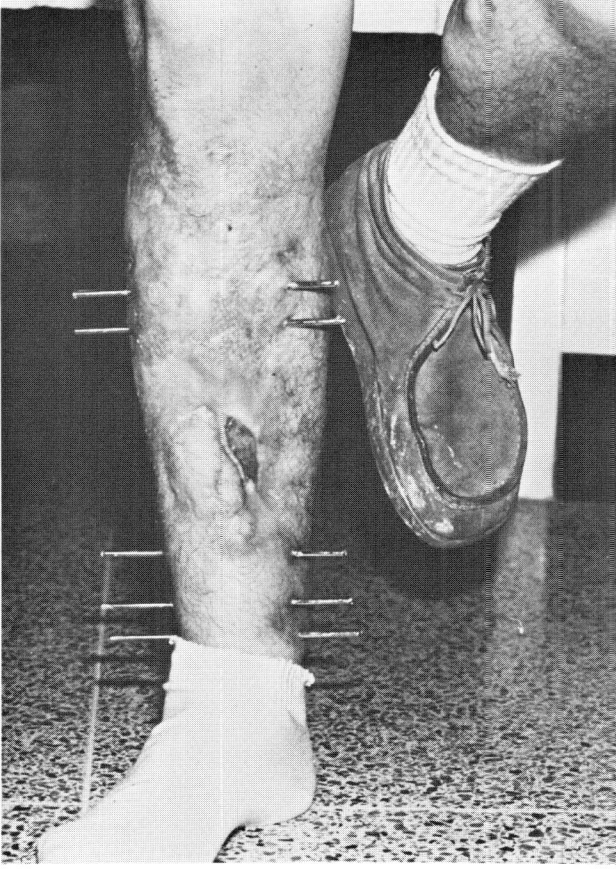

Figure 116: Testing stability before pin removal.

are removed, however, the bone is tested manually to determine stability. If the fracture feels solid after all bars have been removed, it is reasonable to plan pin removal as soon as possible.

If the fracture demonstrates a firm but slightly flexible resistance to bending without displacement at the fracture site, the pins can be left in place after the bars have been removed, and the limb supported with plaster-slab splints. (The splints can be applied over soft padding, which is wrapped around the limb. They should be arranged so that they do not make contact with the pins.) When the patient returns the following week for evaluation, the splints are removed and the fracture tested again. If the firm but flexible resistance is still present, the pins are removed and the patient managed with a molded fracture brace or cast. If, on the other hand, the fracture has become progressively unstable, the uprights should be reapplied and a bone grafting procedure considered.

This approach to the final removal of a fixator has been developed because a false sense of consolidation may be present immediately after the frame is disassembled. The apparent union is due to the presence of very dense fibrous tissue (instead of bone) around the fracture site. Within a week, however, the fibrous tissue will stretch, revealing its true identity. Therefore, it is wise to wait several days for final removal of the pins in order to properly "stress test" the fracture site.

As mentioned earlier, the patient should be made aware of the entire treatment program, before or shortly after the frame is initially applied. This avoids the disappointment, frustration, and impatience that may occur if a frame is reapplied after it has been off the patient for one week.

A technique for unloading the frame that eliminates errors in judgment regarding the "feel" of fracture stiffness has been developed by Jorgensen.[145] At each clinic visit, he replaces one bar in the frame configuration with a measuring gauge that is capable of determining very slight deflections in the configuration of the frame when it is manually stressed. The reading on the gauge is mathematically converted into degrees of deflection at the fracture site, based on the geometry of the frame configuration. The deflection decreases throughout healing as the fracture matures.

When the fracture site deflection is noted to be one degree or less, the patient is permitted to partial weight-bear with the fixator in place. Two weeks thereafter, if stiffness improves again, the patient is encouraged to take full weight on the injured limb. If continued progress is noted after two more weeks have passed, the connecting bars are loosened. The bars are finally removed two weeks later if progressively increasing stiffness is noted with gauge measurements.

With deflection measurements, Jorgensen not only can determine when to remove the frame, but he can also plot changing fracture stiffness as a function of time. In this manner, a curve is derived that can be compared to curves obtained from other patients who have healed in a normal manner. Disturbances in fracture healing can be detected quite early with this method.

Jorgensen's contribution to the problem of assessing fracture healing while a patient is in external skeletal fixation has not been widely employed. His technique and that of Burny, (see below) deserve consideration by anyone treating fractures with external skeletal fixation.

Elastic External Fixation

Franz Burny, of Brussels, promotes the concept of "elastic external fixation."[37] He notes that rigid external fixation retards physiologic fracture callus formation by reducing fracture motion. Burny's system permits slight motion at the fracture site, while achieving alignment control of the limb with external fixation. He utilizes the Hoffmann fixation system but employs a single bar connecting two pin-gripping clamps, one above the fracture and one below it. The frame is supplemented with a second bar for use with

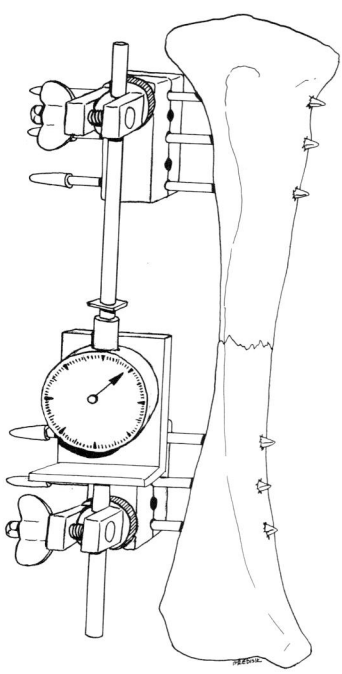

Figure 117: Measurement gauge temporarily substituting for bar to determine fracture stiffness.

very unstable fractures or with fractures of the femur.

Burny's idea is hardly new. The external fixation device invented in 1897 by Clayton Parkhill[214] utilized half-pins connected to a single fixator frame. Lambotte[168, 169] developed a similar system in 1902 Raoul Hoffmann[128, 134] invented the fixator now being employed by Burny. By utilizing double threaded half-pins and a strong pin-gripper, the Hoffmann system appears to have overcome the shortcomings of the previous unilateral frames.

Burny's experience with this system of fracture management is extensive. He has reported results in 1,421 tibial fractures[37] and 100 humeral fractures[41] treated in this manner. According to his studies, the slight flexibility of the unilateral fixator configuration promotes the development of a normally contoured fracture callus. Anatomic reduction is important to Burny's concept, because the bone must bear some of the mechanical load. He reports that 28.8 percent of his cases required open reduction for alignment of the fracture at the time the fixator was applied. An additional 6.8 percent required a "minimum" open reduction (usually enough to get one finger around the bone). He managed 64.4 percent of the cases with closed reduction and application of the fixator. In 5 percent of his tibial series (usually oblique fractures), Burny combined external fixation with one or more screws across the fracture site to maintain alignment.

At the termination of treatment, 25.8 percent of the patients were placed in a plaster cast, but Burny states that he now routinely leaves the frame in place until the fracture is healed. During fracture healing, the patient's single bar is replaced at clinic visits by one that contains a strain gauge.[38] In this manner, Burny is able to measure the stiffness of the fracture as healing progresses. These measurements can be plotted and a curve is generated that has the same configuration to the healing curve that is obtained with Jorgensen's mechanical deflection gauge, described earlier.

A disturbance in fracture healing is readily detected when the changing slope of the curve does not follow the predicted pattern. Bone grafting, or another modification of the treatment protocol, can be planned accordingly. The "elasticity" of Burny's system depends primarily on the flexibility of the transcutaneous pins. Unfortunately, cyclic movement of a pin can promote pin loosening, a problem that can lead to secondary pin tract sepsis.

If Burny's experience with flexible external fixation is confirmed by other clinicians, a valuable tool will have been added to the armamentarium of the fracture surgeon. The use of half-pins and a single connecting bar greatly simplifies the fixator application and reduces the weight and bulk of the frame. More important, a method of fracture management will be available that permits restoration and maintenance of anatomical alignment without the extensive surgical exposure required for plate fixation.

Biocompression

Zbikowski,[294] a Spanish orthopaedist, has reported his preliminary experience with a system he calls "biocompression." He has developed a fixator frame that permits up-and-down pistoning at the fracture site while maintaining overall alignment. He accomplished this by substituting telescoping bars for the compression bars in a Vidal-Adrey quadrilateral frame (Figure 119). Because the bars must telescope freely without binding, Zbikowski has devised a special alignment guide to insure that all four connecting bars are parallel to each other.

He reported his preliminary results at the Seventh International Conference on Hoffmann External Fixation in 1980. He presented nine patients treated for acute tibial fractures (four closed, five open), with an average time to union of 3.95 months. He states that "one outstanding feature is the capacity of the patient to jump on the injured leg with no pain, and with the feeling of a strong solid union, since the moment the external fixation is removed."

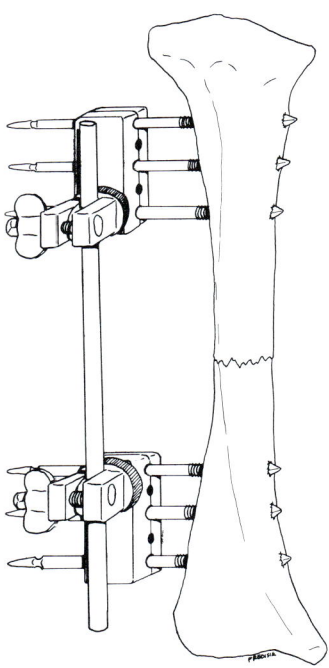

Figure 118: "Elastic" external fixation.

Unfortunately, Zbikowski has noted an increased incidence of pin sepsis with his system, compared with his experience with the standard quadrilateral frame. This is not surprising, considering the role that motion plays in the development of pin sepsis. Nevertheless, Zbikowski's preliminary reports are quite encouraging.

More fixator designs that will permit controlled motion of the fracture site while maintaining overall bone alignment will probably be forthcoming. What is needed, however, is experimental data defining the role of motion in fracture healing. Specifically, the question to be answered is whether flexion at the fracture, telescoping at the fracture site, or shear motion at the fracture site most rapidly produces union. Furthermore, as Chou[49] indicates, "No knowledge is currently available concerning the ideal amount of fixator stiffness required for optimal bone healing." External fixation may become a useful research tool to allow these questions to be answered in a precise and orderly way.

Electrical Stimulation

Electrical stimulation of bone healing is another modality for dealing with retarded union. While electrical stimulation of bone healing is still somewhat controversial, there is an impressive body of clinical and experimental evidence to suggest that electric current can stimulate bone healing, especially if the union has been delayed[16, 24, 114, 216]. As noted in the beginning of this chapter, remodeling within a fracture callus responds to electrical currents generated by the piezoelectric properites of bone crystal and collagen. Rigid external fixation, by reducing fracture motion, decreases the natural stimulation of fracture healing. It is hoped that an externally applied current will substitute for this deficiency, resulting in swift healing of complex fracture problems that require external skeletal fixation for stability.

Stimulation of bone healing with electric current can be accomplished in several ways: (a) the current can be applied directly into the fracture site via implanted[216] or percutaneous electrodes;[24] (b) the current can flow across the fracture site through electrodes at a distance;[144] (c) the electric current can be induced within a fracture site by an external electromagentic field.[16]

Jorgensen[144] has, for years, been applying electric current to the pins of a Vidal quadrilateral frame, using one set of pins as the anode, the other as the cathode. Fortunately, the pins are electrically insulated from each other and from the metal frame of the Hoffmann system.

Early in his experience, Jorgensen observed excessive bone build-up around the cathode pins and osteolysis around the anode pins. He solved this problem by reversing the polarity of the circuit frequently. He noted augmentation of fracture healing when electrical stimulation was used. Jorgensen defined the end-point of healing as one degree of fracture site deflection when

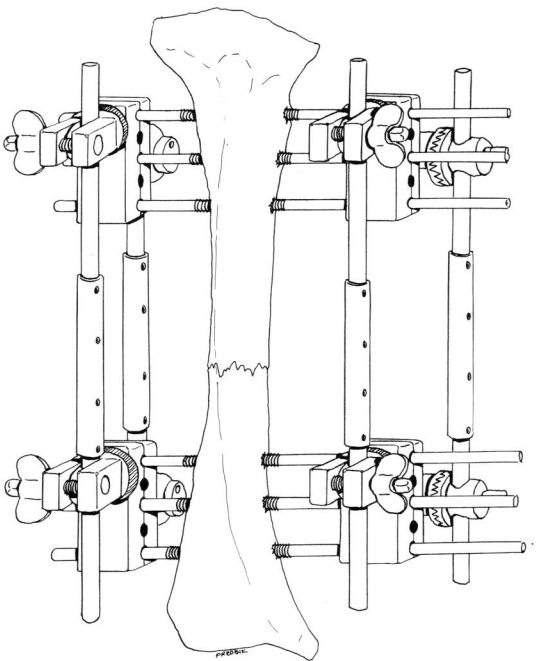

Figure 119: Biocompression—freely telescoping bars in a quadrilateral configuration.

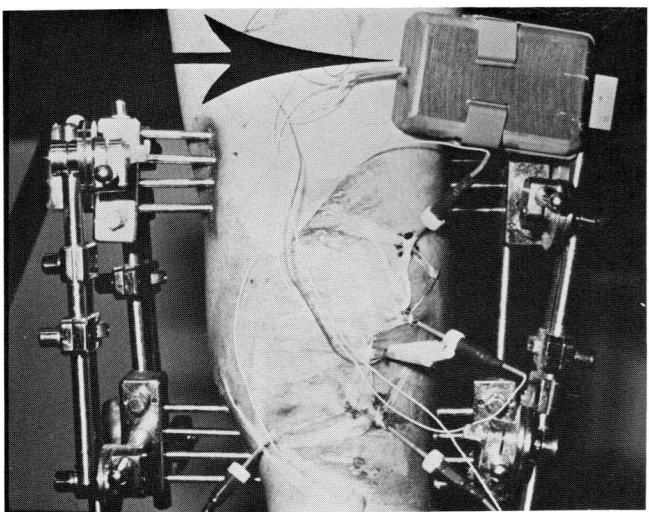

Figure 120: Electrical stimulation pack attached to fixator frame—note percutaneous electrodes.

measured with a gauge placed in the Hoffman frame. Fractures treated in the quadrilateral frame with electrical stimulation "united" in 2.42 months, whereas a similar group of patients treated in a Hoffmann frame but without electrical stimulation united in 3.3 months.

Various branches of electrical stimulation research hold great promise for the future of external skeletal fixation. For example, it may be possible to use electrical stimulation to overcome the problem of retarded bone healing associated with rigid immobilization of highly comminuted fractures. In fact, if the newer methods presented in this chapter prove fruitful, I can foresee a system of fracture management that utilizes external skeletal fixation, with purposefully constrained fracture site motion and electrical stimulation to deal effectively with difficult fracture problems.

Internal Fixation Following External Fixation

In some situations, it may become obvious to the treating physician that a fracture healing in a fixator is not progressing according to plan. This may occur when one or more of the factors known to contribute to retarded bone healing are present. To deal with the problem, the physician may elect to convert the patient from external skeletal fixation to internal fixation, an approach that is reasonably safe. (The practice is widely accepted when limb lengthening procedures are carried out with external skeletal fixation.[283]) In Burny's series of 1,421 tibial fractures treated with external skeletal fixation, 1.2 percent were subsequently treated with a plate and screws, and 1.7 percent were eventually managed with an intramedullary nail.

When planning internal fixation following external skeletal fixation, allow a one-month interval to elapse following removal of the fixator to allow the pin holes to heal completely. Alternatively, one may apply internal fixation immediately after the fixator is removed, if the wound over the plate is left open in the region of the pin holes.

There is a risk of late sepsis if an intramedullary nail is used in a bone that had been previously managed with external skeletal fixation. Krempen[164] describes a case of infection following intermedullary rodding of a fracture that had been managed with external skeletal

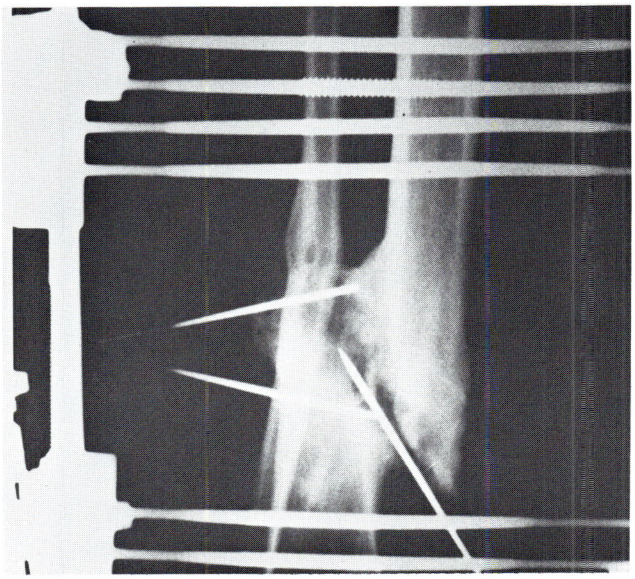

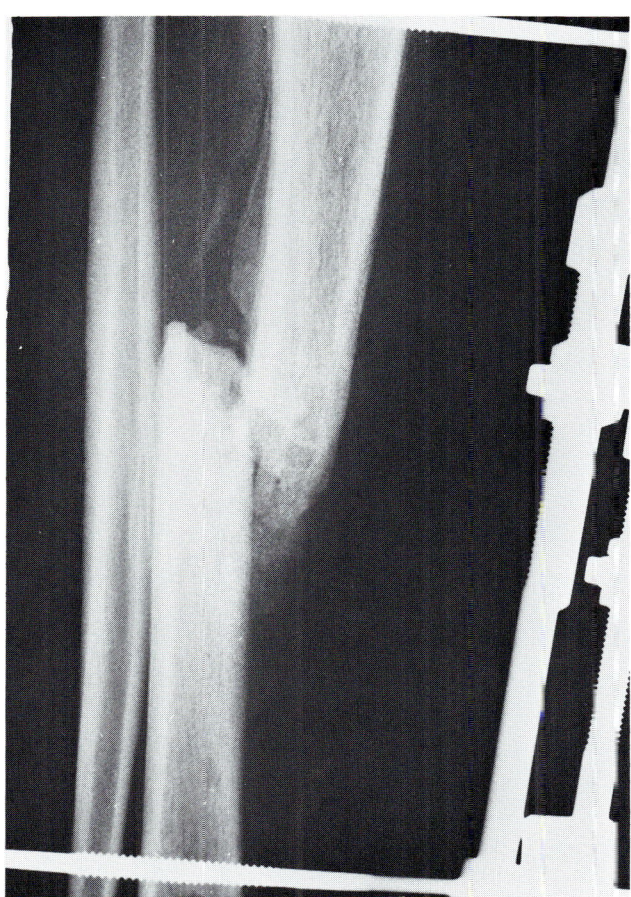

Figure 121: Percutaneous electrical stimulation with fixator in place. Preliminary results are promising for this combination.

Figure 122: Incorrect technique: inadequate reduction. The fracture healed because the patient was sixteen years old.

fixation. Olerud[208] notes two cases in his series that also developed infection following secondary intermedullary nailing, three months and seven months postinjury. Arcq,[11] on the other hand, has experience with eleven intermedullary rod stabilizations after external skeletal fixation, without deep sepsis. Nevertheless, a wise policy is to avoid intermedullary nail stabilization of a bone that has been previously treated with external skeletal fixation.

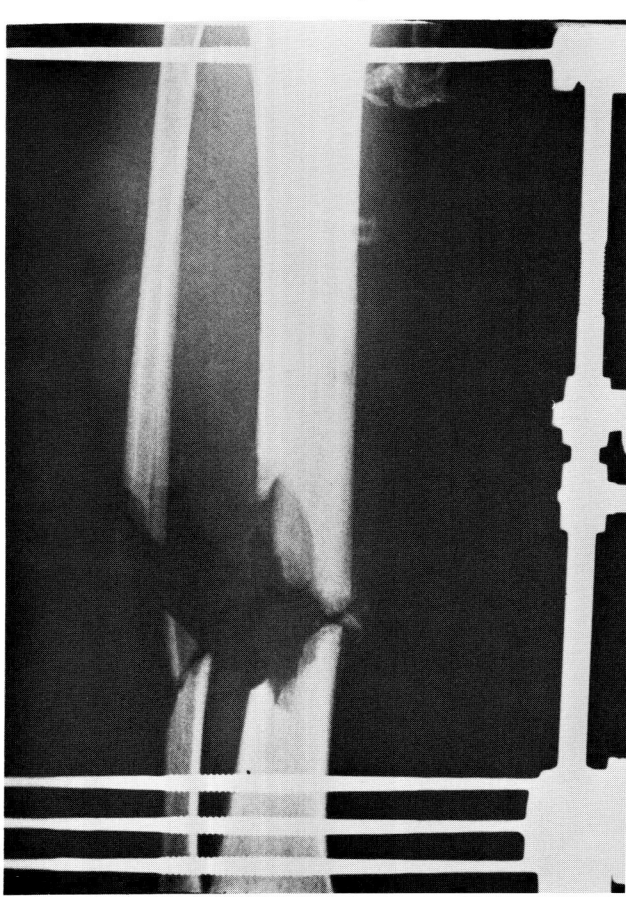

Figure 123: Cortical defect requiring extensive bone grafting. Shortening the limb one inch by squaring off the fracture site will speed healing.

Chapter 6

UNSUCCESSFUL ARTHRODESIS

Introduction

External fixation is an excellent way of achieving a solid joint fusion, especially in circumstances where arthrodesis by other techniques is difficult to achieve. Failure to follow certain technical details, however, can lead to an unsuccessful arthrodesis. In some joints, the rate of successful fusion with compression arthrodesis employing external fixation may be higher than the rate of fusion obtained with other methods. Lance and coworkers,[170] for example reviewed twenty years experience with arthrodesis of the ankle. They noted that compression arthrodesis accomplished with external fixation (Charnley clamps) resulted in the highest rate of union (94%) when compared to other techniques such as anterior sliding bone graft (89%) or fibular bone graft (64%).

Indications

The patient requiring a joint arthrodesis after *removal of a total joint replacement implant* represents a challenge that can tax the ingenuity of any surgeon. In the hip joint, for example, arthrodesis is difficult to achieve because the femoral head and neck have been resected. Arthrodesis under these circumstances is not easy to obtain with any technique. Supplementary bone graft is usually required. In the ankle, shoulder, and elbow joints, arthrodesis is difficult after total joint replacement because of the removal of bone when the components were originally inserted.

In the presence of *chronic joint sepsis*, compression arthrodesis with external fixation is the treatment of choice.[52, 191, 252] In this situation, thorough joint debridement can be combined with the application of a fixator in the same operative session. As Clawson and McKay[59] point out, "arthrodesis in the presence of active infection . . . can be performed as a one-stage procedure if the basic surgical techniques of radical debridement, stable fixation and adequate postoperative drainage of the wounds are followed."

Neuropathic arthropathy is another indication for compression arthrodesis with external fixation. Failure rates as high as 50 percent have been reported with other techniques.[27, 85] Thorough debridement of the synovium, careful bone carpentry, and long-term immobilization are required to obtain a successful fusion.

A joint damaged by *degenerative or posttraumatic osteoarthritis* can also be successfully fused with external skeletal fixation techniques. As noted above, the rate of successful fusion may be higher with compression techniques than with other methods. In the absence of a history of sepsis, however, compression arthrodesis utilizing internal plate fixation is probably preferable to compression and external fixation, provided the surgeon is familiar with the exacting technical details developed by the AO-group in Switzerland. These techniques, including double plating of the knee joint, "cobra" plating of the hip joint, and inverted T-plating of the ankle joint, eliminate the need for the external fixator; yet, they provide compression across the fusion surfaces.

Causes of Failure

An arthrodesis may be unsuccessful for several reasons. First, solid fusion might not have been achieved. Second, a joint may be solidly fused but in a position unsatisfactory for optimal limb function. Third, a joint fused for the relief of pain might continue to be painful in spite of a solid fusion in an excellent position. Finally, an arthrodesis performed to control joint sepsis might continue to drain even if solid union is achieved.

Failure to achieve a solid fusion does not necessarily mean that the arthrodesis procedure was unsuccessful. Fibrous ankylosis may be acceptable to the patient if it is painless. Ratliff[226] noted four (6.7%) cases of fibrous ankylosis out of fifty-nine ankles fused with Charnley's technique. All the patients were pleased with their results; when healed, these ankles could not be distinguished from those in which bony fusion had occurred.

Nevertheless, every effort should be made to obtain a solid fusion. The basic principles of compression arthrodesis were outlined by John Charnley[54] in his book *Compression Arthrodesis*, a classic monograph that should be reviewed by the surgeon before undertaking these procedures.

In joint arthrodesis, several technical considerations are essential to insure a high rate of union. The goals are to obtain two congruous cancellous surfaces that can be pressed together and to maintain the compression until a solid union is achieved. It is difficult, however, to achieve total congruence of both surfaces. Fortunately, rigid contact at one point (which Charnley calls the "beach-head") will rapidly stabilize the entire arthrodesis if no shearing motion is present. Union will progress across the cancellous surfaces once the beach-head has achieved bone union. If, on the other hand, the cut cancellous surfaces constantly move, slide, or shift with respect to each other, union at a contact point will not be achieved; a nonunion may be the result.

In order to insure congruous cancellous surfaces, it is important to have wide exposure of the joint at the time of surgery. Where possible, the surfaces should be cut with a broad hand saw or oscillating saw. In the hip and shoulder, where the joint surfaces are rounded, considerable care is required to completely remove all cartilage and subchondral bone. Congruity and contact of the bone surfaces must be checked with roentgenograms after the fixator is applied. Any error should be corrected before the patient leaves the operating room. Lance et al.,[170] in their review of arthrodesis of the ankle joint, identified forty-four cases where technical errors were clearly evident on the earliest postoperative roentgenograms. Among these technically inadequate procedures, 50 percent ended in nonunion, and one-half of the remaining cases had an unsatisfactory end result for some other reason. The use of intraoperative roentgenograms following application of the fixator, therefore, is an essential part of the procedure.

After the appropriate bone contact and position is achieved, the fixator should be compressed. Excessive compression should be avoided because of the possibility of pressure necrosis of bone at the point of contact. It is reasonable to compress the frame until the transcutaneous pins deflect 2-3°. The frame should be tightened slightly at one- or two-week intervals until it is removed. It is best to prohibit unprotected weight-bearing in the lower extremity applications and limb use in upper extremity applications in order to prevent pin loosening and subsequent pin sepsis.

In general, the fixator should be left in place two to four months for most applications. Hip and shoulder arthrodesis may require four to six months of application. Also, any application where cancellous bone is not present (as after removal of a total knee arthroplasty) may require extensive cancellous bone grafting and long-term fixator application.[114, 194] Follow fixator application with an external support, either a plaster cast or a molded orthosis, until the union is quite mature.

Unsatisfactory limb position is a major cause of patient dissatisfaction following joint arthrodesis. Pain, gait abnormalities, and mechanical disturbances in adjacent joints result from a joint fused in suboptimal position.[3] Unfortunately, the optimal position of fusion has not been clearly established for all joints. Nevertheless, certain guidelines can be used in selecting the appropriate position of fusion according to the patient's age, sex, occupation, and life-style. Detailed consideration of the position of fusion will be made in the subsections of this chapter dealing with specific joints.

In most situations, *persistent pain after arthrodesis* can be readily explained. Nonunion, or union in an unsatisfactory position, accounts for most cases. Postoperative neuromata, chronic swelling, and degenerative changes in adjacent joints will account for most of the rest. There are some patients, however, whose persistent symptoms following arthrodesis defy explanation. Sadly, it seems that the patients with the worst **pain are the ones who fit into the last category. Compensation-liability litigation frequently lurks in** the background. In some of the cases, a review of the roentgenograms leading to the arthrodesis may reveal minimal pathology at best. In the absence of clear-cut pathology, the best way to avoid the problem of unexplainable pain is to carefully screen patients with chronic joint complaints before the arthrodesis procedure.

Wound sepsis can occur following joint arthrodesis, as with any surgical procedure. To reduce the incidence of postoperative infection, employ the usually described methods, including administration of perioperative antibiotics, adequate hemostasis prior to wound closure, and suction drainage to prevent hematoma formation.

The problem of wound sepsis is especially troublesome if the joint arthrodesis was intended to eradicate a chronic joint infection. In this situation, allow the arthrodesis to consolidate before attempting to debride the septic focus. In some cases, drainage will stop when bony ankylosis is achieved. If drainage persists after bone union, it may be necessary to wait at least one year, until the fusion mass is quite mature, before performing a saucerization or sequestrectomy.

TECHNICAL CONSIDERATIONS

Ankle Joint

Discussion

Satisfactory fusion of the ankle joint is sometimes difficult to achieve (failure rates as high as 50% to 80% have been reported). Furthermore, a solid arthrodesis in an unsatisfactory position is known to lead to difficulties in walking.[252] Many techniques for ankle arthrodesis have been described. Surgical procedures utilizing staples, axial pins, sliding and interposition bone grafts, and compression plates have been employed.

As noted earlier, Lance[170] reported that the highest rate of successful fusion of the ankle was obtained with external fixation and compression, when compared with other methods. Verhelst and coworkers[263] also evaluated the success rate for ankle fusion, but they limited their report to experience with the transfibular approach, comparing three different methods of fixation. They noted that compression arthrodesis resulted in a higher fusion rate and a more rapid time to union than fibular on-lay bone graft or fixation with Blount staples. Recently, Morrey and Wiedeman[135] reviewed sixty patients who had ankle arthrodesis performed for chronic posttraumatic arthritis. They noted that the most reliable results were obtained when compression arthrodesis was performed through combined medial and lateral incisions. The combined approach allows removal of both malleoli, which yields

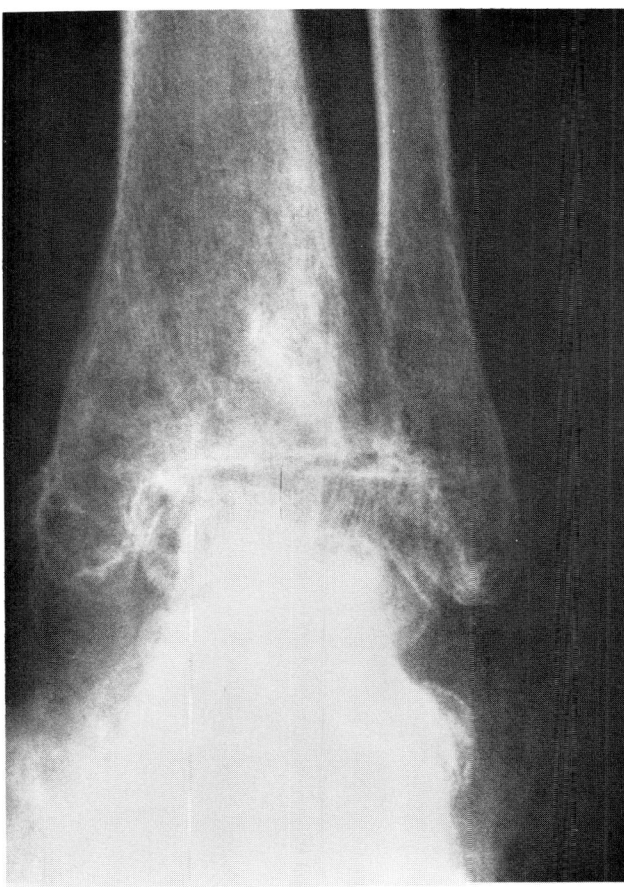

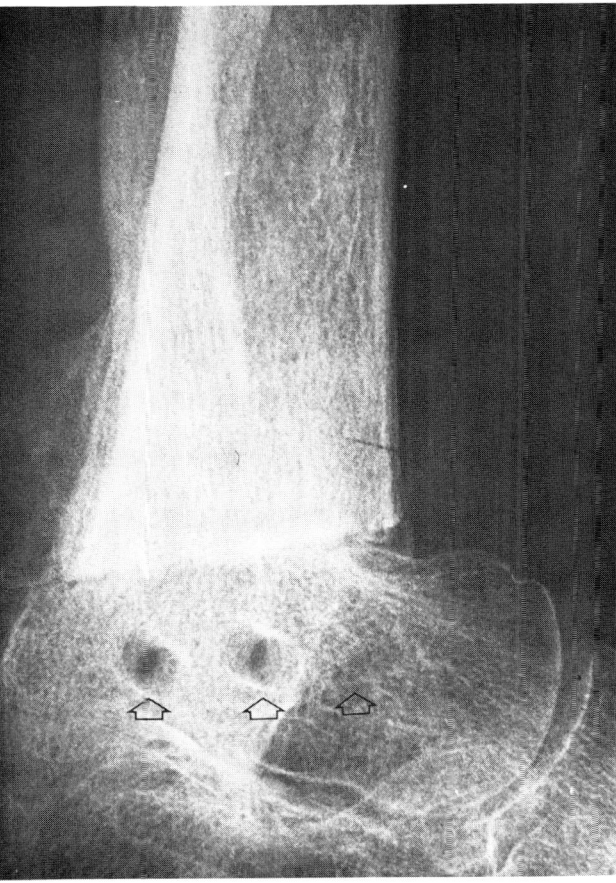

Figure 124: Ankle arthrodesis through anterior incision—malleoli retained. Technical problems include lack of precision with bone cuts; difficulties with pin placement into the talus because of the presence of the malleoli; and a squat bulging ankle.

Figure 125: Ankle arthrodesis with malleoli removed. Two or three pins placed in the talus. Note posterior position of the talus on the tibia.

a surprisingly acceptable cosmetic appearance for the area in contrast to the squat bulging ankle that remains when both malleoli are left in place following ankle fusion. The main advantage of the combined approach, however, is the excellent exposure that can be obtained for the distal tibia and upper talus, enabling the surgeon to protect both the anterior and posteromedial neurovascular bundles while cutting bone.

Frame Configuration

Compression across the ankle can be achieved with a pin or group of pins in the distal tibia connected by an external fixator to a pin or group of pins in either the talus or the calcaneus. At Rancho Los Amigos Hospital, we place two or three pins in the talus in the horizontal plane — parallel to the ankle joint.[114] Placing the pins in the talus rather than the calcaneus avoids compression of the articular cartilage of the subtalar joint.[88, 207] If, however, the subtalar joint is already stiff, we place pins in the calcaneus instead.

Supplementary fixation of the forefoot is desirable when performing an ankle arthrodesis in the presence of sepsis or for any application requiring more than four months in the fixator frame. This prevents forefoot motion, which tends to be transmitted to the hindfoot, resulting in premature pin loosening. A triangular arthrodesis frame can be constructed utilizing pins in the hindfoot, both of these groups being connected to a third group of pins that transfix the metatarsal shafts. Alternatively, half-pins attached to the fixator with an outrigger bar can be inserted into the first metatarsal shaft.

Position of Arthrodesis

The optimum position for ankle arthrodesis is one that is *neutral* with respect to dorsiflexion-plantar

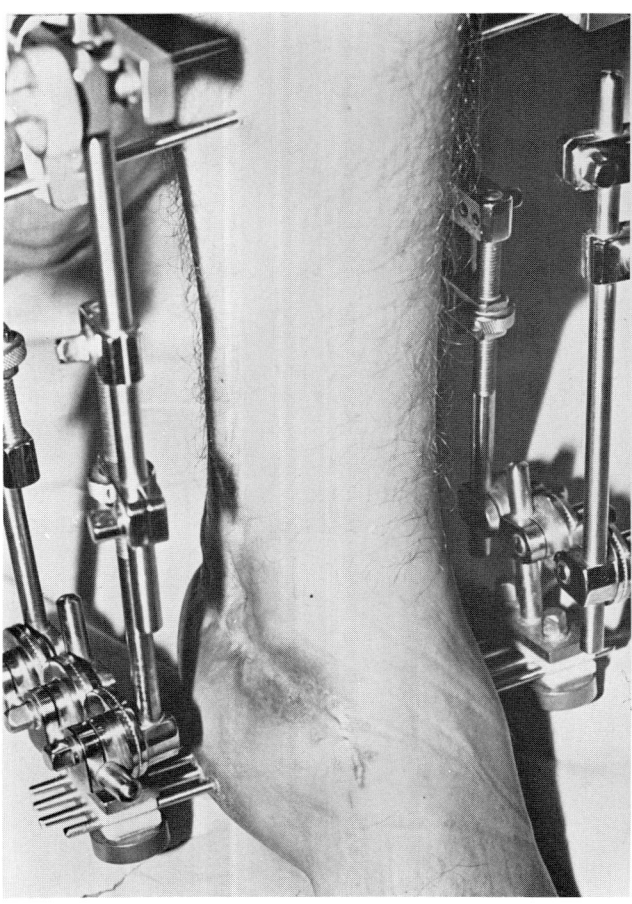

Figure 126: Ankle arthrodesis in the presence of a stiff subtalar joint. Pin placement is in the calcaneus.

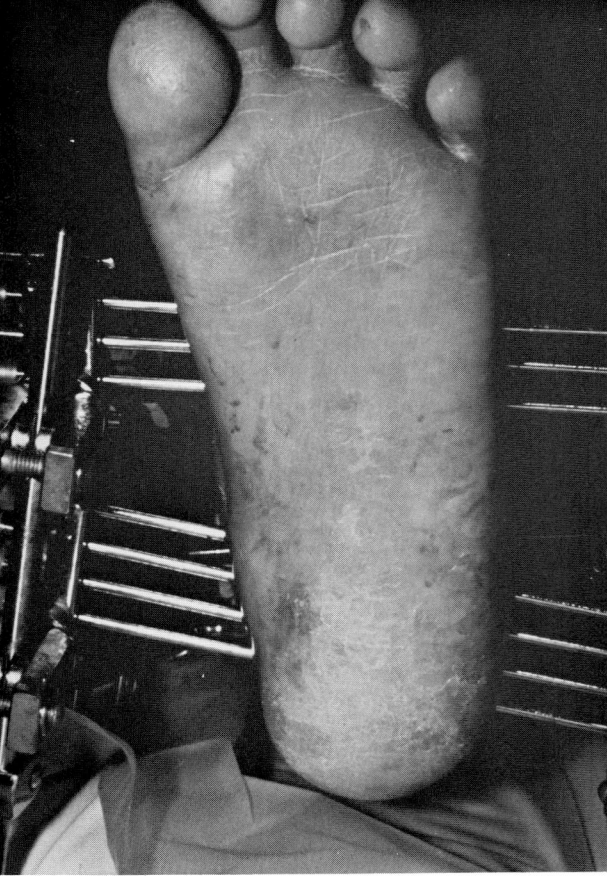

Figure 127: Forefoot stabilization for reconstruction of distal tibia and upper talus.

Unsuccessful Arthrodesis

flexion, and varus-valgus of the hindfoot.[1, 189, 226, 252] In this position, patients can walk well whether barefooted or wearing shoes.

Technique

With the patient in the supine position, tilt the operating table sufficiently to permit visualization of the lateral side of the foot and ankle. Make a longitudinal incision over the distal fibula, curving anteriorly at the tip of the lateral malleolus. Expose the fibula subperiosteally and transect the fibula, slanting lateral to medial, at a level slightly above the tibial plafond. Remove the distal fibula by cutting the talofibular and calcaneofibular ligaments. (Alternatively, the distal fibula can be retained and hinged laterally on the posterior talofibular ligament. If this technique is employed, the distal fibula must be

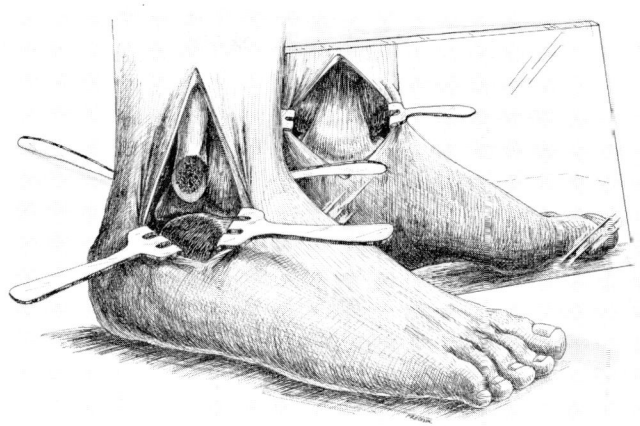

Figure 129: Combined lateral (transfibular) and medial approach for ankle arthrodesis.

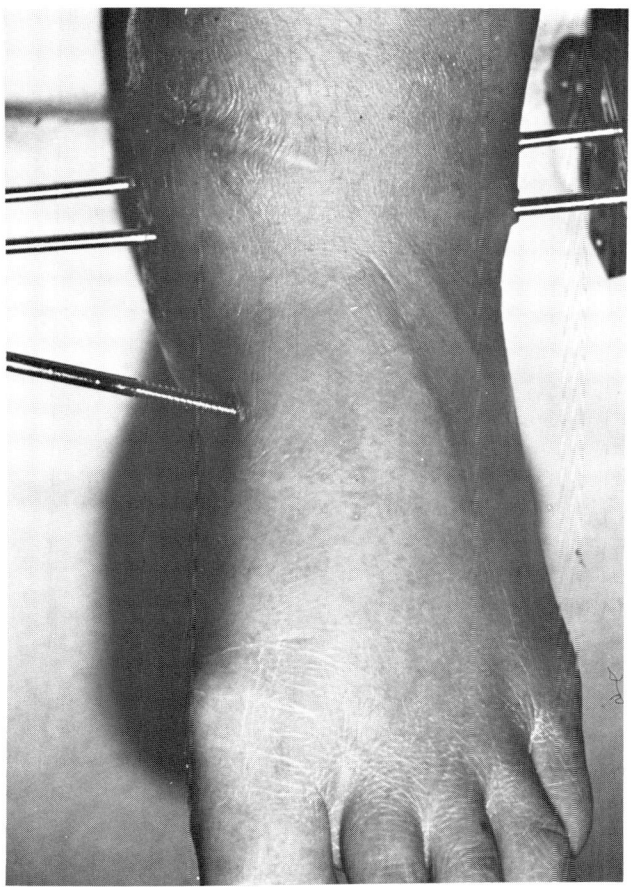

Figure 128: Incorrect technique (temporary forefoot stabilization during ankle arthrodesis): one pin was not enough; loosening and pain developed around the forefoot pin which was removed without further problems.

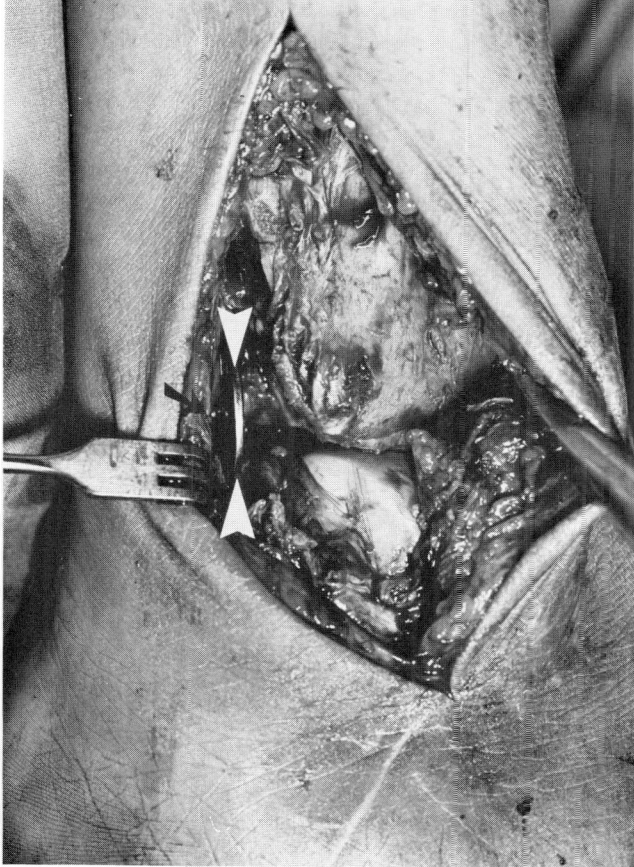

Figure 130: Exposure of ankle joint through lateral wound after distal fibula is removed. Note tip of Bennett retractor (arrows) inserted from posteromedial side, between tibia and neurovascular bundle.

extensively trimmed at the end of the procedure so that it does not interfere with pin placement when it is returned to its original position.) Using the ankle joint as a guide, expose the anterior tibia subperiosteally, extending the dissection to include the anterior and lateral upper talus.

Next, tilt the operating table laterally and make a longitudinal skin incision over the medial malleolus. Expose the medial malleolus subperiosteally, continuing dissection anteriorly and posteriorly around the distal tibia. Carefully develop the interval between the posteromedial neurovascular bundle and the posteromedial corner of the distal tibia while maintaining the foot in the plantar flexed position to relax tension on the soft tissue structures. In this manner, expose the entire distal tibia and upper talus subperiosteally. Place elevators and retractors around the distal tibia and upper talus to protect the soft tissues while the bone cuts are being made.

Make the tibial cut through the lateral wound with an oscillating saw. The plane of the arthrodesis should be perpendicular to the axis of the tibia. If carefully done, a flat piece of bone and articular cartilage consisting of the plafond of the tibia and the medial malleolus can be removed as one piece. If hand tools are used, the cancellous surface tends to be curved rather than flat.

Next, plan the removal of the articular surface of the talus. The success or failure of the entire surgical procedure depends upon this step. A good anatomical eye and wide exposure with adequate hemostasis are necessary. Place the foot in neutral position with respect to both dorsiflexion-plantar flexion and varus-valgus and, with an oscillating saw, make the cut in the talus parallel to the flat surface of the distal tibia. Be sure to protect the soft tissues with retractors and elevators. When the talar cut is complete, remove any remaining cortical lip with a rongeur and smooth out any irregularities with a rasp.

Position the talus posteriorly on the distal tibia to preserve the contour of the heel.[54] This is easily accomplished if the posterior edge of the distal tibial cut is made flush with the posterior edge of the talar cut (Figure 125). Check the position of the foot with roentgenograms at this point during the procedure, because a lateral projection will be impossible to obtain after the fixator is in place. (Ratliff[226] recommends placing a wire or other marker on the sole of the foot to insure a neutral position.) Rasp smooth any high spots or incongruities in the cut surfaces noted on the roentgenograms. If the position of the foot is unsatisfactory, remove additional bone by trimming the distal tibia rather than the upper talus. Remember: excessive equinus or calcaneus of the foot may mean functional difficulties for the patient. A varus position of the foot is also unacceptable.

Apply the external fixator prior to skin closure in order to permit adjustments in position if necessary to insure good bone contact at the arthrodesis site. It is important to replace the skin flaps prior to pin insertion in order to avoid the problem of undue tension on the flaps when the wound is closed. Protrusion of one or more pins through the operative wound itself should not cause concern.

The body of the talus is about one inch high after removal of the articular surface. Utilize an image intensifier to position the pins correctly in the middle of the talus. (If they are too close to the fusion site, they will pull out of the bone.) Avoid putting pins through the subtalar joint.

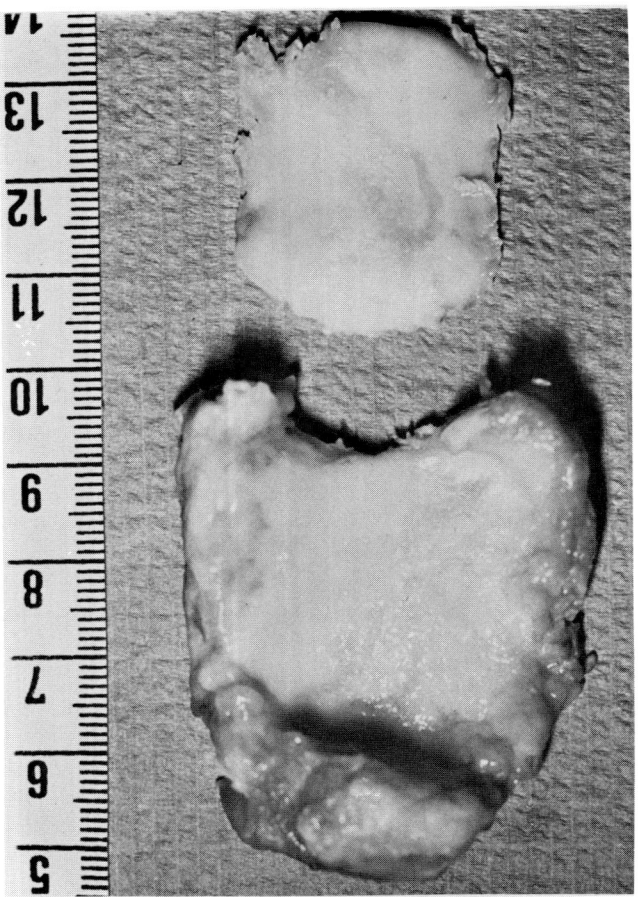

Figure 131: Bone removed for ankle arthrodesis. Tibial plafond and medial malleolus (top); dome of talus (bottom).

Following pin placement but prior to skin closure, apply the rest of the fixator frame. Obtain final alignment and compression, then check position once again with roentgenograms and direct observation. Any residual incongruity or malpositioning must be eliminated before the skin is closed. This can be accomplished by distracting the arthrodesis site with the fixator frame. A flat rasp should be used to make any final correction on the cut surface of the tibia. Great patience is required at this point in the procedure, because rasping is difficult and tedious in the tight confines of the ankle joint, especially when the components of the fixator block easy access to the distal end of the tibia. After making the final correction, the frame should be compressed, and any bone defect filled with cancellous graft obtained from the medial or lateral malleoli or proximal tibial metaphysis. Close the wound over suction drains to prevent the accumulation of blood around the fusion site. Wrap Kling® or Kerlex® gauze dressings around the pins to fill the space between the pin-gripping clamps and the skin.

If a long-term application is planned, supplementary stabilization of the forefoot can be achieved by inserting either half-pins or full pins through the metatarsal bones and connecting them to the frame of the fixator. Alternatively, the entire foot and fixator can be placed in a plaster of Paris cast, or posterior splint.

Postoperative Management

The patient is permitted out of bed as soon as pain tolerance permits. Any pin hole infection must be treated aggressively. This is especially true if the distal (talus) pins are involved with sepsis. Weight-bearing is prohibited because it may result in early pin loosening.

If the arthrodesis is not solid within sixteen weeks, the fixator can be removed and a short-leg walking cast applied. Union may occur very slowly in some cases. External support may be required for as long as one year.

Knee Joint

Discussion

Compression arthrodesis of the knee joint utilizing external skeletal fixation is a concept developed by Key[158] and popularized by Charnley.[52] It has, in recent years, become a popular method of achieving joint fusion. Stewart and Bland[253] report that knee fusion occurred in 15.5 weeks with external fixation, compared to 22.7 weeks with other techniques. Furthermore, union rates of 95-100 percent have been reported for cases where arthrodesis was performed in the absence of sepsis.

A successful knee fusion results in gait impairment—a slight, but tolerable, social handicap. In 1967, Green and coworkers,[112] analyzing the results of 142 knee fusions, found that a majority of the patients avoided public transportation after the surgery but were able to participate in most other activities. The loss of joint mobility is usually well compensated by the gain in stability and relief from pain.

When treating degenerative joint disease of the knee, the use of any arthrodesis technique will probably result in a high rate of fusion. In the presence of neuropathic arthropathy, as noted earlier, failure rates as high as 50 percent have been reported. In this situation, external skeletal fixation is a useful adjunct to the arthrodesis procedure. For cases involving chronic joint sepsis, arthrodesis with external skeletal fixation is the preferred method of treatment. This is especially true if the arthrodesis follows an infected failed total knee replacement arthroplasty.

If the arthrodesis is performed for a failed total knee

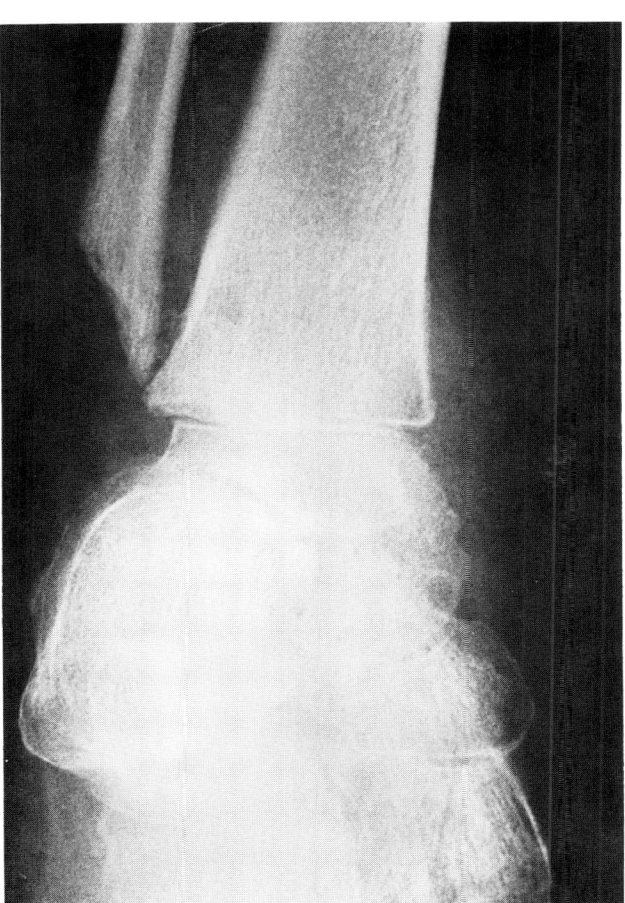

Figure 132: Roentgenogram illustrating level of fibular osteotomy.

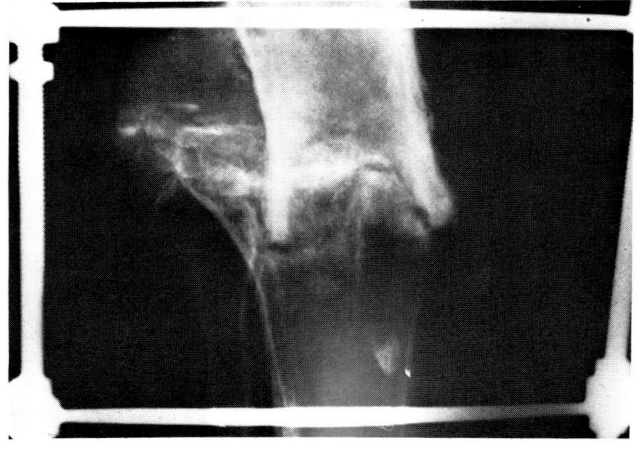

Figure 133: Unsuccessful attempt at fusion of charcot knee following removal of septic hinge total knee prosthesis. Errors: fixation not rigid enough; no supplementary bone graft.

arthroplasty, shortening of the limb can be expected. Fortunately, patients with a fused knee walk better when the limb is one inch short than they do with the limb at its original length, because a slightly shortened limb requires neither circumduction nor hip hiking during the swing phase of gait. Many minimal or partial constrained knee arthroplasty designs permit salvage arthrodesis with approximately one inch of shortening. A recent report from the Mayo Clinic[25] described successful arthrodesis in twenty-nine of thirty-six patients (81%) who had had a minimal or partial constrained arthroplasty prior to arthrodesis.

Removal of a hinged-knee prosthesis, such as the Walldius, Shiers, Guepar, or Herbert, presents a much more difficult problem. Removal of the prosthetic components and loose cement leaves the surgeon with two shells of cortical bone that look like empty ice cream cones. Compression arthrodesis of these bone rims is sometimes difficult to achieve. In fact, results of

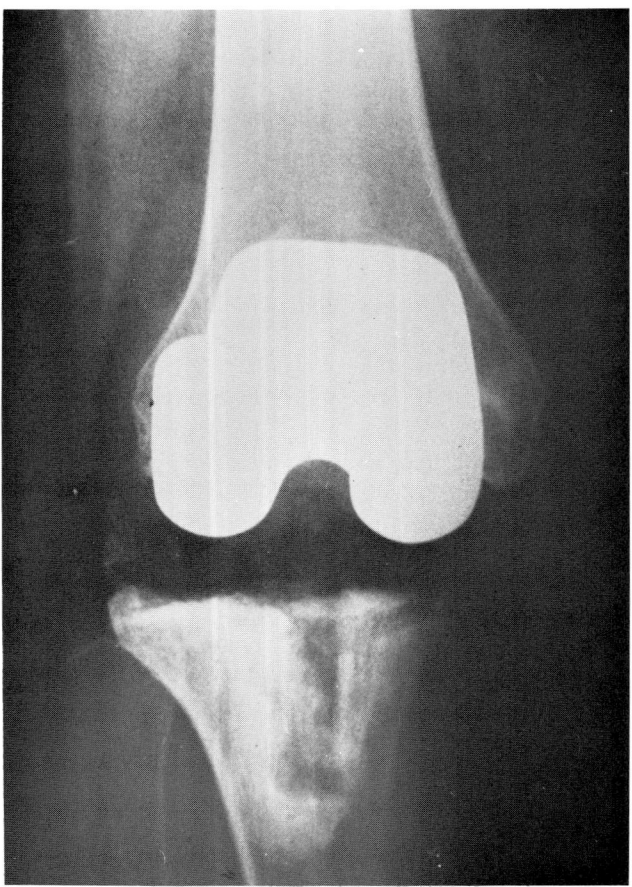

Figure 134: Infected unconstrained total knee prosthesis.

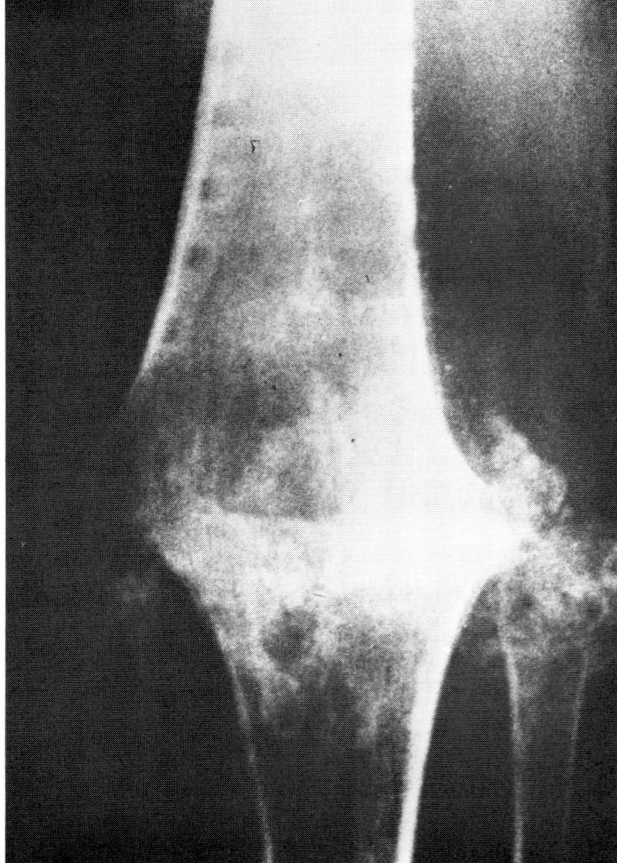

Figure 135: Successful arthrodesis with biplanar configuration and supplementary bone graft.

attempted arthrodesis in the Mayo Clinic series indicate that success was achieved in only five (56%) of nine patients in this group. Other reports indicate a 21 percent successful fusion rate following removal of the Guepar hinge and a 12.5 percent successful fusion rate following salvage of the Shiers hinge.[78] These dismal salvage results suggest that the hinge-type prosthesis has limited indications.[12, 201]

Precise flat bone cuts are important to insure a high rate of union following the arthrodesis. At Rancho Los Amigos Hospital, we make the tibial and femoral bone cuts with cutting guides developed for total knee arthroplasty. These guides produce cuts parallel to each other and perpendicular to the axis of the limb.

Position of Arthrodesis

The recommended position for knee fusion is 20° of flexion in females and 5-10° of flexion in males.[252] Charnley,[54] however, suggests that more than 10° of flexion is excessive. If significant shortening is anticipated, it is best to fuse the knee in full extension, thereby avoiding unnecesary additional reduction in length. The tibia should be externally rotated approximately 10° on the femur. Five degrees of valgus at the knee will align the foot directly under the hip joint when the patient is standing.

Frame Configuration

The surgeon may select any of several frame designs for knee arthrodesis. Charnley clamps can be used but may become unstable because of the lack of rigidity in the fixator configuration. More stable configurations, such as the Vidal-Adrey quadrilateral frame, Kroner ring frame, Day frame or ASIF tubular frame are superior to Charnley clamps.

Most fixator configurations require pin placement in the coronal plane, with parallel pin groups in the femur and in the tibia. This arrangement allows anteroposterior bending of the frame, which can be recognized at surgery as a "jog" of motion in flexion-extension after the fixator is in place and the cut surfaces have been compressed together. Slight motion may not in itself be detrimental to bone union, but in the presence of an infective process may contribute to the persistence of sepsis. Also, where arthrodesis is carried out in the absence of broad cancellous surfaces (as occurs following removal of infected total joint components), motion at the level of the arthrodesis is detrimental to healing of the opposed cortical surfaces. The best way to control undesirable anteroposterior motion is to utilize a fixator configuration that includes supplementary half-pins inserted into the femur and tibia in a second plane.

Technique

Make an anterior parapatellar incision of sufficient length to permit wide exposure. (A scar from previous surgery may dictate the position of the incision.) The patella may be retained and used as a supplementary bone graft if sepsis if not present. If it is retained, mobilize the patella sufficiently to allow it to slide to the side of the joint when the knee is flexed.

Cut the extensor retinaculum, cruciate ligaments, and collateral ligaments, and flex the knee as much as possible. When treating a previously failed arthrodesis, force the knee into flexion by firmly but carefully breaking apart the fibrous tissue present at the site of failed union. If treating a septic total joint replacement, remove all hardware and loose cement. Cement still solidly affixed to bone can be left in place.

With the knee fully flexed, carefully elevate the tissue along the posterior tibial margin with a periosteal

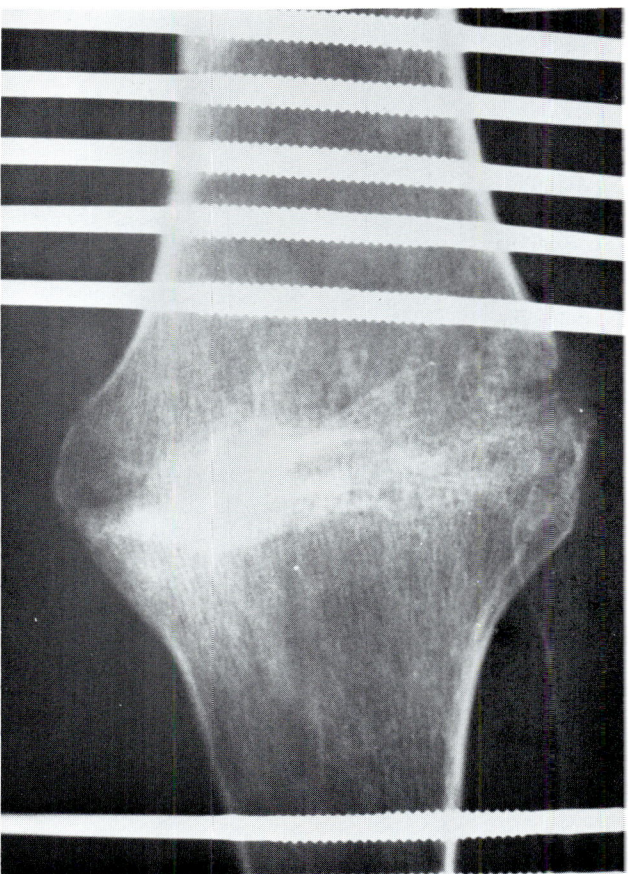

Figure 136: Technically unsatisfactory (but successful) arthrodesis. Initial tibial cut was not perpendicular to the axis of the limb, requiring oblique cut in femur to counterbalance the error.

elevator (or flat retractor) inserted into the interval between the posterior surface of the tibia and the posterior capsule of the knee joint; this will protect the popliteal artery and vein, the common peroneal nerve, and the tibial nerve from injury.

Make the tibial cut by using the tibial cutting guide from a total knee arthroplasty set. Align the tibial cutting guide with the anterior crest of the tibia to permit 5° valgus at the fusion site; then begin the tibial cut with an oscillating saw. After the cut is started, remove the guide and complete the cut with a broad hand saw. When the tibial cut is completed, place the femoral cutting guide on the cut surface of the tibia and extend the knee. A slight amount of knee flexion (5-10°) is desirable, especially in the female patient, but the knee should be fused in full extension if any shortening is present.

Make the initial saw cut through the femoral cutting guide with an oscillating saw, completing the cut with a broad hand saw. While the cuts are being made, protect the posterior neurovascular structures with appropriate retractors. With a rongeur and rasp, trim and flatten any remaining high spots of cortical bone that project above the cut surfaces. Small areas where bone is completely absent are of no great concern. Small areas of extremely dense sclerotic bone can be drilled with a small hand drill, provided these areas do not involve a significant portion of the visible surface of the proximal tibia or distal femur. (The ivory-like subarticular sclerosis that sometimes occurs in degenerative or septic joint disease does not fuse well.)

Replace the skin flaps to their original position prior to pin insertion to avoid skin tension at the pin holes

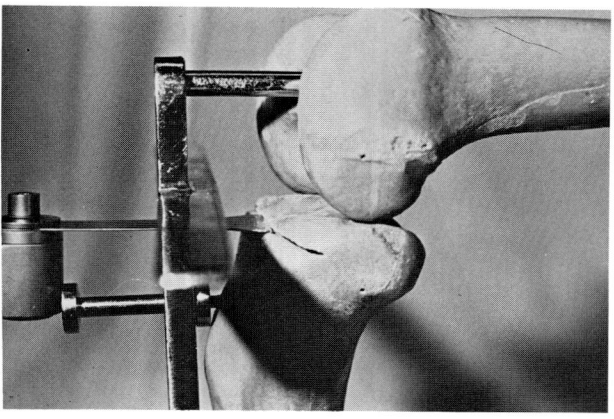

Figure 137: Tibial cutting guard (Insall and Ranawat).

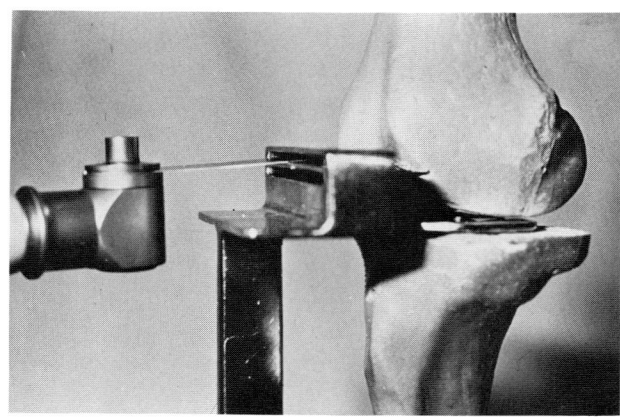

Figure 139: Femoral cutting guard (Insall and Ranawat).

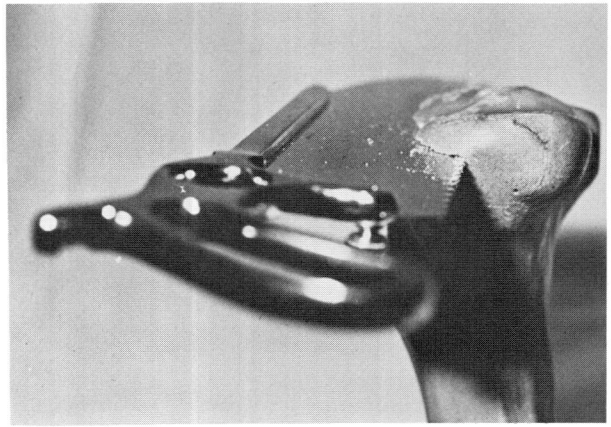

Figure 138: Completion of tibial cut with broad hand saw.

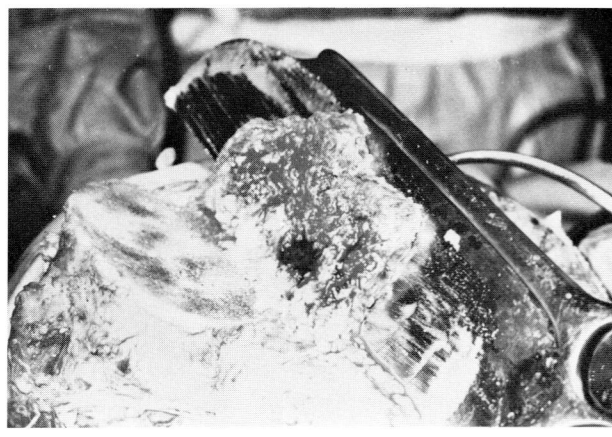

Figure 140: Completion of femoral cut with broad hand saw.

when the wound is finally closed. (The wound is not closed until the entire fixator frame has been assembled and the final position of the cut surfaces of the bone checked both visually and roentgenographically.) Insert the transcutaneous pins and apply the external fixator. Approximate the cut surfaces of the joint and tighten the fixator while palpating the margins of the arthrodesis to insure an even fit. Obtain roentgenograms of both the anteroposterior and lateral projections. If the components of the fixator block visualization of the arthrodesis site, obtain oblique projections and make the appropriate mental reconstruction necessary to evaluate the arthrodesis surfaces.

If the roentgenograms are satisfactory, close the wound in layers with suction drainage tubes in place. If a tourniquet cuff is utilized during the procedure, release it prior to wound closure and obtain adequate hemostasis. Bleeding from the genicular vessels can occur. If the patella has been retained, use it as a supplementary bone graft by removing the posterior articular surface and attaching it to the femur and tibia with bone screws, spanning the arthrodesis.

Postoperative Management

The patient is permitted out of bed when tolerated. Fusion is usually solid in ten to sixteen weeks but may occur in six weeks or less. When dealing with a joint infection or neuropathic arthropathy, fusion may take twice as long as normal. Test the consolidation by removing the upright bars (leaving the pins in place) and manually stressing the arthrodesis site. If it feels solid, leave the upright bars off but keep the pins in place for one week to test the strength of the arthrodesis. Apply anterior and posterior slab-splints to the limb with Ace® Bandages for protection and support. If at the end of one week the fusion remains solid, remove the transcutaneous pins under a light general anesthetic. Apply a molded orthosis or cast after the fixator is removed to permit the fusion to mature and to reduce the likelihood of fracture through a pin hole.

If the fusion is not solid when the bars are removed, replace them for another month. If a pin tract infection develops while the fixator is in place, the septic pins should be removed if the infective process cannot be brought under control. If multiple pins have become infected or are painful, the entire fixator can be removed and the patient managed with a snugfitting orthosis or cast until fusion is solid.

Hip Joint

Discussion

Experience with arthrodesis of the hip utilizing external fixation is limited.[191] One reason for this is that the excellent results obtained with total hip arthroplasty have largely eliminated the necessity for the former procedure in the primary management of degenerative hip disease. However, primary arthrodesis still retains an important role in the treatment of hip problems in active younger patients. In the pediatric and adolescent age groups, excellent functional results have been obtained following hip fusion.[102] Young and middle-aged adults can also expect a satisfactory outcome from a hip that is solidly fused in good position.[176]

Where compression arthrodesis with external fixation is desirable, it can be accomplished with a fixator frame that permits compression parallel to the axis of the femoral neck. We have successfully utilized the Hoffmann system fixator to obtain arthrodesis of the hip in three patients, employing a technique that

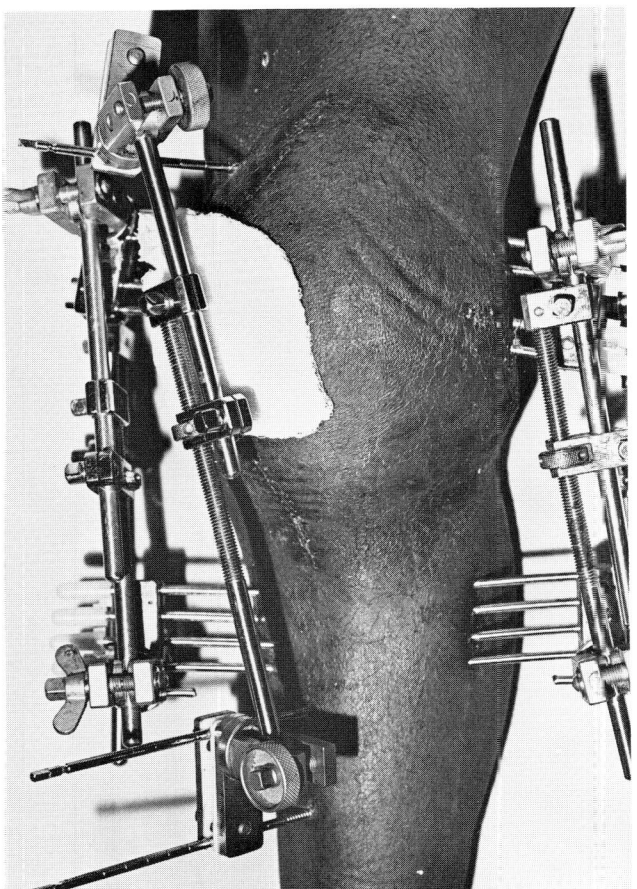

Figure 141: Biplanar fixator configuration for knee arthrodesis.

utilizes compression bars that connect a femoral mounting to a pelvic mounting. [It should be noted that Vidal (personal communication), on the other hand, reports a significant failure rate with this configuration.] Because experience with external fixation arthrodesis of the hip is so limited, internal fixation techniques should be used if sepsis is not present.

The large number of hip replacements being performed will, with the passage of time, result in a significant number of patients presenting for treatment of failed (or infected) total hip replacement. The accepted method of dealing with this problem is to remove components and cement and leave the patient with a flail joint. In some cases, components can be reinserted at a later date, but the failure rate from the secondary procedure is high. The lack of stability following head and neck resection may be a considerable hardship for many patients. Several procedures have been devised to deal with this problem. Judet[149] and coworkers,[179] utilizing a procedure called "iliotrochanteric coaptation," attempt to obtain a mechanically stable pseudarthrosis between the upper femur and the acetabulum after removal of infected total joint replacement components and cement. They report that a stable coaptation between the proximal femur and pelvis is more acceptable to the patients than a girdlestone-type resection, which permits telescoping of the upper femur into the soft tissues lateral to the iliac wing.

Judet employs external skeletal fixation to hold the upper femur into the acetabulum until a pseudocapsule forms. Several measures in this procedure are essential to prevent the upper femur from sliding out of the acetabulum. First, the muscles attached to the apex of the greater trochanter are removed and the trochanter is "squared off." Second, the gluteus medius must remain attached to the lateral portion of the greater trochanter. Third, the iliopsoas is detached from the lesser trochanter, thereby preventing adduction from displacing the upper femur from the acetabulum. Last, the acetabulum must be prepared for a stable coaption by squaring off the superior weight-bearing area and removing the condyles on both sides of the acetabular notch. If this step is not performed, the femur will lever out of the acetabulum as soon as the patient's leg is adducted. Judet has performed this procedure on more than 400 patients. He has designed a simple external fixator frame that allows the upper femur to be held to the acetabulum until a fibrous tissue envelope forms, but any fixator frame that allows attachment of the upper femur to the pelvis may be utilized. The fixator is left in place for six weeks.

Mears[191] has described a technique utilizing external skeletal fixation to aid in obtaining a stable femoropelvic arthrodesis following removal of infected total hip components. This difficult procedure is performed in two stages. The first stage consists of thorough debridement of the joint replacement components and loose methyl methacrylate cement. A rigid external skeletal fixation frame is applied at the same time. The frame configuration utilized by Mears involves multiple half-pins inserted into the crest of the ilium and additional full pins inserted from the anterior inferior iliac spine to the posterior inferior iliac spine. A special pin alignment guide helps the surgeon to transfix the pelvis without damage to visceral or vascular structures. A rigid upper femoral mounting is connected to the pelvic mounting with bars. When a clean granulating wound is obtained, a vertical osteotomy through the acetabulum is performed and the upper femur is held firmly against the cancellous surface of the pelvic osteotomy. Supplementary bone graft is utilized. If successful, the procedure permits a stable pelvic-femoral arthrodesis without loss of limb length.

Frame Configuration

A pelvic mounting is created by inserting three or four pins in each iliac crest and connecting the pin groups with a bar across the front of the lower abdomen. It is important to allow enough clearance for the patient to sit up without his abdomen striking the bar. Next, an upper femoral mounting is applied with two groups of half-pins inserted into the proximal femur according to the pin placement positions described in Chapter 4. Finally, the femoral mounting is connected to the pelvic mounting with a compression bar parallel to the axis of the neck of the femur. A second compression bar is placed posterior to the plane of the hip joint to prevent the hip from flexing when the anterior compression bar is tightened. This lateral stabilizing bar attaches from the most posterior femoral pin group to a posterior extension of the pelvic mounting. Additional connecting bars are sometimes added to stabilize the configuration.

At Rancho Los Amigos Hospital we utilize the Hoffmann system to fabricate a frame designed to prevent shortening of the limb following removal of total hip arthroplasty components, prior to reinsertion after an appropriate interval. The configuration utilized consists of a pelvic mounting, incorporating both iliac crests, connected to a distal femoral mounting; it is kept in place five weeks. This frame configuration is similar to the frame configuration described above for arthrodesis of the hip joint except that the femoral mounting is distal rather than proximal. In this manner, transcutaneous pins are kept away from the region of the femur into which a femoral stem prosthesis will be inserted.

Position of Arthrodesis

Various positions for hip fusion have been advocated — the basis for the difference of opinion being the trade-off between the optimum position for sitting comfort and the best position for maximum limb length.[174, 176] Increasing hip flexion allows greater ease of sitting (lessening low back strain), but it also shortens the effective length of the limb, increasing gait abnormalities. For most individuals, the ideal position of flexion is between 20° and 30° with 5° external rotation and 0°-5° of abduction.

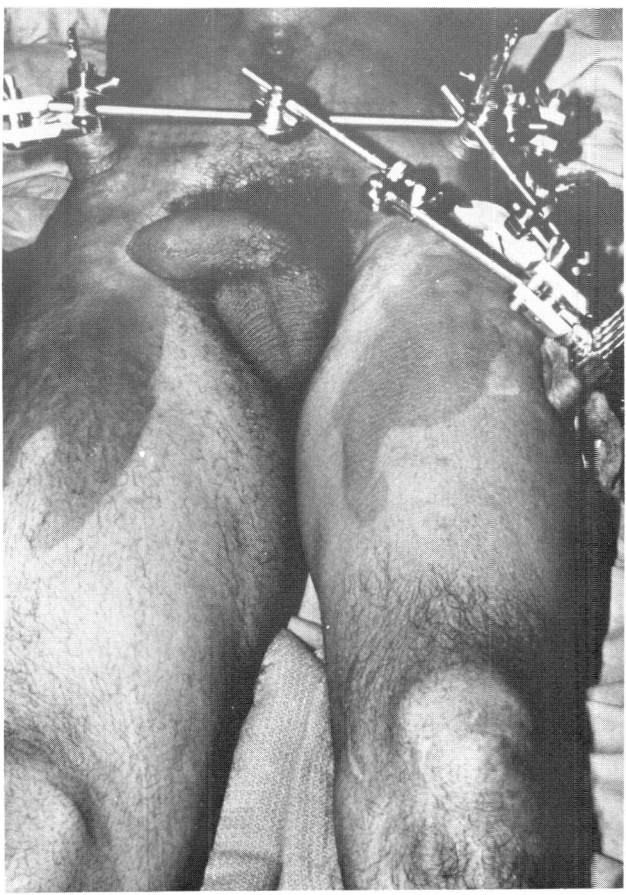

Figure 142: Upper femoral frame connected to pelvic frame for hip stabilization or arthrodesis. The axis of compression is in line with the femoral neck. The hip will adduct and flex unless a compression bar is placed posterior to the level of the hip joint.

Technique

With the patient supine, approach the hip joint through an anterior iliofemoral incision. Dislocate the hip joint and remove all cartilage from the femoral head and acetabulum.

Gouges and shapers designed for cup arthroplasty can be used to denude and contour the femoral head and acetabulum. In the absence of sepsis, a supplementary cortical-cancellous bone graft spanning the hip joint can be utilized. Cut an appropriate trough into the femoral head extending across the acetabulum into the pelvis. Davis' technique employs a muscle pedicle graft consisting of the anterior iliac crest attached to the tensor fascia lata muscle and anterior fibers of the gluteus medius and gluteus minimus.[25] An intertrochanteric osteotomy of the femur, as advocated by Lipscomb and McCaslin[176] and by Stewart,[25] is unnecessary when external skeletal fixation is used. Close the wound over suction drains and apply an external skeletal fixator that spans the hip joint.

Postoperative Management

The patient can be out of bed as soon as he is comfortable. The fixator frame is left in place for at least six months. Pin hole infections from the pelvic area are quite common but rarely lead to chronic pin hole osteomyelitis. The problem can best be managed by a one-week course of enforced bedrest and oral or parenteral antibiotics selected on the basis of the pin tract cultures.

Shoulder Joint

Discussion

Arthrodesis of the shoulder is not a common operation. The classic indications for the procedure, tuberculosis and paralysis secondary to poliomyelitis,[53] are now uncommon. Degenerative and posttraumatic joint disease are rarely as disabling to the shoulder as they are to joints of the lower limbs. The availability of total shoulder arthroplasty further reduces the need for shoulder fusion. Thus, primary arthrodesis of the shoulder is rarely indicated, except for chronic sepsis and failed joint replacement arthroplasty. Degenerative or posttraumatic arthritis of the shoulder in a young active individual may also benefit from fusion of the shoulder in a satisfactory position.

Arthrodesis of the shoulder can be accomplished with the help of external skeletal fixation, since the fixator not only holds the humeral head to the glenoid but also permits compression across the arthrodesis surfaces. Furthermore, fixation can be obtained without the use of a shoulder spica cast. Unfortunately, considerable torque will be present at the fusion site and at the pin holes due to the weight of the arm. For this reason, it is wise to stabilize the arthrodesis with internal screw fixation (if sepsis is not present) and employ the external fixator as a substitute for a shoulder spica cast. In the presence of sepsis, external skeletal fixation can be employed without internal fixation; however, the fixation across the shoulder joint is less secure.

Frame Configuration

The fixator frame must be capable of allowing compression perpendicular to the plane of the arthrodesis surfaces. The only reasonable place for pin insertion is into the scapular spine since the scapula is thin. Needless to say, half-pin insertion into the spine of the scapula must be done carefully to avoid penetration of the anterior surface of the bone. Roentgenograms of pin placement are difficult to interpret in this situation. The proximal humeral mounting is applied in accordance with the guidelines in Chapter 3. Compression anterior and posterior to the axis of the glenohumeral joint will insure a snug fit.

Position of Arthrodesis

There is no universal agreement concerning the optimal position of shoulder fusion. A study done by the Research Committee of the American Orthopaedic Association described the optimal functional position as 45°-50° abduction of the humerus on the scapula, 15°-25° forward flexion, and external rotation of the upward flexed forearm 25°-30° above the horizontal. This corresponds to a position of 25° internal rotation when the arm is at the side of the body. Rowe[233] argues for considerably less abduction of the shoulder. He believes that undue scapulothoracic pain and stress on the acromioclavicular joint follow humeral abduction of 50° and recommends abducting the arm 20° from the neutral position. Clinical investigators who have reviewed the matter agree on one thing: humeral rotation is critical. Approximately 25° internal rotation of the humerus when the arm is at the side of the body corresponds to a position of approximately 25° upward tilting of the forearm above the horizontal level when the humerus is flexed 90° (pointing straight ahead). This position brings the hand to the mouth and permits hair grooming without difficulty. In this regard, humeral rotation is more important than either abduction or forward flexion.

Operative Technique

The objective of the surgical approach is to obtain a wide enough exposure to allow complete decortication of the humeral head, glenoid, and undersurface of the acromion process.

Place the patient on the operating table in the lateral decubitus position with the shoulder to be fused uppermost. Make a curved incision following the origin of the deltoid muscle, starting at the clavicle and extending along the acromion process to the spine of the scapula. Elevate the deltoid origin from the bone and fold the muscle distalward. Take care not to injure the axillary nerve that lies on the undersurface of the muscle (approximately two inches from its origin on the acromion process). Elevate the trapezius muscle from the upper surface of the acromion process. Enter the subacromial bursa and transect the tendons of the supraspinatus and infraspinatus muscles to allow

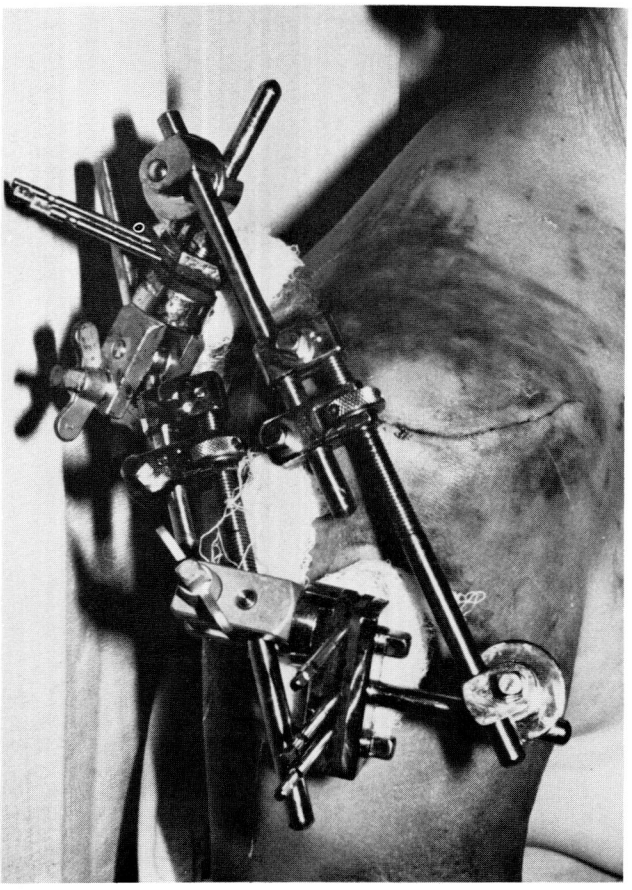

Figure 143: Shoulder arthrodesis frame connecting proximal humerus to the spine of the scapula.

complete exposure of the humeral head. Exposure of the glenoid cavity can be obtained by transecting the subscapularis muscle and opening the capsule of the shoulder joint. Pass a gauze tape around the neck of the humerus and retract it laterally to obtain excellent visualization of the entire humeral head and glenoid cavity. Remove articular cartilage and subchondral bone from the humeral head, glenoid fossa, and undersurface of the acromion process.

Place the limb in the planned position of arthrodesis and obtain anteroposterior roentgenograms. Because the vertebral and axillary border of the scapula will be poorly visualized, the spine of the scapula can be used as a guide in determining the proper amount of glenohumeral abduction. Remember, humeral rotation is critical.

After the proper position of the humeral head has been determined, make a flat cut in the superior surface with an osteotome. Next, make a partial osteotomy through the superior cortex of the acromion process, permitting it to be folded downward to the humeral head. If sepsis is not present, place screws from the humeral head into the glenoid and from the acromion process into the humeral head.

Apply the external skeletal fixator with a mounting configuration appropriate to the components used. Compress the arthrodesis surfaces and check the final position of the limb once again. Make sure the hand can be brought to the face and head. Close the wound in the usual way.

Postoperative Management

The patient is permitted out of bed as soon as the wound is stabilized. The arm should be supported in an abduction brace either when internal fixation has not been used or where the internal fixation is not securely holding the humeral head to the glenoid. If any pin sites become infected, they should be treated aggressively. The fixator is left in place for three to four months, or until fusion is solid.

Elbow Joint

Discussion

Arthrodesis of the elbow is rarely required for pain relief and may not even be necessary to control a septic process: resection arthoplasty can also be performed.[116] External fixation is useful in resection arthroplasty because it permits rigid immobilization and seems to promote control of the infective process until a fibrous nonunion develops.

A fixator immobilizing the elbow joint at 90° for six

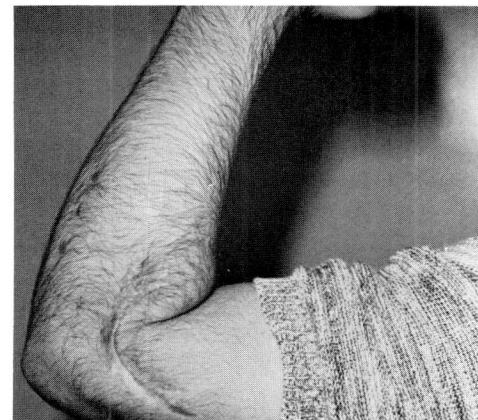

Figure 144: Range of elbow flexion after resection arthroplasty.

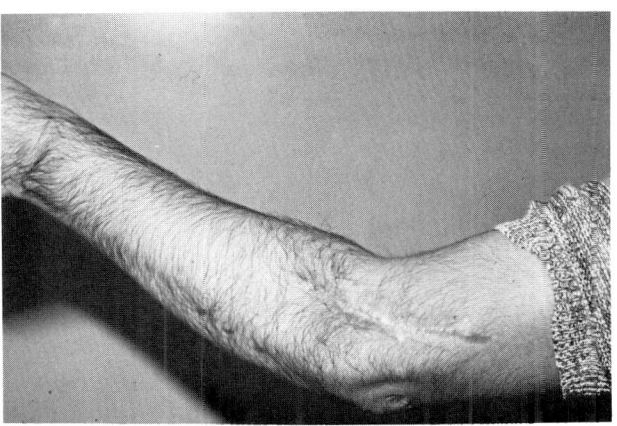

Figure 145: Range of elbow extension.

weeks following debridement will permit the elbow to heal by secondary intention; gentle exercises may be started thereafter. Elbow arthrodesis may be necessary following trauma or resection of a total elbow arthroplasty if there is loss of bone substance sufficient to preclude resection arthroplasty.

Frame Configuration

A unilateral fixator configuration connected to the ulna and humerus with half-pins will be sufficient to stabilize a resection arthroplasty. A more complex triangular frame, utilizing full pins and bilateral supporting bars, will be necessary if the goal is solid fusion.

Position of Arthrodesis

The proper position for arthrodesis of the elbow is 90°[252]. Following resection arthroplasty, an arc of

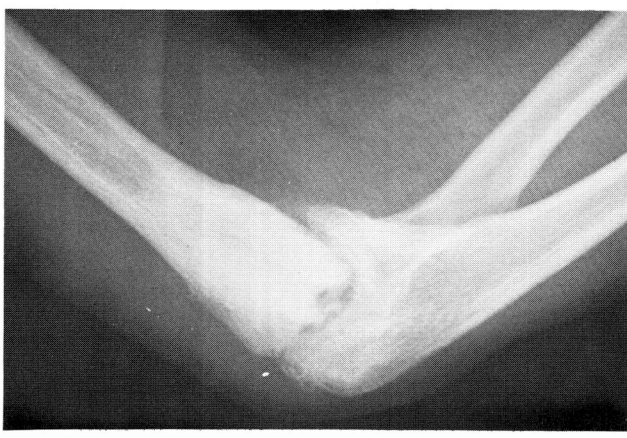

Figure 146: Successful resection arthroplasty for chronic fungal infection of elbow.

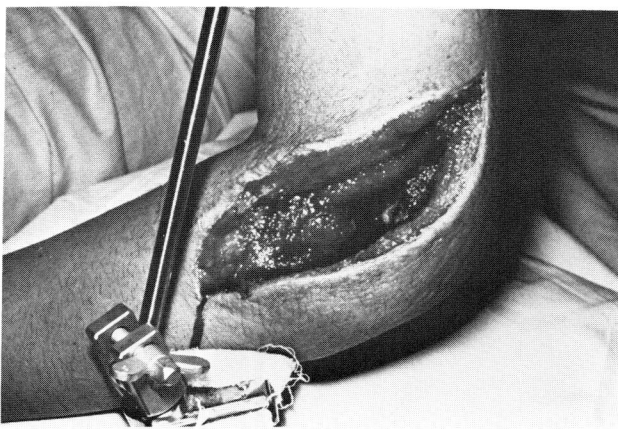

Figure 147: Unilateral fixator will stabilize elbow for resection arthroplasty; wound packed open after debridement.

elbow motion of approximately 60° can be anticipated if the elbow is held at 90° while the wound heals.

Technique

Approach the elbow joint through a lateral Kocher incision. Remove all septic and necrotic bone, cartilage, and soft tissues. Employ standard debridement techniques with appropriate precautions regarding the neurovascular structures, especially the ulnar nerve. At the completion of the debridement, pack the wound open with sterile gauze dressings soaked in antibiotic irrigating solution. Apply the external skeletal fixator configuration appropriate to the procedure. If a solid fusion is desired, supplementary bone graft will be necessary, especially if there is a significant defect present.

Postoperative Management

For resection arthroplasty, the fixator is left in place for six weeks while the wound is permitted to heal by secondary intention. Appropriate dressing changes once or twice a day will promote rapid growth of granulation tissue. Gentle active and passive exercises are commenced as soon as the fixator is removed. If arthrodesis is to be achieved, the fixator should be left in place until this occurs; periodic tightening of the compressor bars will promote union.

Wrist Joint

Discussion

Many satisfactory techniques that utilize bone grafts and/or internal fixation are available for wrist arthrodesis. External fixation is only necessary if there is uncontrollable sepsis or a significant loss of bone substance, necessitating extensive grafting.

Frame Configuration

Any frame configuration that permits stable fixation connecting the radius to the metacarpals of the hand is satisfactory. Ideally, pronation-supination of the forearm should be preserved. For this reason, the fixator should not be fixed to the ulna.

Position of Arthrodesis

The optimum position for unilateral wrist fusion is 10°-15° dorsiflexion, with the axis of the third metacarpal aligned with the axis of the radius.[252]

Technique

Approach the wrist through a dorsal curvilinear incision. Retract the extensor tendons and expose the wrist joint in the usual manner. If a radial-metacarpal bridge graft is planned, make an appropriate trough in the dorsal aspect of the carpal bones, extending the trough proximally to the distal radius and distally to the base of the second and third metacarpals. Fashion a Y-shaped cortical cancellous bone graft from the outer table of the ileum, and insert it into the trough. If the graft has been properly tailored, dorsiflexion of the wrist will lock it in place. In the presence of sepsis, it is safer to use cancellous bone even though union will take longer.

When the graft is in place, apply the external fixator in accordance with the pin placement recommendations in Chapter 4.

Postoperative Care

The patient is permitted out of bed as soon as he is comfortable. A sling should be worn to reduce postoperative edema and swelling. The fixator should be left in place until union is solid, but may be removed if pin sepsis becomes a problem. The limb can be supported, if necessary, in a plaster cast or custom-molded orthosis.

Chapter 7

PERSISTENT WOUND INFECTION

Introduction

External fixators are frequently applied to acute open fractures to reduce the likelihood of serious wound sepsis. It is not unusual, however, to discover purulent drainage discharging from such a wound one or two weeks after application of the frame. A persistent wound infection, like an unsuccessful arthrodesis, is not a complication of external fixation per se, but if it occurs, it will lead to an unsatisfactory fixator experience for the surgeon and the patient.

There are different opinions about the management of open fractures.[47,124,152,260] Basic disagreement focuses on two considerations: (1) the use of internal fixation and (2) the place (if any) of primary wound closure. Advocates of immediate internal fixation of open fractures claim that rigid stabilization reduces the likelihood of wound sepsis.[51,230,295] Proponents of more conservative methods of fracture management note that the additional surgical dissection necessary for internal fixation extends tissue injury unnecessarily.[29,80,280,281] External fixation should appeal to both groups, because it permits rigid stabilization of bone and soft tissue without additional tissue dissection.[89,153,208,270]

INCIDENCE OF WOUND SEPSIS

Type of Wound

The probability of an open fracture becoming septic depends on many factors, the nature of the original injury being the most important. For the purpose of analysis, open fractures were classified into three categories by Gustilo and Anderson:[119]

Type I—An open fracture with a clean wound less than 1.0 cm long.

Type II—An open fracture with a laceration more than 1.0 cm long without extensive soft tissue damage, flaps, or avulsion.

Type III—Either an open segmental fracture or an open fracture with extensive soft tissue damage.

In Gustilo and Anderson's[119] study of 352 open fractures treated in Minneapolis, the overall infection rate was 2.4 percent. The infection rate according to the type of injury was 0.0 percent for Type I fractures; 1.1 percent for Type II fractures; and 9.9 percent for Type III fractures. When Rosenthal and coworkers[231] analyzed the incidence of wound sepsis according to the type of injury, they arrived at similar findings: 0.0 percent for Type I fractures; 0.0 percent for Type II fractures; 19.0 percent for Type III fractures.

Type of Treatment

The incidence of wound sepsis is affected by treatment. Patzakis and colleagues[217] have demonstrated the importance of antibiotics in the management of open fractures by randomly assigning patients to groups receiving either: (1) no antibiotics; (2) penicillin and streptomycin; or (3) cephalothin. The incidence of infection was 13.9 percent in the no antibiotic group; 9.7 percent in the penicillin/streptomycin group, and 2.3 percent in the cephalothin group.

It is difficult to compare reported infection rates for different methods of open fracture management. Even within an institution, a comparison of results is not valid unless patients are randomly assigned to treatment groups (as in the Patzakis study.) Rosenthal's series, cited earlier, disclosed a 10 percent rate of infection following cast treatment of open fractures, compared to an 18.7 percent infection rate following internal fixation. However, the selection of cast or fixation treatment may have been determined by the appearance of the wound when treatment was initiated.

Van der Linden and Larsson[260] reported a study designed to evaluate the results of plate fixation versus plaster cast treatment of tibial shaft fractures. The patients were randomly assigned to the two treatment groups. Most fractures were closed, but there were six open fractures in each group. No complications developed in the cast group, but four of the six patients in the AO plated group developed complications. Karlstrom and Olerud[152] summarized many reports pertaining to the treatment of open tibial fractures. They concluded that a distinction must be made between *rigid* and *nonrigid* internal fixation before any comparison can be made. The reported incidence of wound sepsis following *rigid* fixation of open fractures ranged from 3 to 8 percent, while the incidence of wound infections following *nonrigid* fixation ranged from 1 to 20 percent. In general, however, rigid fixation led to a lower incidence of sepsis. They noted the incidence of infection following the cast treatment of open fractures ranged from 5 to 17 percent. Karlstrom and Olerud also described their own experience with *external fixation* for nineteen Type II and thirty-two Type III tibial fractures. The overall infection rate was 0.0 percent in both groups. (However, two patients did develop infections after secondary intramedullary nailing.)

Edwards,[87] describing his experience with forty-four open Type III tibial fractures managed with external fixation, noted that 23 percent of the tibias developed posttraumatic osteomyelitis. This complication was usually associated with skin coverage of a segment of necrotic bone, a practice no longer employed at his institution. The injuries in Edwards' series were unusually severe: 73 percent of the cases had either loss of bone or major fracture comminution, 70 percent were grossly contaminated, and 50 percent had sustained soft tissue loss. Many of his patients might have been treated by primary amputation at another institution.

No study has been reported comparing the infection rate following external fixation to the infection rate obtained with other methods of treatment. Until such a prospective evaluation is completed, the effectiveness of external fixation in reducing wound sepsis in open fractures will not be clearly defined.

THE SELECTION OF EXTERNAL FIXATION

In the following sections, it is presumed that the surgeon has already made a decision to utilize external fixation as an adjunct to open fracture management. At the present time, many trauma centers around the world utilize external fixation for the serious open (Type III) fractures, especially if there is absence of bone substance, extensive soft tissue damage, or gross comminution.[88, 119, 208, 227, 294] A fixator might also be applied to a less serious, or even closed, fracture if stabilization of the bone permits a polytrauma victim to be more rapidly mobilized. Fixators are also applied to comminuted closed fractures. In Europe, Burny[57] applies a unilateral frame to almost every tibial fracture, including those which are closed and minimally displaced. More data are needed before his approach is universally applied.

PREVENTION OF WOUND SEPSIS

Initial Management

Antibiotics

Antibiotics are important in the management of open fractures. Cephalosporins are widely used for initial antibiotic treatment. Obtain a culture at the time of initial debridement, because there is a correlation between the organisms identified initially and those which subsequently produce wound sepsis.[119, 217] If resistant organisms are identified in the primary culture, the antibiotics can be changed accordingly.

The recommended duration of treatment is five to seven days if no sepsis ensues. My antibiotic regimen for open fractures is based on the possibility of a polymicrobic etiology to early wound sepsis. For this reason, I administer an anti-staphylococcal agent (nafcillin or a cephalosporin): and an anti-gram-negative aminoglycoside (tobramycin). An antianaerobic antibiotic (clindamycin) is added if the wound is contaminated. The latter two antibiotics are given for the first forty-eight hours following trauma, provided renal function is normal. The nafcillin is discontinued after five days if no evidence of sepsis develops.

Debridement

SOFT TISSUE: Thorough irrigation of the wound and debridement of devitalized soft tissue are important in preventing severe wound sepsis.[29, 33] A Water-Pic® will aid in removal of microorganisms from the wound. With a fresh skin-soft-tissue-bone injury, it is difficult to determine the amount of tissue damage by visual inspection. Marginally viable tissue may survive, while tissue that initially appears fairly healthy may become necrotic. Gustilo and Anderson[119] noted that soft tissue necrosis was generally greater than expected when the wound was reevaluated forty-eight to seventy-two hours after initial debridement. Because of this problem, staged serial debridement of soft tissue should be planned as part of the management of a severely traumatized limb.[89]

BONE: Debridement of bone should be very conservative during the initial phases of open fracture management.[29, 211] At the time of primary debridement, any small fragments of bone that are completely devoid of soft tissue attachments should be removed, while fragments attached to soft tissue should be returned to their anatomic position. Be ultra-conservative when

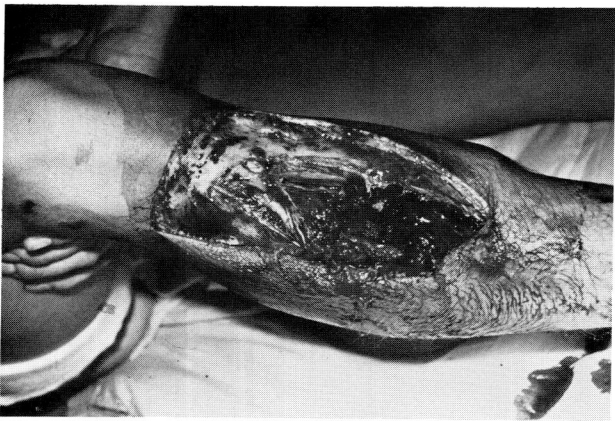

Figure 148: Extensive soft tissue injury with a transverse fracture that would not ordinarily be treated with a fixator in the absence of the skin trauma. Aggressive soft tissue management will permit early conversion to cast or orthrosis.

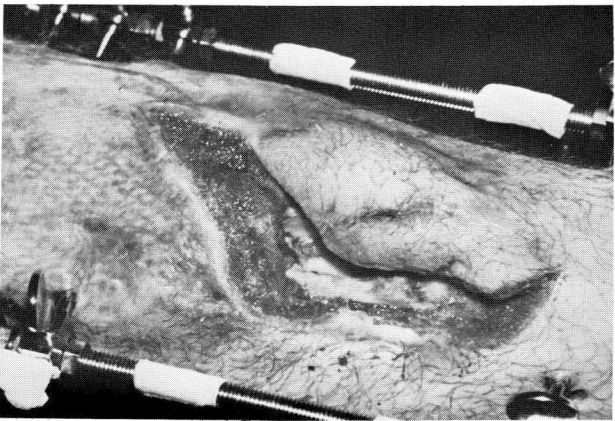

Figure 149: Granulation tissue advancement stopping at the edge of cortical bone necrosis. The granulation tissue will "heap up" and cover the dead bone, eventually creating a sequestrum.

deciding whether to discard a large fragment of bone. If any soft tissue attachment is present, replace the fragment to its original anatomical position, utilizing components of the fixator to hold it securely in place if necessary. If the fragment is completely devoid of soft tissue attachment, it still may be worth replacing, especially if it involves the cancellous end of a bone. A large cortical segment not attached to soft tissue can be replaced to its original position but may have to be removed later if it becomes the focus of a persistent wound infection. This occurs when soft tissues cover contaminated necrotic bone. The sequestrum must be removed if evidence of serious wound sepsis develops.

Fixator Configuration

The fixator frame must adequately stabilize the fracture fragments. Provisions for intermediate segment control may be necessary. Avoid, if possible, inserting pins into the fracture site or into a fracture hematoma. Try to place pins through healthy skin and soft tissue, consistent with adequate stabilization. Restore the skin flaps and wound edges to their normal position prior to pin insertion, but the pins should not be inserted in a manner that will create undue tension on the skin flap. When dealing with a Type III compound fracture of the tibia, there is a tendency for the weight of the calf musculature to draw the skin margins posteriorly, especially if the limb is suspended from an overhead frame. As the wound gapes anteriorly, the circulation of the wound edges is compromised because of the ill-effects of traction on the small cutaneous blood vessels. Tension on the pin holes also occurs. These problems can be prevented by supporting the calf while the limb is suspended by wrapping the limb circumferentially with gauze bandages, utilizing the fixator itself for support. When the wound dressing is changed, however, the calf tends to sag away again, separating the anterior margins. This difficulty can be overcome by providing continuous support of the calf during dressing changes. A calf sling, independent of the dressings, can be made and connected to the fixator frame. (Software manufacturers offer ready-made calf slings padded with synthetic sheep skin.) Alternatively, a rigid calf support that attaches to the frame of the fixator with articulations can be devised.[88]

Wound Closure

PRIMARY CLOSURE: Wound closure after initial debridement is fraught with danger.[29, 106, 119, 268] Primary closure of the wound creates tension on the wound edges that increases as the limb swells. This tension may cause tissue ischemia, resulting in extension of the effects of trauma. Furthermore, primary wound closure restricts drainage of potentially contaminated fluids from the wound area. Also, the danger of gas gangrene is always present when a wound is closed primarily. These potential complications can be largely prevented if the wound is left open.

Advocates of primary wound closure believe it is an acceptable procedure for wounds of lesser magnitude that do not involve extensive soft tissue damage, large skin flaps, or avulsions, are not caused by gunshot wounds, and are not more than eight hours old. Patzakis and coworkers[217] closed all of the 310 open fractures in their series not caused by gunshot wounds. Two patients (who received no antibiotics) developed gas gangrene. Gustilo and Anderson[119] closed all Type I and Type II wounds primarily and utilized delayed or secondary closure for Type III wounds. There were no cases of gas gangrene in their series of 1,025 open fractures. Nevertheless, Gustilo warns: *"If there is the slightest doubt in the surgeon's mind as to whether there has been adequate debridement of the wound after an open fracture, the wound should not be closed regardless of the type of open fracture. For the surgeon who manages only an occasional open fracture, the safe rule is not to close the wound."*

The problem with leaving a wound open, according to Reckling and Roberts,[227] is that avascular structures, such as cortical bone denuded of periosteum, tendon, and fascia, are prone to dry out and become the necrotic focus of a wound infection. They recommend coverage of bone and other tissues of low vascularity, while leaving muscle and soft tissue open, thereby reducing the likelihood of gas gangrene.

For the same reason, Ger[106] (and others [122, 206]) has devised muscle flaps that can be transferred over exposed bone to provide the necessary coverage. Ger believes his procedure is safer than extensive relaxing incisions, local skin transposition flaps, or immediate skin grafting for initial wound management. He states that the average civilian injury does not damage muscle bellies as severely as it damages skin, subcutaneous tissue, and bone. (If the viability of the muscle to be transposed is in doubt, the procedure can not be performed.) Unfortunately, the open fracture likely to require external fixation may also be severe enough to preclude the additional dissection necessary for an immediate muscle transposition flap. As noted earlier, the extent of soft tissue injury in Type III fractures is frequently greater than originally anticipated.

DELAYED CLOSURE: Delayed primary closure, performed five to seven days following injury, can be accomplished with little risk if the wound is stable.[33] After initial wound debridement and application of a skeletal fixator, pack the wound with gauze bandages moistened with a physiologic irrigating solution. Hold the wound edges loosely in position with large retention sutures over the gauze packs. Delayed primary closure can be easily accomplished with the external fixator in place, if pin placement does not interfere with the plan. If the soft tissues have been contaminated with considerable foreign material, employ an open irrigation system in place of wet dressings.[275] If tissue of marginal viability has been left in the wound, repeat debridement forty-eight to seventy-two hours later should be done.[6]

Subsequent Management

Serial Debridement

SOFT TISSUE: For the first wound inspection, the patient may require a light anesthetic with an agent such as ketamine. Tissue viability may be assessed according to the classic principles of color, consistency, and contractability. If the local tissues are clean and healthy, the wound can be closed if no dead space is created by the procedure. Drains placed under the skin flaps will insure egress of contaminated fluids.

If necrotic tissue is present, it must be debrided. If there is soft tissue of marginal viability, pack the wound open and reevaluate it a few days later. This repeated serial debridement process will eventually result in a healthy wound area. If the wound cannot be rendered free of all tissue of questionable viability, or if any evidence of wound sepsis is present, delayed primary closure cannot be contemplated. Plans for secondary coverage or healing by secondary intention should be made.

BONE: After initial conservative debridement of bone, other bone may become necrotic with the passage of time. Nonviable bone has several characteristics that make it easy to identify: it is whiter than living bone although it will turn brown if it dries out; it is more brittle when cut with an osteotome or rongeur; and it does not bleed when cut (the last being a reliable test of bone viability). If early skin coverage is planned, all nonviable bone should be debrided. In this manner, cutaneous, musculocutaneous, and muscle flaps can be transferred without fear of creating deep wound sepsis, the focus of which is necrotic bone. Edwards and his coworkers,[87] treating Type III tibial fractures, found that covering devitalized bone with a muscle or musculocutaneous flap resulted in a 50 percent infected pseudarthrosis rate.

Removing loose fragments of nonviable bone is simple. Necrotic bone on the surface of viable bone, however, should be debrided carefully by shaving off the outer layers of the cortex with an osteotome down to healthy bleeding bone. If the debridement procedure progresses through the full thickness of the cortex, supplementary cancellous bone grafting should be planned as a subsequent procedure. Occasionally, the debridement may progress not only across the near cortex but across the far cortex as well, resulting in a segmental defect that greatly complicates subsequent treatment. In view of the difficulties encountered reconstructing a segment of absent cortical bone it is reasonable to shorten the limb if the segment defect

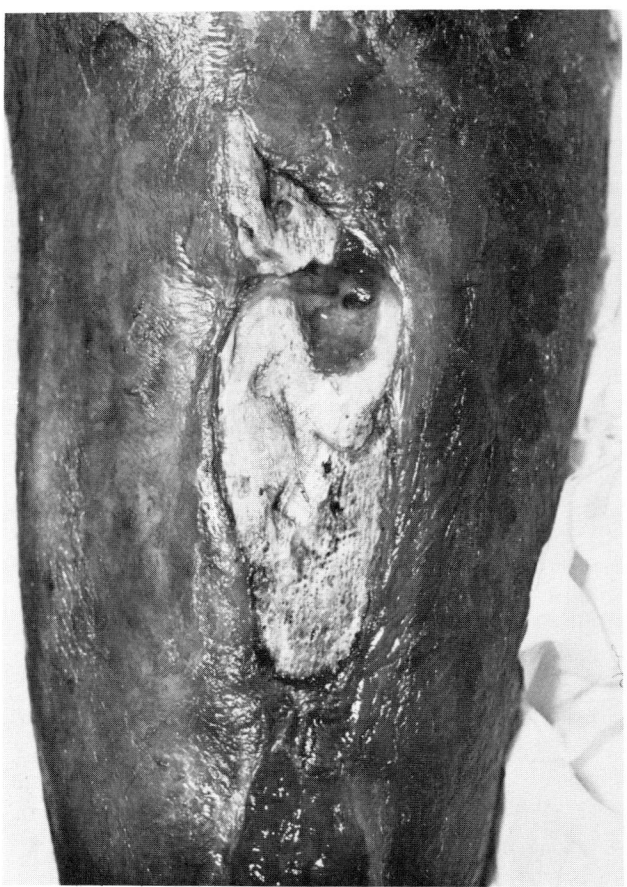

Figure 150: Fixators stabilized nonunion. Large necrotic segment of bone is in continuity with distal shaft.

created by debridement is one inch or less in length. If the defect is more than one inch long, the fixator should be kept on the patient until reconstruction of skeletal tissue has been completed. (See below "Bone Defect.")

Soft Tissue Coverage

HEALING BY SECONDARY INTENTION: The wound may be permitted to heal by granulation tissue and secondary epithelialization from the wound edges. This method is safe, reliable, and unlikely to lead to the unfortunate situation that occurs when devitalized bone is covered by a viable skin flap.

The advancing front of granulation tissue will not adhere to necrotic bone. If the area of bone necrosis is only a few cell layers deep, the advancing edge of granulation tissue can dissolve the superficial necrotic

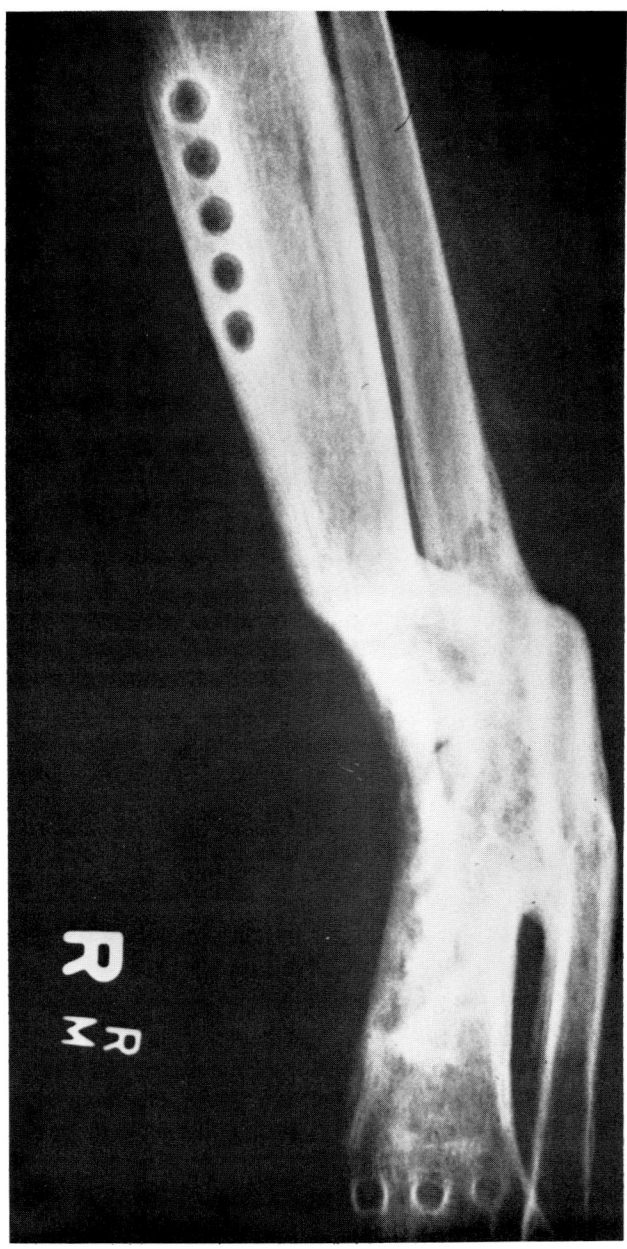

Figure 151: Resection and bone grafting, creating tibia-fibula synostosis. Note unsatisfactory placement of proximal pins into anterior crest of tibia, which led to premature loosening.

material and adhere to viable bone beneath it. The process can, nevertheless, be hastened by removing the superficial outer layer of dead bone from the exposed surface. The freshly exposed surface must be kept moist thereafter.

When the advancing front of granulation tissue

Persistent Wound Infection

stops at the margin of full-thickness bone necrosis, the remaining granulation tissue behind it continues to proliferate, causing the advancing front to "heap up." The heaped up edge can eventually cover the area of necrotic bone but does not adhere to it, leaving a space between dead bone and granulation tissue. This space becomes an abscess cavity as the granulation tissue continues to close over the necrotic bone until only a small hole (the sinus tract) is left. Occasionally, the granulation tissue will close over the necrotic area completely, blocking the egress of contaminated fluids, and resulting in deep abscess formation. Systemic signs of sepsis, including fever, malaise, and chills, accompanied by increasing local wound pain, will be present until the abscess is decompressed either surgically or spontaneously.

The sinus tract, in the presence of underlying necrotic bone, is a natural mechanism to permit the discharge of fluid formed by granulation tissue surface

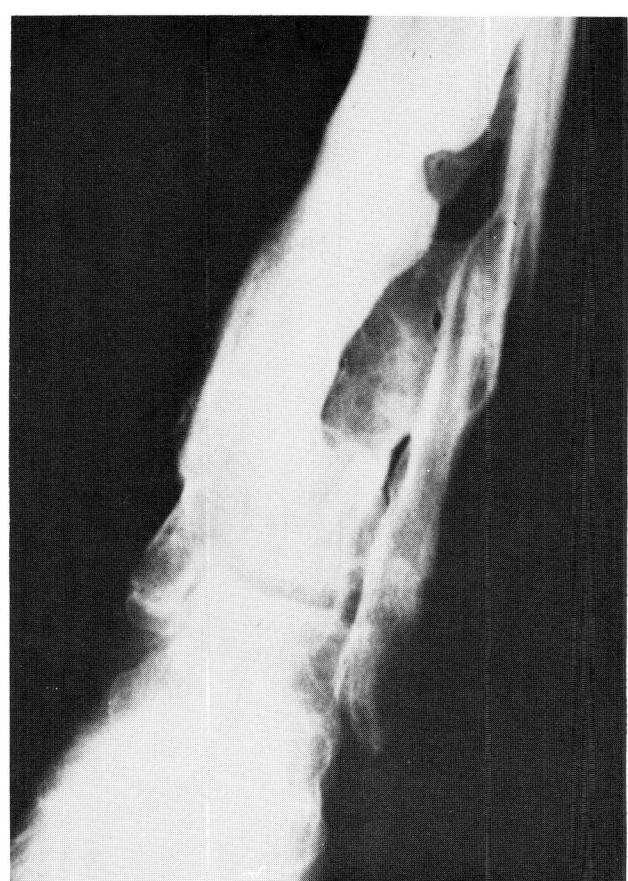

Figure 152: Posterolateral bone graft following resection and one-half inch shortening of the distal tibia and fibula.

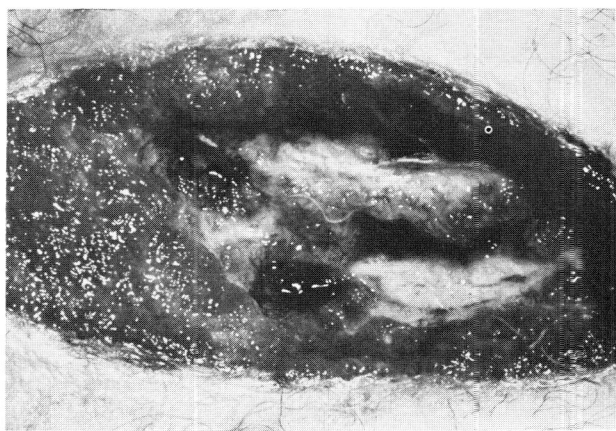

Figure 153: Denuded bone rapidly covering with granulation tissue. Additional debridement not necessary. Skin or flap coverage of this healthy bone is acceptable.

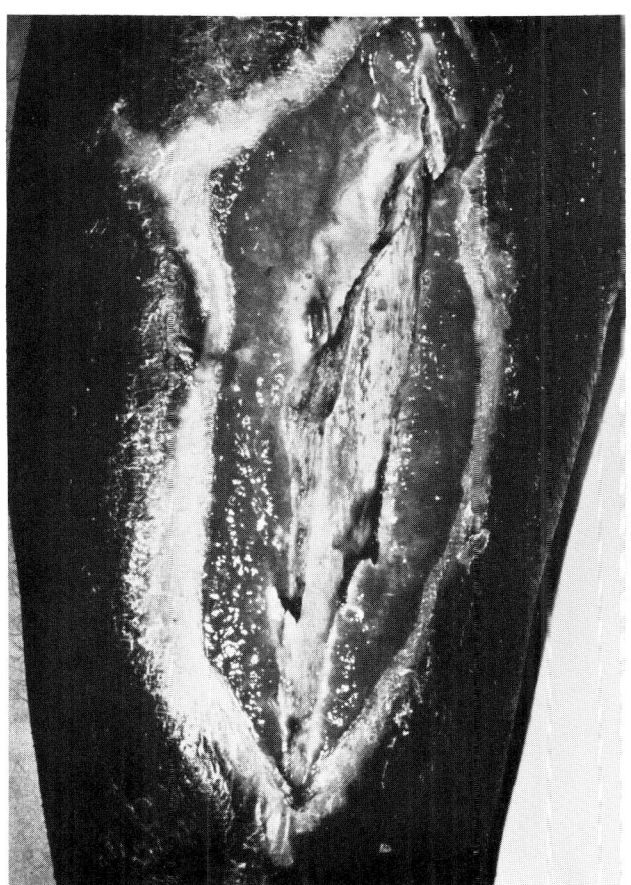

Figure 154: Denuded segment of bone being engulfed—but not covered—by proliferating granulation tissue.

facing, but not adherent to, necrotic bone. The fluid is contaminated by microorganisms that occupy the abscess space. The dead bone acts like a foreign body; granulation tissue will not adhere to it. The space (abscess) cannot be obliterated. The draining sinus tract and its underlying necrotic bone are not dangerous to the patient as long as free egress of contaminated or purulent material occurs.

If the necrotic bone is in continuity with viable bone, it will be replaced, in time, by creeping substitution. The transformation from nonviable to viable bone allows obliteration of the abscess space, because granulation tissue will adhere to the living osseous surface. In this manner, resolution of the infective process can be anticipated. This substitution process is unlikely to occur if the necrotic segment of bone is *not* in continuity with the living bone. For this reason, loose necrotic bone should be removed whenever it is associated with wound sepsis.

SKIN GRAFT: In cases where exposed areas consist only of healthy muscle and healthy granulation tissue, split-thickness skin grafts can be applied. Free skin grafts are unlikely to survive over bone that has been denuded of its periosteum, or over large exposed tendons.[106] When periosteum (or peritenon) is present, the likelihood of graft acceptance varies inversely with the thickness of the skin.[106] Unfortunately, successful coverage of bone with a thin split-thickness skin graft results in an adherent patch of skin that is easily traumatized, resulting in recurrent breakdown.

FLAPS: If the exposed bone in a wound is viable, it can be covered with a cross-leg flap, a transposition flap, a muscle flap, or a musculocutaneous flap. The best type of skin coverage for a particular case depends upon anatomic considerations, the condition of adjacent skin and soft tissues, and the patient's medical status. It is wise to consult with a surgeon knowledgeable in reconstructive techniques when planning skin

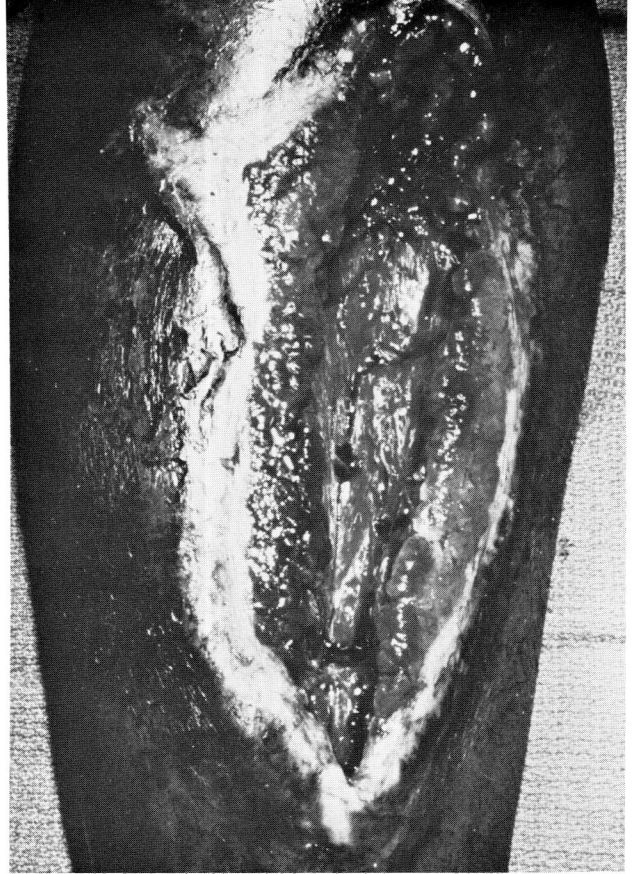

Figure 155: Debridement of necrotic segment in Figure 154. Prompt coverage with full thickness skin will permit closed bone grafting.

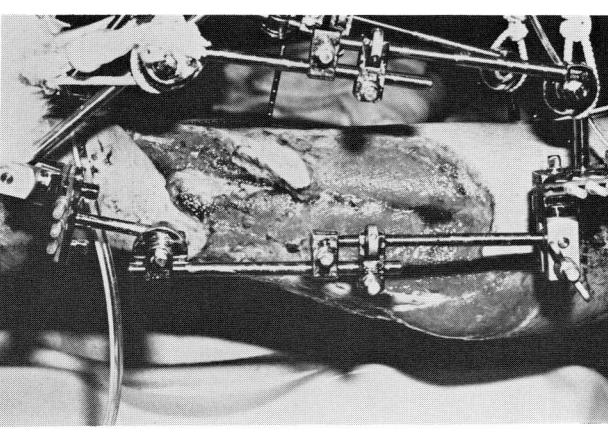

Figure 156: Motorboat propeller injury. Fracture configuration permitted compression (bone graft unnecessary).

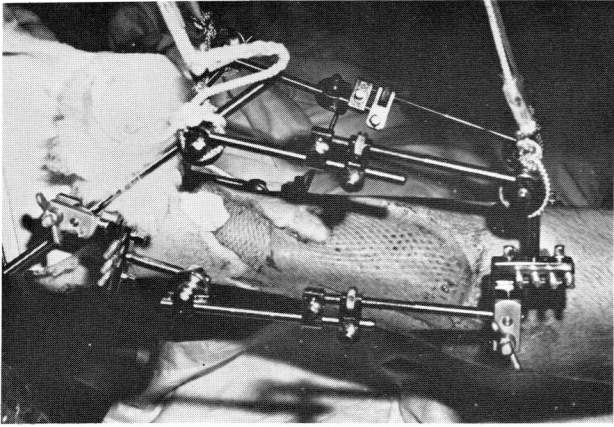

Figure 157: Mesh graft applied to a clean granulating base.

coverage for severely traumatized limbs.

Muscle and musculocutaneous flaps have become popular methods of obtaining tissue coverage of exposed bone. Numerous flaps have been described.[125, 206] For lower limb problems, the most useful muscle flaps include the medial or lateral head of the gastrocnemius muscle, the soleus muscle, and the flexor digitorum longus, among others. Flaps utilizing superficial muscles, such as the gastrocnemius, can be transposed with their overlying skin as a musculocutaneous flap. Muscle flaps from deeper areas, such as the soleus, will require an overlying split-thickness skin graft, which can be done either at the time the flap is transposed or as a separate procedure.

Free flaps, transferred by anastomosing the vascular supply onto local arteries and veins, have also been described.[73, 74] This rapidly expanding field of reconstructive surgery offers great promise to the problem of tissue loss following trauma. Free flaps may consist

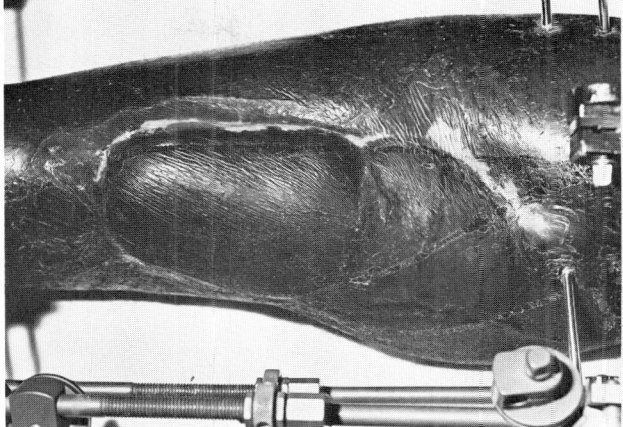

Figure 160: Cross leg flap. The flap will be elevated along the margin and a cortical cancellous graft inserted for segmental reconstruction.

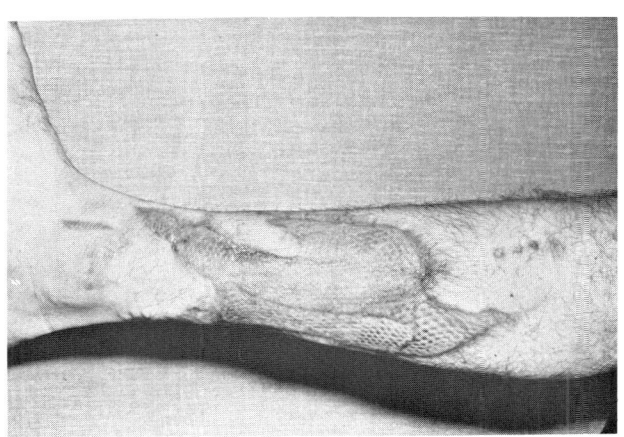

Figure 158: Union in fixator due to primary bone healing.

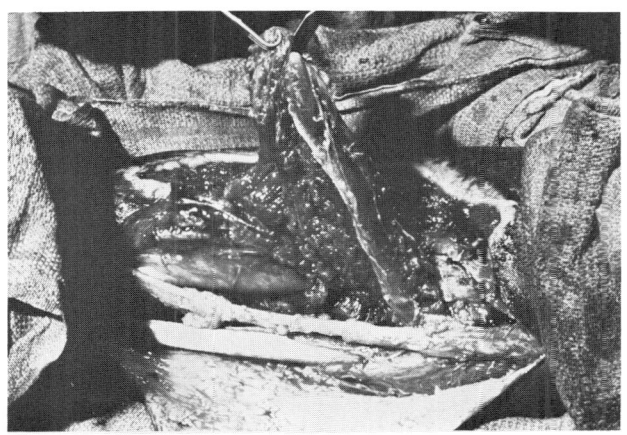

Figure 161: Soleus muscle flap.

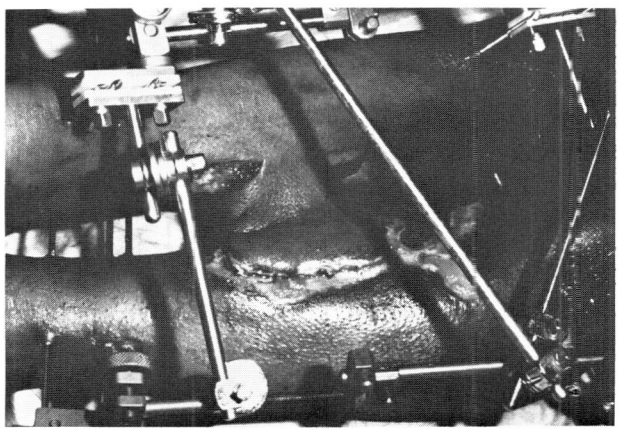

Figure 159: Cross leg flap (partially detached). Fluorescein dye was used to assess the flap during detachment.

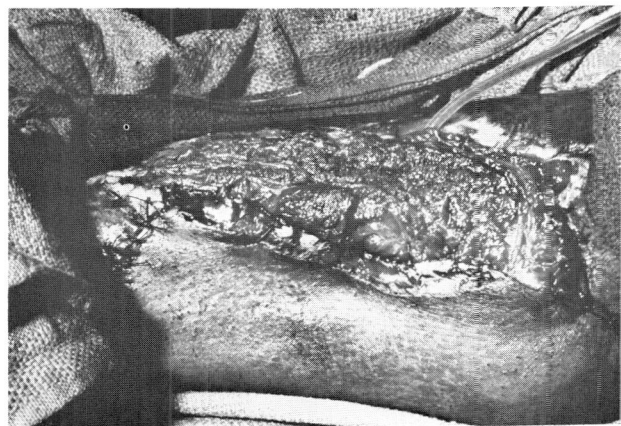

Figure 162: The flap can be covered with split thickness skin immediately, or after five days, if there is any question of muscle viability.

either of skin and subcutaneous tissue, or of skin, subcutaneous tissue, and muscle. (The muscle serves as a nonfunctioning unit to protect essential musculocutaneous arteries.) Free groin flaps,[206] deltopectoral flaps,[122] and latissimus dorsi flaps[197] have been especially useful in limb reconstruction.

Recent work with composite free tissue transfers is also promising.[32] In this situation, the composite tissue serves as a "functional" reconstruction. A free osteocutaneous transfer,[255] for example, provides living bone to be incorporated into a skeletal defect. A free musculocutaneous transfer permits the transferred muscle's motor nerve to be anastomosed to the stump of a motor nerve in the recipient area, providing a functional muscle. Experience with microvascular anastomosis techniques is a prerequisite for performing these procedures.[84]

DEALING WITH A BONE DEFECT: The considerations noted above apply primarily to limb injuries in which the soft tissue damage is not associated with extensive loss of bone substance. The absence of osseous tissue requires that the fixator be left in place until bone continuity is restored. In fact, one of external fixation's most valuable roles in fracture management is for treatment of wounds associated with a segmental bone defect, whether due to traumatic loss or surgical debridement.

Bone grafting should be delayed until healthy tissue capable of sustaining the graft is present.[87, 268] On occasion, a cancellous bone graft can be applied at the time of initial debridement, if tissue in the recipient area is of good quality.[275] In general, however, trauma sufficiently severe to result in segmental bone loss or to require extensive bone debridement is usually severe enough to preclude primary bone grafting.

The type of bone grafting technique to be employed depends upon whether or not there is adequate skin coverage. If the bone defect is accompanied by a clean open wound with loss of skin and subcutaneous tissue, the area can be covered with an appropriate soft tissue flap or transfer. When the flap is healed, an incision along its free margin will permit the flap to be elevated, exposing the osseous defect. Fresh autogenous or cancellous or cortical-cancellous bone can be packed in the defect.

One of the goals of wound management, in fact, is to obtain sufficient full-thickness skin coverage to permit closed bone grafting of skeletal defects. The fragile skin that follows healing by secondary intention is not as suitable for supporting a cortical-cancellous graft as is full-thickness skin. If skin coverage is incomplete, only fresh autogenous cancellous bone should be used.

An alternate way of dealing with the absence of cortical bone is to apply a cancellous bone graft into a

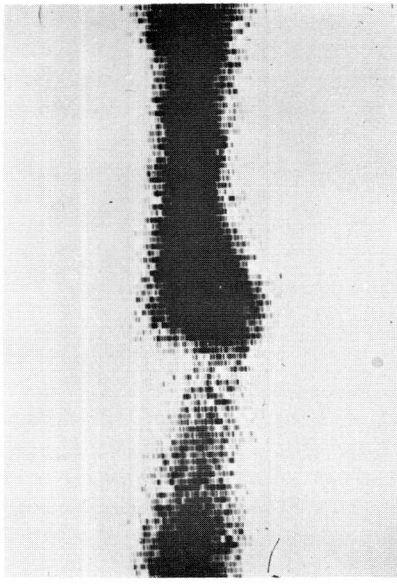

Figure 163: Bone scan demonstrating typical "cold spot" indicating full thickness bone necrosis.

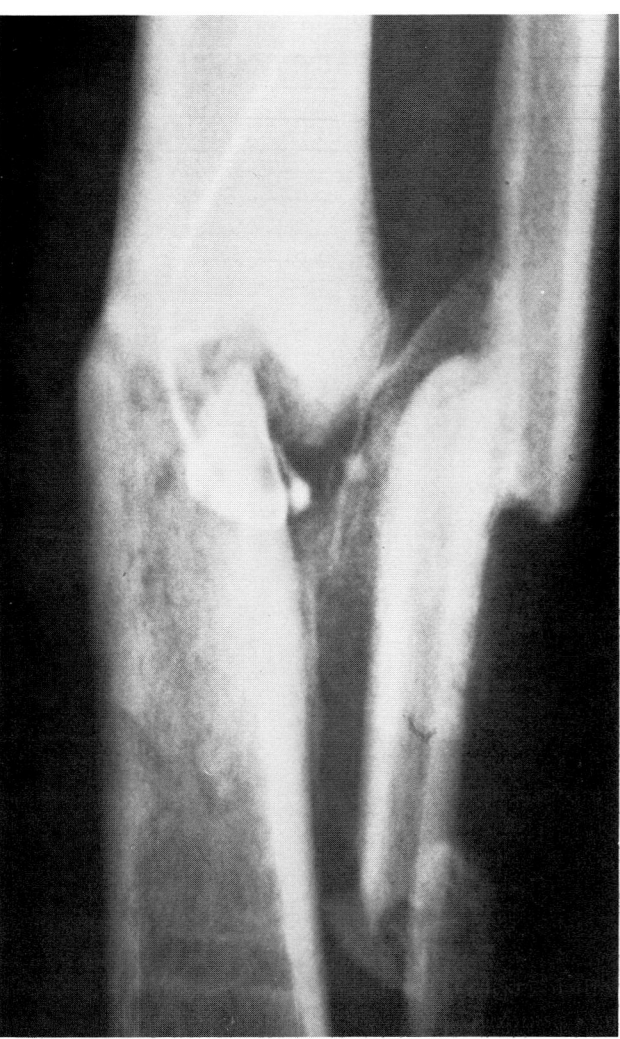

Figure 164: Characteristic pencil-in-cup roentgenographic appearance of necrotic end of fracture fragment held in place with external fixation.

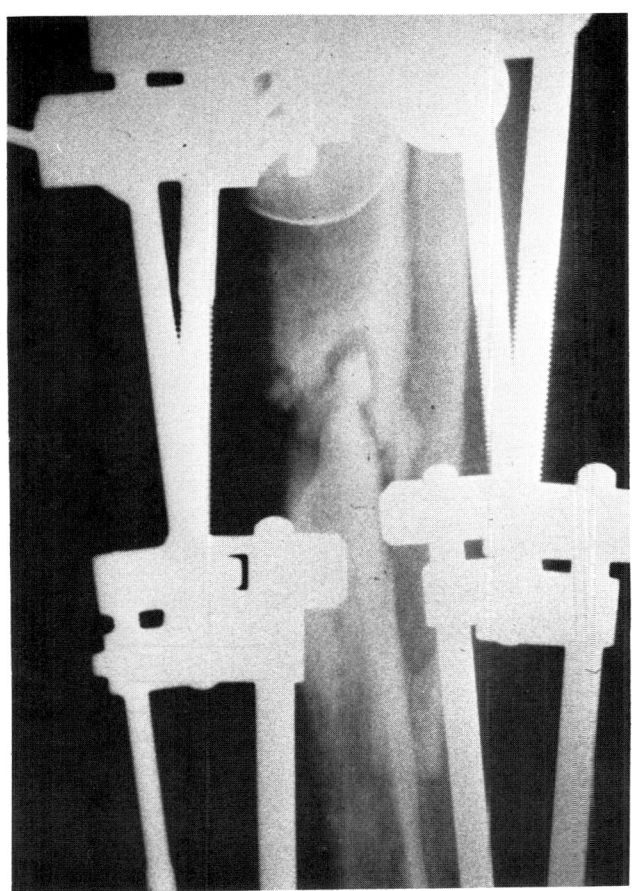

Figure 165: Pencil-in-cup appearance of necrotic cortical bone in contact with viable bone. The fracture callus attached to the proximal fragment is trying to bridge the dead section. Resorption of bone from the viable end is due to micromotion.

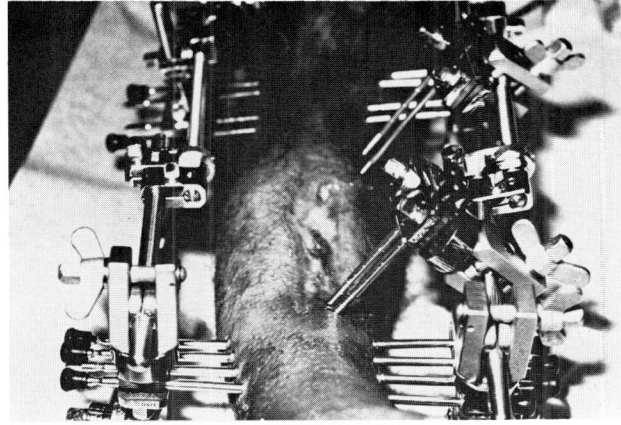

Figure 166: Biplanar fixation is helpful for long-term application where there is absolute loss of bone substance, whether due to trauma or surgical debridement. This configuration decreases pin-bone interface motion, which leads to pin loosening.

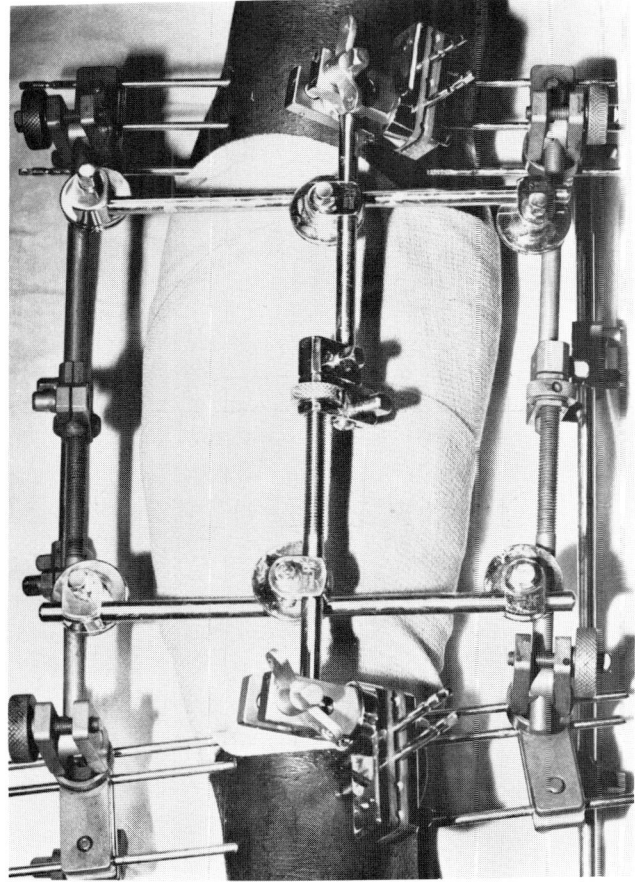

Figure 167: Biplanar configuration utilizing anterior bar. Many fixator systems permit biplanar configuration with the addition of supplementary bars and pin grippers.

cleanly granulating but open wound cavity.[116, 192, 270] This technique, hardly new to orthopaedic surgery, has been revived by Papineau[213] of Montreal, and popularized by Roy-Camille[234] of Paris. A fresh autogenous cancellous bone graft is applied when the wound is lined with a continuous bed of healthy granulation tissue. It is important to have enough cancellous bone to create a bulky wad of grafted bone at the level of the skeletal defect—wider, in fact, than the bone itself.

Cover the graft with a gauze dressing soaked in Ringer's solution. (Continuous irrigation of the dressing prevents the graft from drying out.) Change the dressing three times a day. The central portion of the graft will become necrotic and secondarily infected, but free drainage through the graft will prevent abscess formation and sequestration. Granulation tissue will grow into the graft followed by secondary epithelialization. Early skin grafting is unnecessary and even dangerous because it prevents egress of contaminated fluid and pus. Any radiolucent defect in the graft mass will require regrafting, repeated every four to six weeks, if necessary, until the graft matures.

Extremely rigid external fixation is required to prevent micromotion from producing a cartilaginous pseudarthrosis in the middle of the graft mass.[88,116] Use of a biplanar frame configuration or a rigid ring-type fixator will usually prevent this problem.

Roy-Camille and his coworkers[234] have reported their results with sixty-four cases treated in this manner. Seventeen involved bone gaps less than 1.0 cm (a crater-like defect) and nineteen involved bone segment defects ranging in length from 1.5 to 14.0 cm (usually created by extensive debridement). Thirty-four (94.4%) of the thirty-six patients with bone defects were solidly united at follow-up—although eleven of the thirty-six patients required one or more regrafting procedures.

The bone formed by this procedure takes years to mature, requiring prolonged orthotic protection. For this reason, when dealing with the severely traumatized limb, it is wiser to obtain full-thickness skin coverage of any bone defect, followed by a closed cancellous or corticocancellous grafting procedure. Open cancellous bone grafting should be reserved for use as a salvage procedure until more data are available regarding its effectiveness and potential hazards.

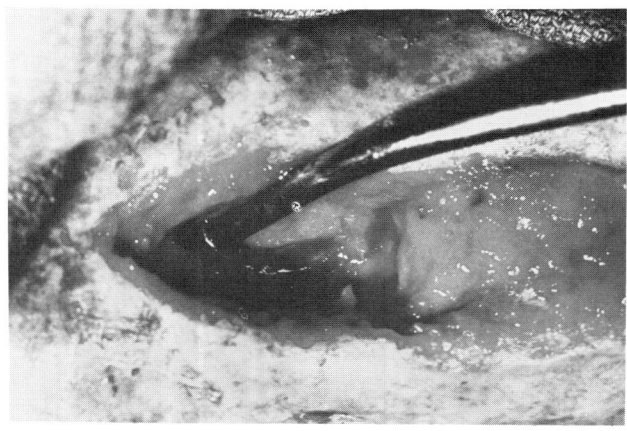

Figure 168A: Open cancellous bone graft. Clean granulating bed without necrotic tissue or purulence is essential for a good graft "take."

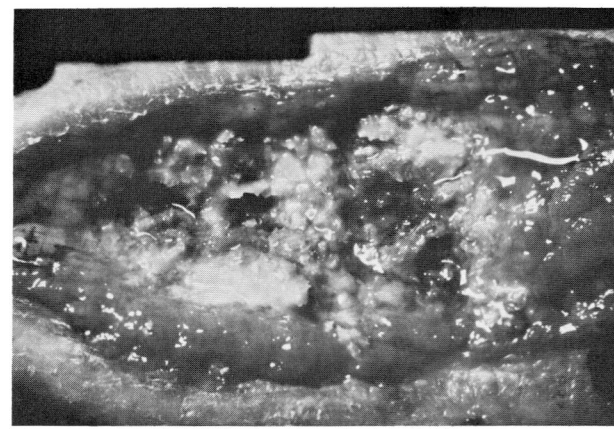

Figure 168C: Cancellous bone graft—appearance of graft at ten days. Note purulent discharge from the center of the graft.

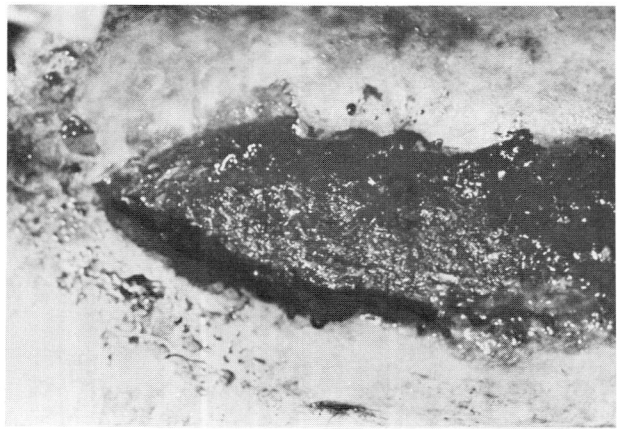

Figure 168B: Cancellous bone graft—appearance of graft at the time of surgery.

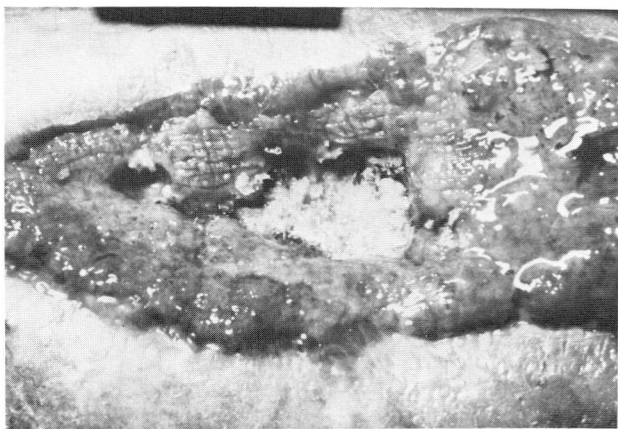

Figure 168D: Granulating tissue closing over the graft mass. A sequestrum would form if the bone were cortical rather than cancellous.

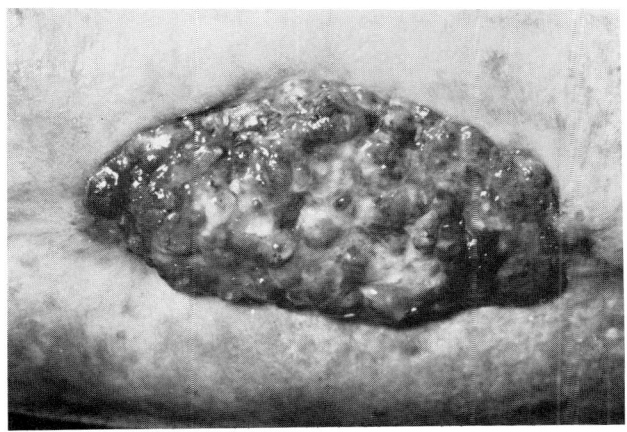 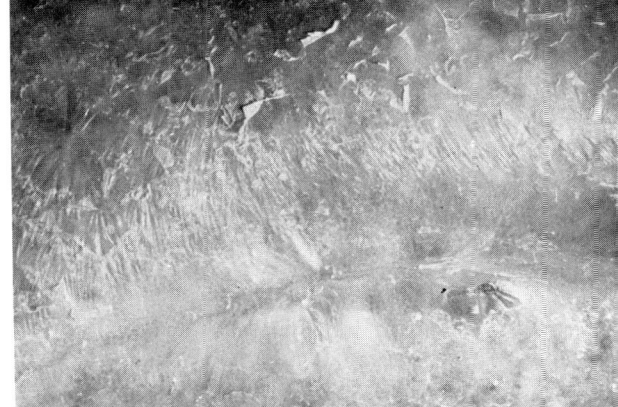

Figure 168E: Graft covered by granulation tissue. I suspect that there is less loss of potential bone forming cells with a closed (rather than an open) graft.

Figure 168F: Final appearance of grafted area. Healing by secondary intention.

Chapter 8

FIXATOR PROBLEMS

Introduction

It is the patient who must tolerate wearing an external skeletal fixator from day to day. Patient-related problems caused by the fixator are pressure necrosis of the skin and undue or excessive pain. Pin breakage may occur while a fixator is in place, causing distress to the patient and his surgeon. Disruption of the patient's life-style and psychosocial problems associated with external skeletal fixation, generally related to long-term application combined with protracted hospitalization, may occur.

PRESSURE NECROSIS

Figure 169 illustrates an area of pressure necrosis that occurred when the thigh of the patient came into continuous contact with the pin-gripping clamp of a fixator. The condition was aggravated by the fact that the pin ends had been cut off next to the clamp, so that the clamp could not be moved away from the skin until the entire fixator was removed. The same situation occurred with two other patients with whom I have had experience. In one of these cases, the patient's overweight condition may have contributed to the problem.

The amount of clearance between the skin and the fixator, the pin-grippers, or bars varies from area to area.

In the upper extremity, and in lower extremity applications where there is bone immediately under the subcutaneous tissues (as in the pretibial region of the leg) one fingerbreadth between the skin and the fixator is sufficient. Two fingerbreadths are required over most soft tissue areas in the lower extremity where there is muscle between subcutaneous tissue and bone. If limb swelling is anticipated, additional clearance will be necessary. Three fingerbreadths are required in the lateral aspect of the thigh in an obese patient, because the tissue will bulge laterally when the patient is lying down. When applying an external fixator to the pelvis, 10 to 15 cm clearance must be left for the abdomen, so the patient can sit up.[117] It is important to fill the space between the skin and the fixator with a bulky gauze wrap, so that excessive motion between the skin and the pins is prevented.

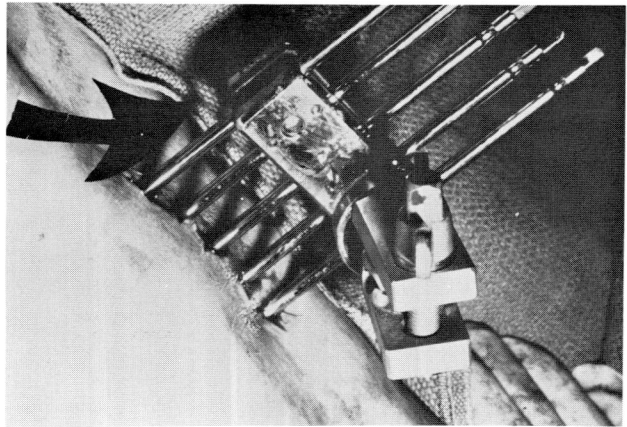

Figure 169: Pressure necrosis caused by fixator against the skin.

Figure 170: Two (or three) fingerbreadths' clearance over mobile soft tissue areas.

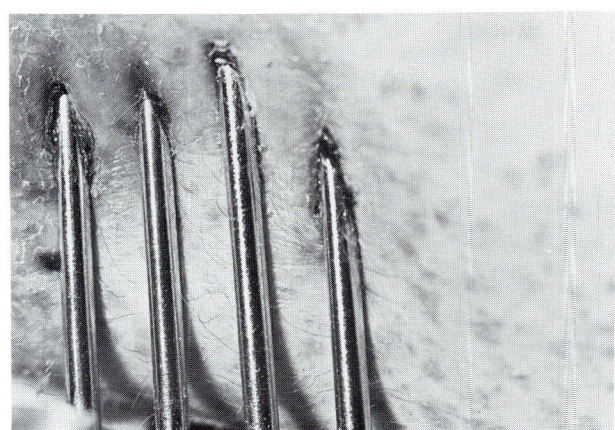

Figure 171: Pressure of skin against the pin in a supine (obese) patient. Bulky wrap helps prevent this problem.

Figure 172: Pin end covers made from miniature picture stand leg tips (hardware store).

BROKEN COMPONENTS

On rare occasions, pins break while the fixator is on a patient. Charnley,[52] in his 1938 communication describing compression arthrodesis of the knee, noted three broken 4.0 mm Steinmann pins out of thirty that were applied. (Charnley's fixator bent the pins significantly during compression arthrodesis. He solved the problem of pin breakage by pre-bending the pins after heating them.)

Burny,[37] in his review of 1,421 applications of external skeletal fixation for tibial fractures, noted that pin breakage occurred in 5.2 percent of the patients. Pins of the upper group broke in 0.9 percent of the cases, while pins in the lower group broke in 1.5 percent. Breakage of pins in both upper and lower groups occurred in 2.2 percent while other types of breakage occurred in 0.6 percent. Burny noted that pin breakage occurred when the pins were used more than once. His incidence of pin breakage was greatly reduced after he began using pins for only one application. Burny indicated that austenitic steel tends to break at the level of a thread or groove when nonparallel pins are forced into the parallel grooves of a pin-gripping clamp, causing a high level of static stress. Ambulation resulted in dynamic stresses superimposed on the static stress already present.

Also of note is the observation of Chao et al.[49] that static stresses on the pins of a fixator applied without compression are seventy times greater than the stresses on the pins of a fixator applied with compression. The reason for this is the contribution to overall fixator stability made by the bone being compressed.

Thus, to reduce the likelihood of pin breakage, one should not reuse pins. Also, pins should be inserted parallel to each other if they are to fit into a pin-gripping clamp with parallel grooves. Finally, compression of the fracture site should be achieved if at all possible.

DISRUPTION OF LIFE-STYLE

When applying an external fixator, the surgeon should consider the logistical problems associated with wearing the frame. The sharp ends of pins should be covered with plastic protectors.[172] Some manufacturers provide protectors with their fixation systems, but they are easy to fabricate from IV tubing.

The pins should be positioned so that they do not interfere with the function of the opposite limb. When applying the fixator to the upper extremity, care should be taken not to apply a frame that will prevent the arm from adducting to the side of the body. In femoral applications, the frame should not occupy the area medial to the upper thigh for reasons of personal hygiene and comfort. In general, fixators should not be applied around the posterior thigh area, which would force the patient to lie only in the prone position for the entire time the fixator frame is in place. If posterior pelvic application is required for stabilization of a complex pelvic fracture, special modification of the patient's bed will be necessary to allow the patient to lie in the supine position.

Fixator frames have a tendency to loosen while they are on the patient. At each clinic or office visit, the patient's frame should be checked and tightened if necessary. Loosening tends to occur where the fixator components meet at a right angle. Compression or distraction, if necessary, should be carried out in a systematic fashion, i.e., symmetrical length adjustments done at each visit. It is important to check for pin loosening, especially if evidence of pin tract sepsis or pain is present when the patient is evaluated. A loose pin should be removed as soon as it is discovered.

Fixators may destroy clothing and bed linens. Patients develop clever strategies to overcome these problems, including modifying their clothes with ties and zippers along the seams to permit easy access to the frame. The surgeon or a member of the hospital staff should advise patients of techniques they can employ to make the presence of the fixator less inconvenient.

UNDUE PAIN

Pain following application of an external skeletal fixator is to be expected during the postoperative period and thereafter. The pain is usually well tolerated by the patient. However, excessive or undue pain requires evaluation and management.

Pin Insertion Pain

Orthopaedic surgeons frequently insert transcutaneous pins into limbs without general anesthesia. Local infiltration of the skin and soft tissues down to the periosteum with lidocaine is an acceptable method for inserting single pins for skeletal traction. Insertion of multiple pins is another matter; one is unlikely to achieve sufficient local anesthesia at enough sites to eliminate the pain of multiple pin insertion. Excessive motion of the limb during insertion because of pain may cause the fracture fragments to move, adding to the patient's discomfort. For these reasons, I recommend general or regional anesthesia when applying external fixation.

Postoperative Pain

External fixation, like any surgical procedure, can be expected to produce pain postoperatively. The pain is usually appropriate to the nature of the problem for which the fixator was initially applied; for example, one can anticipate as much postoperative pain from the application of a tibial external fixator as would result from the use of internal fixation for the same injury.

The patient's personality and pain tolerance threshold also determine the level of pain experienced and, consequently, the quantity of analgesics necessary for pain control. Patients with considerable drug experience seem to require more narcotic medication for relief than do other patients.

Pain around the pin sites following surgery, while significant, is usually overshadowed by the operative site symptoms. However, if pain around the pin predominates among the early postoperative complaints, the pin sites should be inspected after the bulky wrap around the pin groups has been removed. Occasionally, pressure from these dressings against the skin can produce discomfort, not unlike the pressure from a snug-fitting cast. The patient often complains of "burning" and can usually specify the group of pins causing the discomfort.

Next, the pin holes should be inspected. Often, tension will be noted around a pin where none had been present when surgery was completed. Skin and soft tissue tension occurs with shifting of a mobile soft tissue area impaled by transcutaneous pins. As with a bulky wrap that is too tight, tension on the skin at the pin hole produces pain or a burning sensation. The lateral thigh tissue, for example, can shift position when a patient is moved from the operating table to a soft bed, creating skin tension.

The calf muscles can sag posteriorly when the patient's limb is suspended from a bed frame, resulting in tension on the skin around the pin holes. Wrapping the calf musculature, or application of a sling to support the calf, may help. Positioning the patient in a way that avoids undue soft tissue tension may also be beneficial. If the soft tissue support or positional change does not relieve the pin hole tension, the skin must be infiltrated with a local anesthetic and released with a sharp pointed scalpel blade. Occasionally, the skin will be trapped and squeezed between two pins; this tissue should also be released. If skin tension is not released promptly, the burning pain of pressure or tension ischemia will eventually stop: when the skin become necrotic! The result will be an infection of necrotic skin around the pin hole.

Tension on bone is also painful, probably because of the sensitivity of the periosteum. The tension occurs when a fixator is used for compression or distraction, or when nonparallel pins must be bent to fit into a pin-gripper. (Kawamura has noted intolerable pain when intraoperative limb lengthening exceeds 3 percent of the initial length of the bone.) Generally, the pain lessens within a few days. (I do not know if the diminution of deep pain is associated with necrosis of bone or periosteum as occurs when the skin pain stops.) It seems that a pin that creates pain problems from the beginning of fixator application continues to be a problem throughout the fixator experience and occasionally even after the fixator is removed.

Pain While the Fixator is in Place

Ordinarily, the pain associated with a fixator diminishes to a tolerable level within one week after frame application. However, it is not unusual for patients, including those who are quite stoical, to describe a *continuous dull ache* requiring codeine or a similar analgesic medication for control during the entire time the fixator is in place. In some individuals, these symptoms vary with activity levels, being greatest when the patient is ambulatory and relieved by rest. When the patient is permitted to ambulate in the fixator without supplementary aids (such as crutches), the problem is worse. For this reason (and to prevent pin loosening), supplementary ambulation aids are recommended whenever a lower extremity fixator is applied. The aids should be continued until the fixator is removed.

Excessive or undue pain may develop during the time the fixator is in place. This symptom is most distressing to the patient and should be closely investigated. At times, the patient may describe pain starting at a particular pin and radiating proximally or distally, suggesting nerve compression. The sensation may be continuous or intermittent and may be related to the position of the limb. Usually, the patient can identify the pin causing the problem. If not, each pin can be tapped or wiggled slightly with the limb in different positions and the offending pin identified when it reproduces the patient's symptoms. Any pin that produces significant radiating pain should be removed because the involved pin may be putting pressure on a sensitive nerve.

If a pin to be removed is surrounded by exquisitely sensitive tissue, pin removal can be accomplished with the patient sedated or anesthetized. Intravenous diazepam (10 mg) and meperidine (50 mg), administered while monitoring the patient, works well in this situation. One ampule of Narcan® can be kept nearby to be administered, if necessary, to rapidly reverse the depressant effect of the meperidine. It may be convenient to remove the painful pin in the emergency room of a hospital because of the availability of adequate resuscitation equipment and trained personnel in case a problem develops. Alternatively, the tissue around the sensitive pin can be anesthetized with a local anesthetic and the pin removed in the office or clinic.

Pin tract sepsis can also cause excessive pain at the pin site. During outpatient visits, the patient may complain of deep pain which may or may not be associated with tenderness around the pins. Frequently, the pain will precede the onset of visible erythema or swelling. Occasionally, the patient will describe a history of fever or malaise corresponding to the time the pain started. I advise patients to contact me if such symptoms develop. The appropriate course of action consists of bedrest and elevation of the limb at home, and oral antistaphylococcal antibiotics (dicloxacillin 500 mg every 6 hours). If the septic process continues beyond this stage, to the development of cellulitis that is not responsive to oral antibiotics, the patient should be admitted to the hospital, placed at bedrest, and administered parenteral antistaphylococcal antibiotics (nafcillin, 2 gm every 4 hours) after swabs for culture have been obtained from the pin hole. It is helpful to release the skin around the pin with a sharp-pointed scalpel blade, on the assumption that a deep abscess may be present near the bone. With early aggressive treatment, the pin hole may respond quickly and the infection subside. If not, the pin producing the infection should be removed.

Figure 173: Deep bone pain with slight redness developing while fixator is in place. Pin securely fastened to bone.

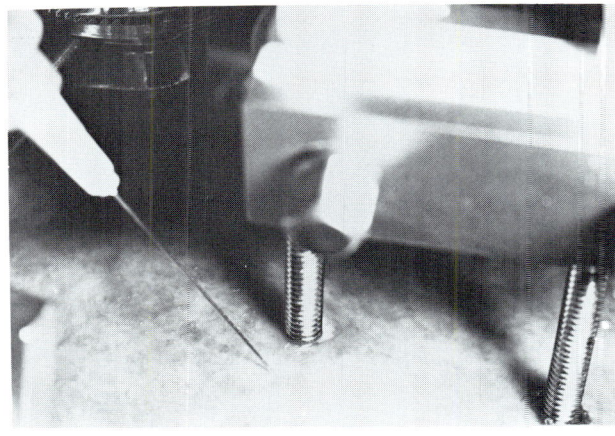

Figure 174: Local infiltration with Lidocaine.

A patient may present in the office or clinic complaining of pain associated with purulent drainage from a pin site but without a history of redness or swelling. The combination of purulent drainage and deep pain requires pin removal.

In reviewing our experience with chronic pin hole drainage following pin removal,[116] we noted four patients developed chronic pin hole osteomyelitis. They all had significant pain and tenderness localized to the bone, associated with persistent purulent drainage while the pin was in place. In each of these cases, I decided to adopt a "wait-and-see" approach in the hope that the drainage would stop or that the pain would go away (instead of removing the offending pin). The result in these four cases was a chronic pin hole osteomyelitis requiring additional surgery long after the fixator had been removed.

Figure 175: Release of soft tissue on presumptive diagnosis of deep abscess.

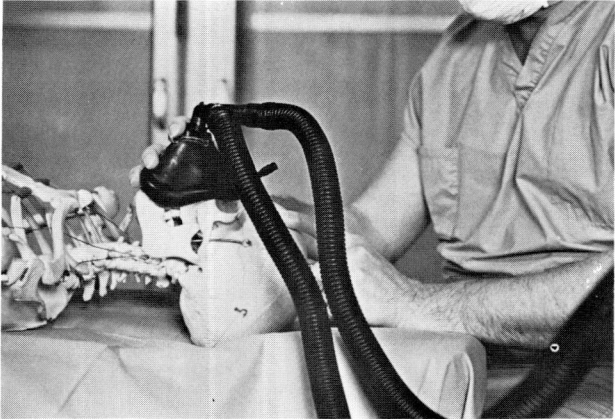

Figure 176: General anesthesia is recommended for removal of multiple full pins.

Pain on Pin Removal

Pin removal can be an unpleasant experience for the patient. This is especially true if the pins are bent. (Full pins, of course, cause more distress than half-pins.) Therefore, although I formerly removed all pins in the office or clinic, I now utilize (and recommend) general anesthesia or heavy sedation for removal of multiple full pins. Patients who have had a very unpleasant pin removal experience seem to remember the event years later, even when other aspects of the fixator application have long been forgotten. Multiple half-pins can usually be removed without an anesthetic in the office or clinic, although Cooney[67] recommends anesthetic sedation for upper extremity half-pin removal as well as for full-pin removal.

Persistent Pain after Fixator Removal

Pain following pin removal usually falls into one of the following categories: (1) bone pain associated with bone hole sepsis; (2) neurogenic pain resulting from persistent nerve irritation; and (3) pain associated with a healing fracture. Other causes of postfixator pain include the usual problems which occur in the post-trauma period. These include joint pain associated with restriction of motion and malalignment of the joint surfaces secondary to fracture deformity.

Bone Pain

Persistent bone pain, localized over the pin hole and lasting more than three or four days after an external fixator is removed, is unusual. If it does occur and is associated with persistent inflammation or drainage, one can expect a chronic pin hole infection to develop. Pain at the pin site can also occur after the pin hole has sealed over and the limb is quiescent. The patient may describe episodes of recurrent pain, sometimes accompanied by redness and swelling, which may subside spontaneously or after a brief course of oral antibiotics. This has occurred twice in our experience and is evidence of a low-grade infection that does not drain freely to the surface. Interestingly, no evidence of a sequestrum was present on radiographs in either case. One responded to curettage of the bone hole, and one did not.

Neurogenic Pain

Chronic pain from nerve irritation should subside with the passage of time. Chronic persistent dysesthesia has occurred with only one patient of mine and was more of an annoyance to him than a significant disability.

Pain Associated with Fracture Healing

Pain associated with fracture healing, localized to the site of injury, is similar to that seen with other modalities of treatment and usually subsides as the fracture consolidates. Solid union in a position of malalignment sometimes causes a dull ache that lasts for many years.

PSYCHOSOCIAL PROBLEMS

The long-term application of an external skeletal fixator can be an unpleasant experience for a patient. In reviewing the various complications of external fixation in a series of thirty-seven patients who had fixators applied as part of the treatment of an infected pseudarthrosis or chronic pyarthrosis, we were surprised to discover that more than one-half (53%) of our patients had one or more significant fixator problems.[116] These problems included pin tract infection, chronic pin hole difficulties, transient paresthesias, and undue pain. Furthermore, many patients, towards the end of their treatment program, were anxious to have the fixator removed, even if no complications or problems had developed while the fixator was in place. We were distressed at the high percentage of patients who were dissatisfied during their fixator experience, especially in view of the enthusiasm for external skeletal fixation being expressed in several recent reports.[67, 164, 269] (Recall, however, that external fixation has had at least one previous wave of popularity among orthopaedic surgeons in the past, only to be followed by a trough of disenchantment.) I should point out that the patients in our series had relatively long-term applications of external skeletal fixation. Patients with septic nonunions of long bone, for example, were in the fixator for an average of six months. When we reviewed our short-term applications, on the other hand, we noted good patient acceptance of external fixation. We concluded that the long-term application (greater than four months) of an external fixator could be a stressful experience for the patients.

Professor Jacques Vidal and his associates[272] have recently reviewed the socioeconomic and psychological effects of long-term treatment in an external skeletal fixator, reporting their experiences with thirty-nine patients who were in external skeletal fixation for long-term applications. The total time of hospitalization, including both acute and rehabilitation institutions, averaged eighteen months, with a maximum time being three years. During this period, the patients were subjected to an average of eight surgical interventions. Vidal's findings related to psychosocial stresses reported by this group of patients are summarized below.

Types of Problems

Problems During Hospitalization

During hospitalization, the patient's concerns and fears were evaluated and tabulated. Thirty-eight percent of the patients complained frequently about the duration of hospitalization; 44 percent felt a sense of estrangement from their families; 45 percent became concerned about their ability to make a living. The latter concern occurred principally among patients in commercial occupations. One-tenth of the group reported that they were haunted by the fear of amputation. Interestingly, this fear was only expressed by patients with septic pseudarthrosis of the tibia, not by those with septic pseudarthrosis of the femur. In summary, the patient's complaints involved both external and internal stresses; they were plagued with anxieties and discomfort related to the duration of hospitalization and its associated separation from family and livelihood, as well as by fear of loss of limb. In reaction to these various problems, 75 percent of the patients had suicidal thoughts at one time or another.

In an effort to gain relief from their problems while hospitalized, 25 percent of the patients experienced a significant increase in consumption of alcohol 36 percent began to smoke cigarettes, and two of the thirty-nine patients became drug abusers. Furthermore, 25 percent of the overall group became deeply religious during their period of confinement. The majority of these patients, however, never entered church again after their hospital treatment was completed.

Problems After Consolidation

The patients were reviewed again after consolidation of their fractures and were noted to have undergone significant changes in their lives. At the time of their initial injury, sixteen of the thirty-nine patients had been single and twenty-three were married. By the time of consolidation, four of the twenty-three married patients had become divorced or separated.

Twenty-one patients had returned to their normal lives, while eight had changed their jobs. Five percent of the patients were on relief and 11 percent began collecting pensions. Fourteen percent were sufficiently incapacitated to preclude their returning to work. These were mostly female patients.

The psychological condition of the patients following completion of treatment was also studied. Although 58 percent of the patients had returned to a normal life 64 percent of these were noted to have significant character changes. They tended to be more aggressive, more irritable, and more anxious. For two of the thirty-nine patients, the personality changes were sufficiently serious to require psychiatric treatment. Eighteen percent (for the most part female) of their patients were disturbed by the esthetic appearance of their limbs.

Problems at Termination of Treatment

The end results of the treatment were analyzed subjectively by the patients and objectively by Vidal and his coworkers. Sixty-eight percent of the patients reported that they were satisfied with the treatment, the proportion being higher (74%) among the younger patients. Satisfaction with treatment was also more characteristic of the more intellectual patients.

At the time the treatment program was terminated, the patients were asked whether they felt that the salvage of their limbs justified the protracted course of treatment. Specifically, they were asked whether they would have gone through the course of treatment had they known it would have taken so long. Of the entire group of thirty-nine patients, only one would have selected amputation rather than endure the course of treatment if he had the choice to make again. Objectively, the results of the treatment were not as encouraging. Vidal and his coworkers rated the end results based on criteria established by the French Workmen's Compensation Board. The final permanent partial disability rating for a resolved septic pseudarthrosis of the tibia was 55 percent, while the average final partial disability rating for a healed pseudarthrosis of the femur was 65 percent. This compared with the official French disability rating for below-knee amputation of 60 percent, and a disability rating for an above-knee amputation of 70-80 percent. Thus, viewed objectively, patients endured protracted hospitalization, psychological duress, disruption of their personal lives, and significant personality changes for what amounted to preservation of a very dysfunctional limb. Vidal wonders what are the limits of treatment that a patient can endure, and what are the limits that society can support to achieve such limited goals? It should be noted that Professor Vidal's study concentrated on patients with the most severe lower extremity problems: septic pseudarthrosis of long bone. These patients are frequently referred to reconstructive centers, like Professor Vidal's clinic, when they face the possibility of amputation. In many cases, the limbs are already quite dysfunctional before reconstruction is started. Ankle stiffness, limb shortening and angulation, and sensory and circulatory impairment are all common problems seen in this group of patients.

Cooney,[67] analyzing his experiences with external skeletal fixation for treating fractures of the forearm, noted that 86 percent of his patients were satisfied with the use of external skeletal fixation, while 3 percent were unsatisfied, and 11 percent had mixed feelings. Not surprisingly, the patients in the latter two groups had a much higher percentage of problems with pain, deformity, or weakness of grip compared to the patients who were satisfied with their treatment.

Helping the Patient

Our experience at Rancho Los Amigos Hospital with patients who have septic pseudarthrosis correlates very well with Vidal's experience at Montpelier. We have noted, however, that many patients with septic pseudarthrosis appear to have significant psychological problems when they first arrive at our facility. Most have undergone multiple surgical procedures in an attempt to obtain bone union. In some cases, the patients have restricted ankle or knee motion, fibrosis and scarring of muscle, and considerable shortening of the limb before their treatment with external fixation commences. These factors are associated with an end result that, although it includes a solid union, is associated with considerable permanent limb dysfunction.

When we embark on such a treatment program, we tell patients that we cannot guarantee them a successful outcome. We try to maintain a positive attitude, but when we advise the patients of the possible pain and potential problems and complications associated with external skeletal fixation, we add to their already considerable worries. We have had ample opportunity to observe reactions similar to those described by Vidal for his patients: depression (sometimes suicidal), estrangement from family, loss of vocation, alcoholism, drug addiction, and fear of amputation. These problems can effectively be managed if the patient and his

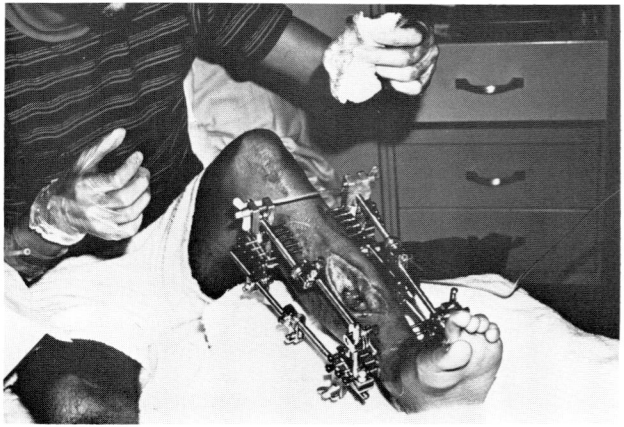

Figure 177: Patient should participate in pin and wound care.

family are advised of them before they occur. Professional psychological counseling is important. Every prospective long-term fixator patient is evaluated by a psychologist at Rancho Los Amigos Hospital. This option is not available to the surgeon who treats trauma victims, but such counseling should be done beforehand when possible or else soon after the fixator application if any problems appear to be developing.

The surgeon has the responsibility to prepare the patient for fixator application. It is worthwhile to tell the patient that pin tract infections are *likely* to occur. They should be told that one or more pins will probably have to be changed during the course of therapy, and that the procedure will probably require general anesthesia. These odds may be a bit higher than the actual likelihood of another anesthetic for pin management, but no harm is done by preparing the patient for the worst and hoping for the best. The patient should also be advised of the strong likelihood of a bone graft.

As with other therapeutic programs, the patient should be advised of alternative methods of treatment. The surgeon should describe a realistic plan to deal with the limb in the event that the fixator does not accomplish its intended goal. I believe that an amputation is preferable to an extremely dysfunctional limb. As a matter of fact, I am reluctant to embark on a prolonged fixator application to salvage a limb that has significant sensory or motor impairment. Marked sensory loss and profound motor weakness are problems that will not resolve with the application of a fixator. I am in agreement with Rosenthal and coworkers[231] who state that a patient would be better off with an amputation and a well-fitting prosthesis than he would with a paralyzed, shortened, insensitive limb. If, on the other hand, the limb is worth saving, we make

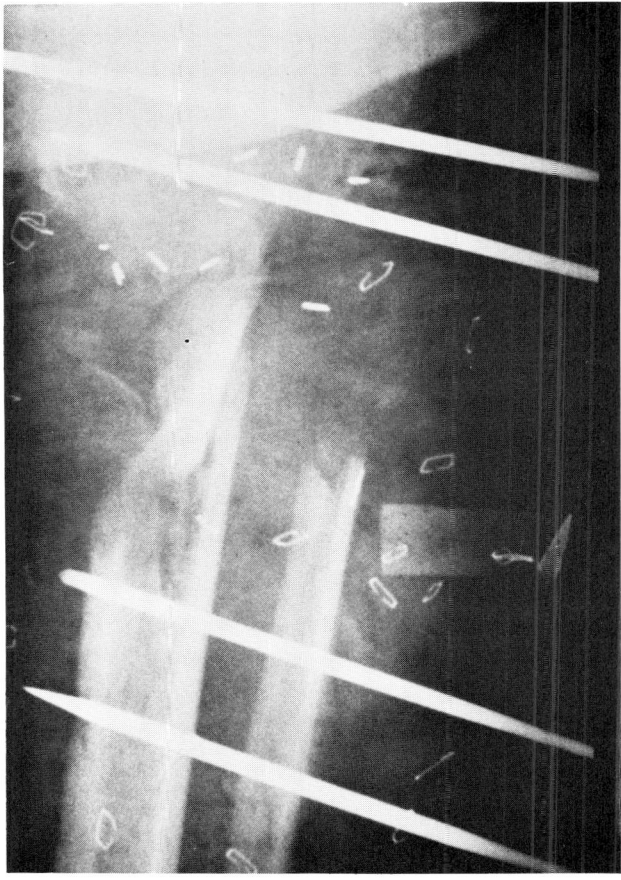

Figure 178: Accident at Indianapolis Speedway. Note absence of proximal fibula and defect in proximal tibia.

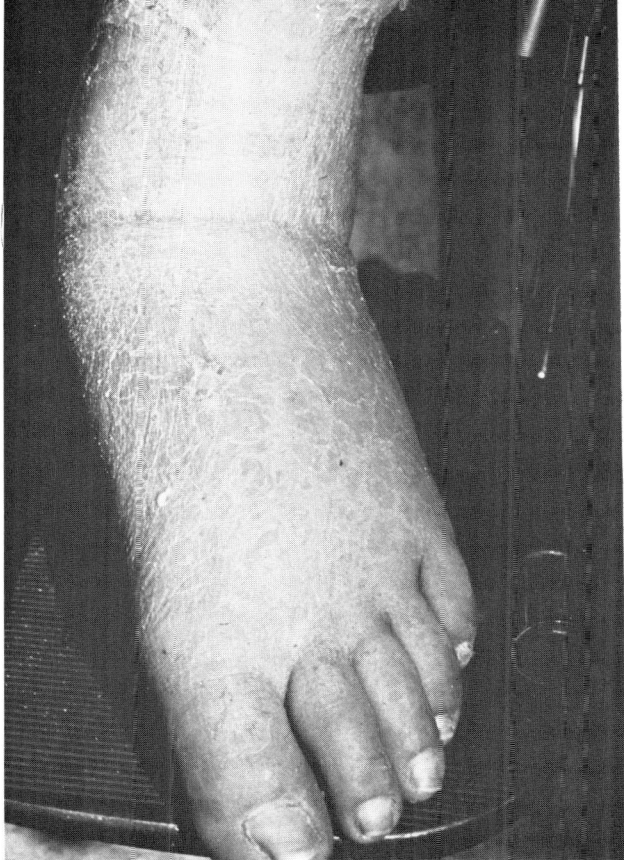

Figure 179: Same patient as Figure 178. Anesthestic functionless foot. An amputation is preferable to the protracted course necessary to reconstruct the osseous defect in such a limb.

every effort to do so. Multiple surgeries, including several bone-grafting procedures, are worthy endeavors if the end result is a functional limb.

In concluding this presentation of the psychosocial effects of long-term external skeletal fixation, Vidal observed that all but one of his thirty-nine patients felt that the prolonged hospitalization and the course in the external fixator was ultimately worthwhile. This feeling of satisfaction was present in spite of the severe disability that remained after healing was complete. Our Rancho experience mirrors Vidal's findings. The patients were pleased to have their limbs salvaged, and this feeling overrode the distress produced by the prolonged hospitalization, multiple surgeries, and general disruption of their personal lives.

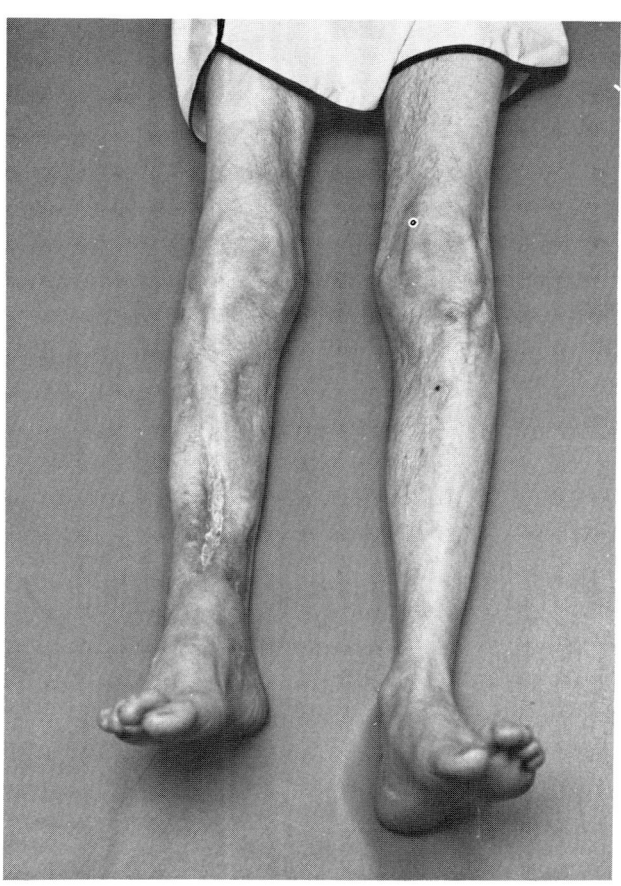

[79]Figure 180: "The limb was saved but the man on top was lost." This patient, a truck driver, became an alcoholic during treatment of an infected pseudarthrosis of the tibia.

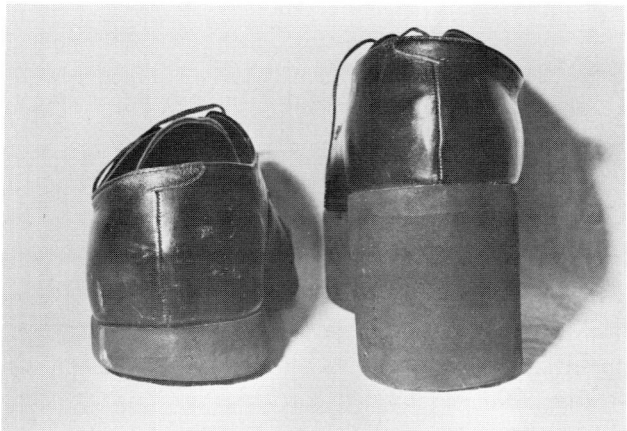

[79]Figure 181: The patient's shoes (Fig. 180).

Chapter 9

POST FIXATION COMPLICATIONS

Certain complications of external fixation become evident after the fixator is removed. These problems include chronic pin hole osteomyelitis, fractures through the pin holes, and refracture through the site of original injury.

PIN HOLE OSTEOMYELITIS

Incidence

Pin hole osteomyelitis is one of the most serious complications of external skeletal fixation. In the preantibiotic era, Siris[248] noted a 22.5 percent incidence of pin hole osteomyelitis when the Anderson and Haynes devices were used. Approximately one-half of his patients with pin hole osteomyelitis developed septisemia and died.

With the advent of antibiotics, the incidence of serious osteomyelitis has diminished. Naden[202] noted 0.4 percent of the patients in his series of 950 treated with the Anderson apparatus developed small sequestra. Three of these healed spontaneously, and one required curettage. Fellander[93] treated forty-nine patients with the Vidal quadrilateral frame. One developed long-lasting drainage from a pin tract. In Burny's[41] series of one-hundred humeral fractures treated with external skeletal fixation, one required curettage of a pin hole after the frame was removed. Lawyer[171] divided thirty-one open tibial fractures into two groups for the purpose of analysis. There were twenty-two Type I/II fractures and nine Type III fractures. One significant postfixation pin tract infection occurred in the Type I/II group, cured with a single debridement. Three chronic pin tract infections occurred in the Type III group, two being controlled with a single debridement, and one requiring three curettage procedures. Krempen[164, 165] noted a 3 percent chronic pin hole osteomyelitis rate in his series employing the Anderson apparatus.

A review of the literature leads to the interesting observation about chronic pin hole osteomyelitis that it rarely occurs in the cancellous ends of bone or in the iliac wing of the pelvis, in spite of a significant incidence of pin drainage while the pins are in place.

Occasionally, a disastrous complication accompanies pin hole osteomyelitis. Figure 182 demonstrates a pin hole infection that continued to erode bone until it produced a full thickness defect of the tibial shaft. The fracture for which the fixator had been applied was healed. The patient requested an amputation of his leg because he could not face the additional surgical procedures necessary to control the infective process. Mears (personal communication) has a similar case, wherein the pin hole osteomyelitis did not drain externally. The patient developed an extensive diaphyseal erosion, which resulted in a pathological fracture. Mears treated the problem by reapplication of a fixator, segmental resection of the infected bone, and extensive cancellous bone grafting.

Pathophysiology

It is interesting to speculate on the origin of the ring sequestrum. A ring sequestrum is a circle of dead bone—formerly surrounding the pin—which lies within a radiolucent defect of the bone. Schatzker et al.[238] have studied the histology of screw holes in bone, as well as the effect screw motion has on adjacent osseous tissue. (Presumably their observations also apply to pin holes.) They observed that bone cells surrounding a screw hole die after screw insertion. The osteocyte necrosis extends into the bone about 1.0 mm in all directions. The necrosis is due to transection (and thrombosis) of the haversian blood vessels that nourished the osteocytes. This cylinder of dead bone is in continuity with living bone surrounding it and would be indistinguishable from the living bone were it not for the empty lacunae.

Replacement of the dead bone by adjacent living bone—creeping substitution—begins shortly after screw

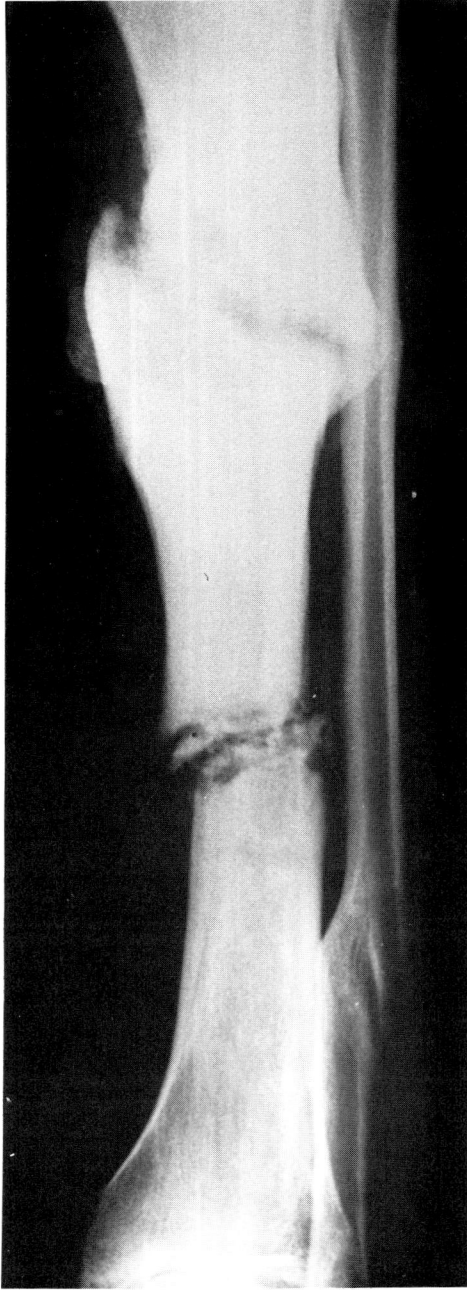

Figure 182A: Disastrous chronic pin tract infection, which resulted in a pathologic fracture after healing of the primary injury.

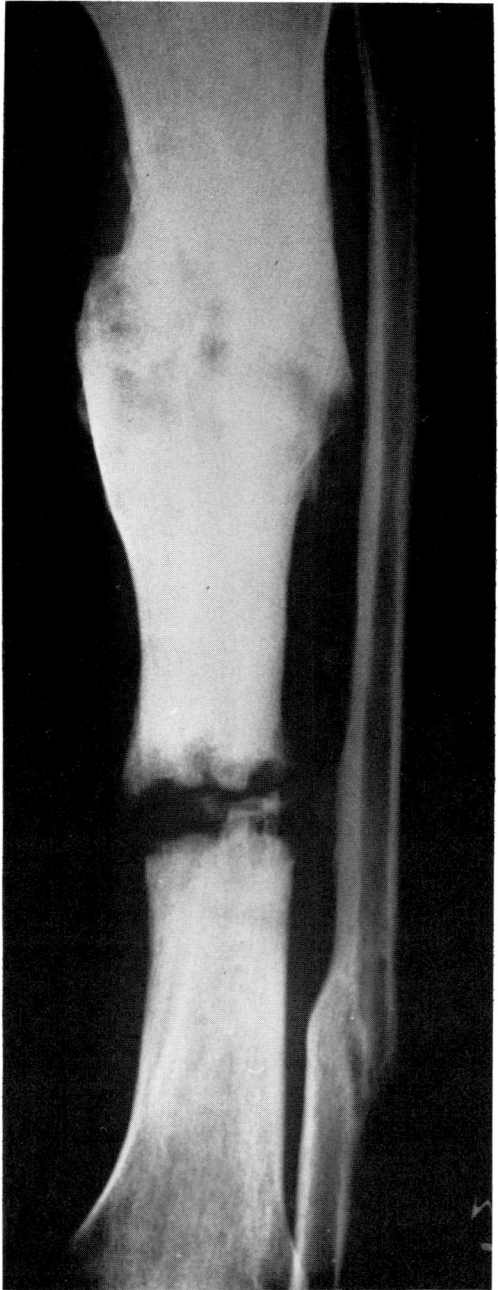

Figure 182B: Further progression while in a cast. The patient requested an amputation.

insertion and is completed in about six weeks, provided the screw is not moving within the screw hole. The process of creeping substitution occurs because there is simultaneous osteoclastic resorption of dead bone and osteoblastic formation of new bone—beginning, no doubt, where necrotic bone and viable bone meet.

If the screw is moving within the screw hole, creeping substitution does not take place. Instead, osteoclastic resorption of dead bone is accompanied by proliferation of fibrous granulation tissue rather than by generation of viable bone. Fibrous tissue formation is the natural response to the presence of motion within

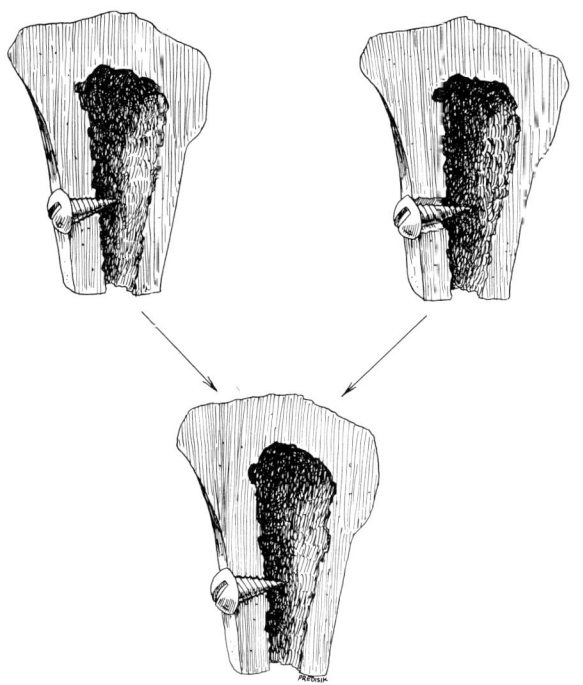

Figure 183: Shatzker et al.'s experiments: In the absence of motion, bone will grow into the threads of a loose screw in six weeks.

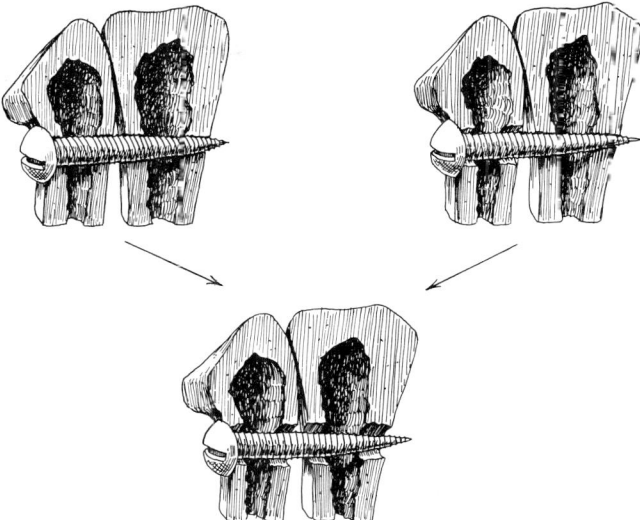

Figure 184: In the presence of motion, bone resorption and replacement with strain tolerant granulation tissue will occur.

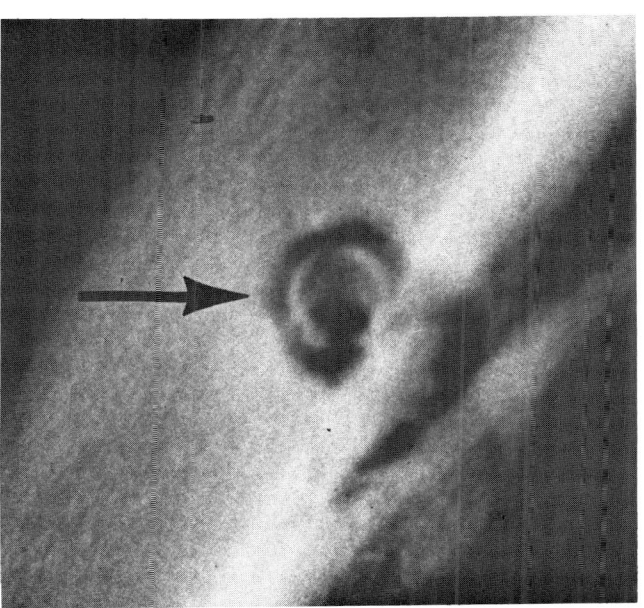

Figure 185: The classic ring sequestrum.

the screw hole. Cyclic screw movement produces high strain (local deformation) in the walls of the screw hole. Nature will wisely place tissue that is strain-tolerant adjacent to a moving screw, and fibrous granulation tissue is one of the most strain-tolerant tissues in the body. (It can be stretched to twice its length before rupturing, while bone, on the other hand, will rupture when stretched only 2 percent of its length.[218]) Fibrous tissue replacement continues until all the strain has been relieved, i.e., the screw is completely loose, surrounded by a (radiolucent) zone of fibrous granulation tissue. With *external fixation* the local strain at the bone hole is presumably higher than with internal fixation. After all, the transcutaneous pin is an enormous lever arm compared to a bone screw.

The local deformation (strain) in the bone around the pin is relieved in the usual manner: bone resorption and replacement with strain-tolerant fibrous granulation tissue. Bone resorption must start where the bone is not necrotic because viable osteoclasts are necessary for the process. It is reasonable to assume that osteoclastic resorption weakens the region between living and dead bone slightly while, at the same time, the enormous cyclic tissue strain (resulting from movement of the transcutaneous pin) favors granulation tissue proliferation rather than new bone formation in the resorbed areas. The bone can easily fracture through the weakened area between the living and dead regions. Following separation, the ring of dead bone becomes, in a mechanical sense, part of the pin, which is still firmly screwed into it.

The pin/dead-bone act as a single unit, wobbling within the bone hole, and causing further osteoclastic resorption at points of contact. The process continues until all the strain is relieved and the pin-dead bone

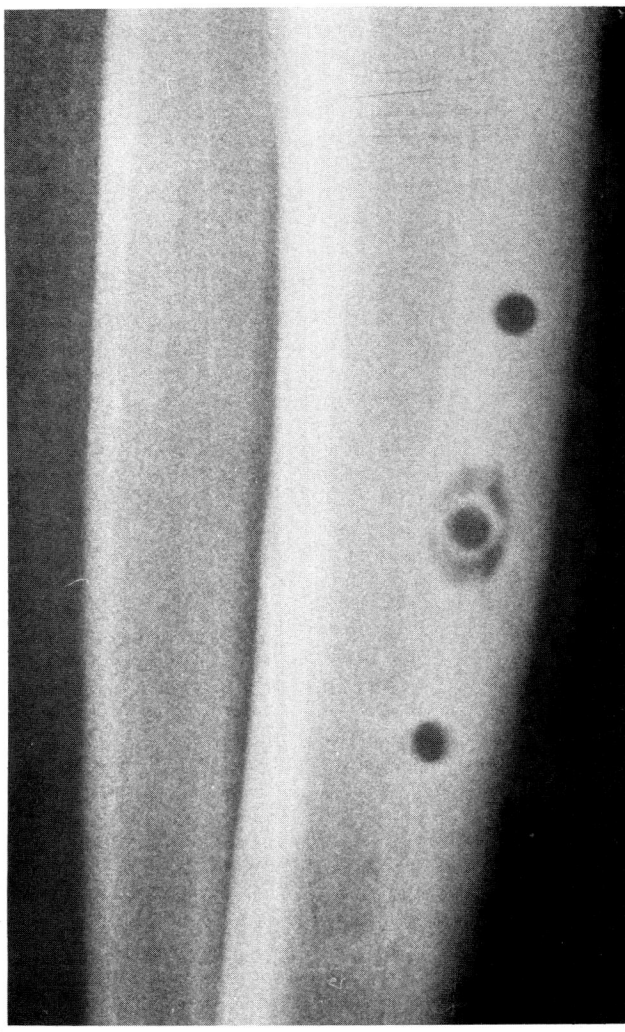

Figure 186: Ring sequestrum following ankle arthrodesis. I do not know why this pin hole developed a sequestrum while the others did not. I suspect that the middle pin became too hot.

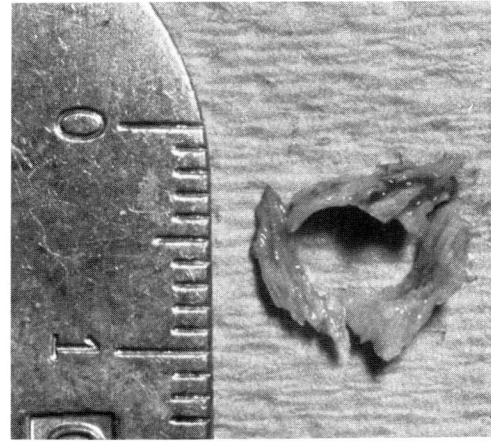

Figure 187: A ring of dead bone relieves strain (local deformation) by fracturing—into three pieces in this case.

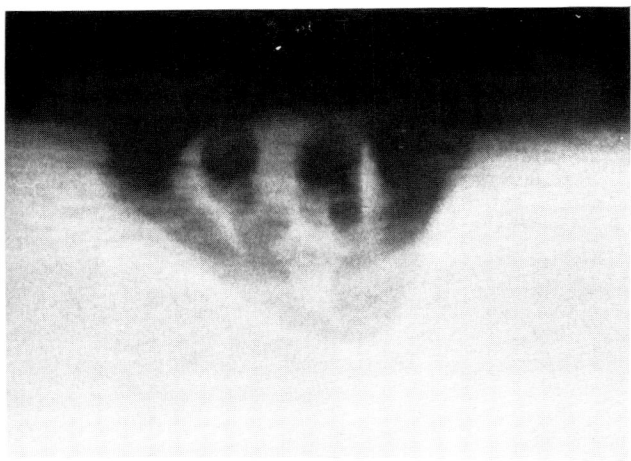

Figure 188: A double ring sequestrum. Note the characteristic location in the anterior cortex of the tibia.

collar floats in strain-tolerant soft granulation tissue. Thus, the ring-like structure surrounded by a zone of radiolucency is a phenomenon of pin loosening, secondary to the cyclic motion of the pin itself. (Most of the time, by the way, the ring actually breaks apart, this being the only way dead bone can "relieve" high strain.)

The presence of a separated ring of dead bone around the pin does not necessarily lead to pin hole osteomyelitis. Pin hole microorganisms probably extend inward from the soft tissues because the strain-relieving granulation tissue becomes an inward extension of the tissue around the pin as it passes through soft tissue. The bone hole granulation tissue comes to be lined with a continuous membrane in regions where it makes contact with the collar of dead bone. It is difficult to say where in the pin hole outright infection arises from the bacterial contamination that is normally present. I believe it can start either in the soft tissue or in the intraosseous portion of the pin hole, depending on the local environment and the presence of necrotic material, motion, and other factors.

With soft tissue pin tract sepsis, removal of the pin leads to resolution of the infection because elimination of the contaminated foreign body permits complete soft tissue healing. In the case of pin hole osteomyelitis, however, fragments of dead bone are left behind to continue to act as contaminated foreign bodies. Presumably, the dead bone is not absorbed for two reasons: first, it is not subjected to strain (deformation) because

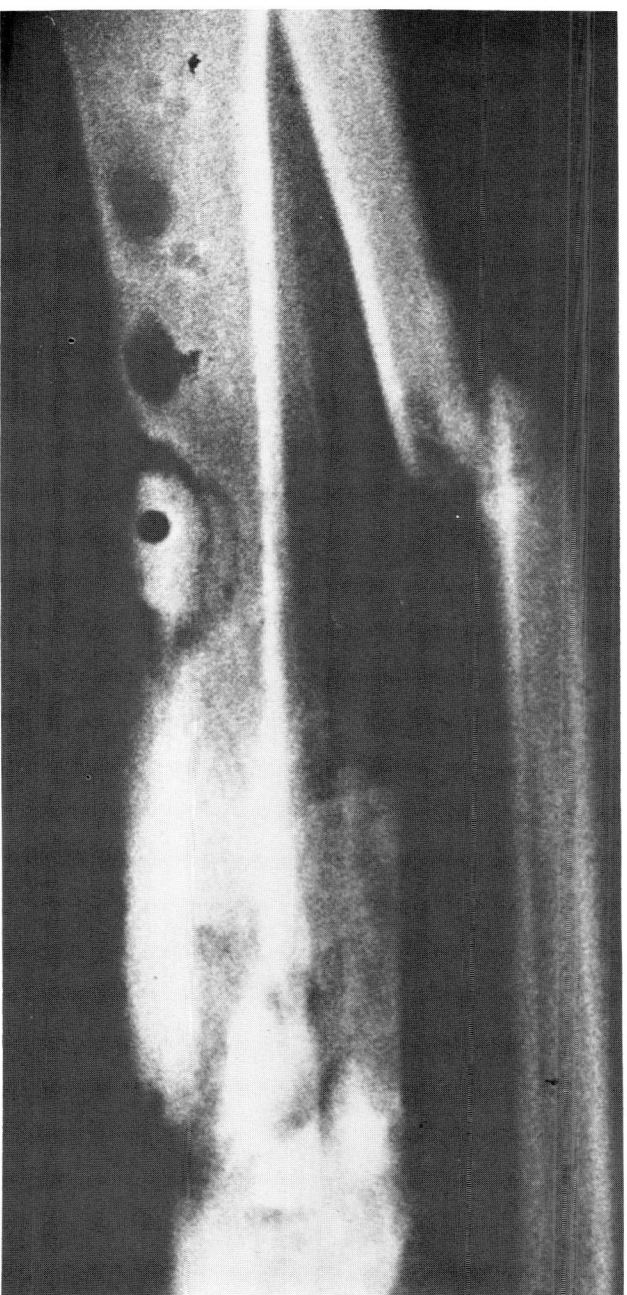

Figure 189: A giant ring sequestrum. I suspect that thermal damage caused the extensive bone necrosis, which preceded separation of the sequestrum from viable bone. If a pin emerges from a limb too hot to be comfortably held in the surgeon's fingertips, it should be removed, cooled off, and reinserted elsewhere into a predrilled hole.

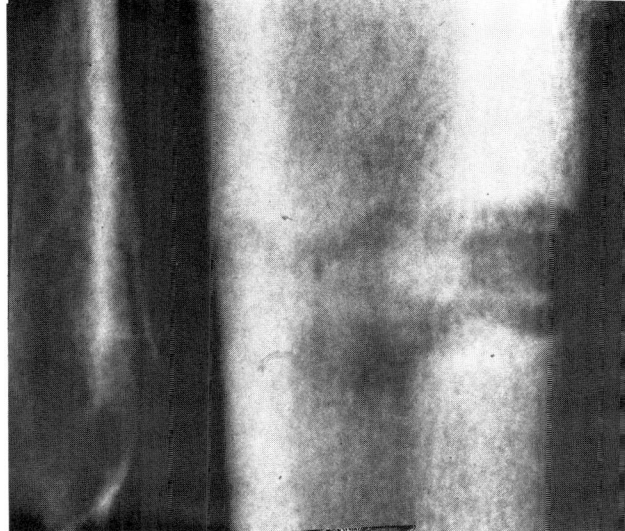

Figure 190: Lateral view (AP roentgenogram) of the ring sequestrum in Figure 187. If this structure is seen when the pin is still in the limb, the pin hole should be gently curetted at the time of pin removal.

it is floating in strain-tolerant granulation tissue; therefore it does not send piezoelectric pulses to osteoclasts. Second, osteoclastic resorption is probably inhibited in the anaerobic low pH environment of pus.

The above considerations do not explain why ring sequestra and chronic pin hole osteomyelitis are more common in cortical than in cancellous bone. Cortical bone, for one thing, is more likely to suffer thermal damage during pin insertion, extending the area of bone necrosis. Cancellous bone, even if dead, might be more easily dissolved by invading osteoclasts, a phenomenon that explains why cancellous bone grafts rarely form sequestra. For these reasons, pin loosening in cortical bone is more likely to result in sequestrum formation than is pin loosening in cancellous bone.

Because fragments of dead bone are so frequently associated with persistent pin hole sepsis, it is wise to gently curette any pin hole following removal of an infected pin. This can be accomplished, even if the patient is awake, with a small curette.

At times, a separated ring of bone, noted on roentgenograms, is not associated with a septic pin hole. In this situation, the dead bone will eventually be replaced by creeping substitution. Periodic roentgenographic reevaluation is desirable in this situation because an indolent but progressive pin hole osteomyelitis may be present.

Management

Persistent or recurrent drainage after pin removal

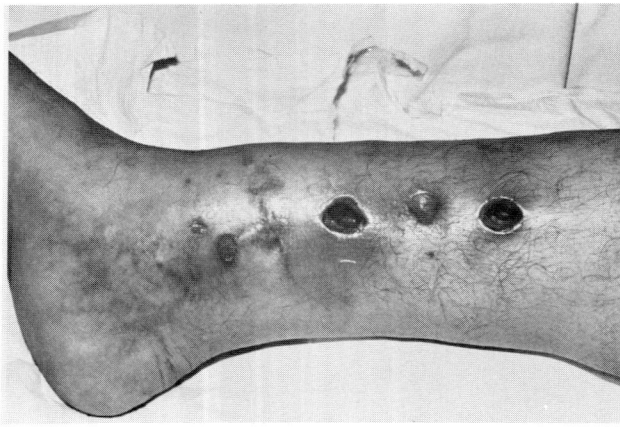

Figure 191: Open management of pin hole after debridement of ring sequestrum.

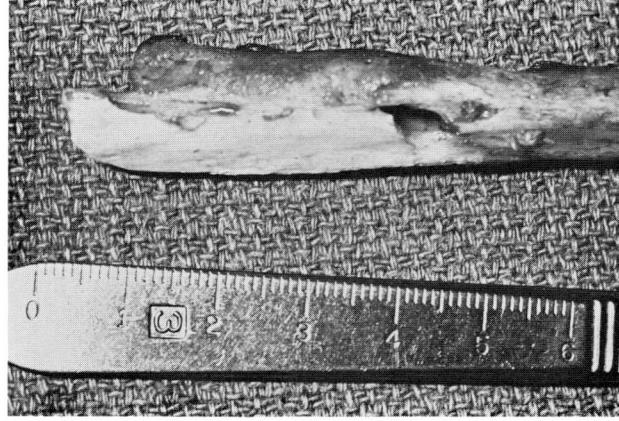

Figure 192: Pin hole of ulna that continued to drain in spite of three curettements. Note the large segment of dead bone (lower left) making up one wall of the pin hole.

usually requires surgical treatment. The bone hole should be curetted completely to remove foreign material, dead bone, infected granulation tissue, and the walls of the pin hole. Curettage should continue until bleeding bone is reached in all directions. The cavity created in this manner should be packed with a fine mesh gauze dressing, which is changed three times a day until the wound is completely healed. Appropriate parenteral antibiotics should be administered during hospitalization. It is wise to continue parenteral antibiotics for at least two weeks, following with oral antibiotics, until the bone is covered by granulation tissue.

Simple curettage of the lesional area may not resolve the problem. Repeat curettage may also fail, leaving the patient (and the surgeon) with a small but persistent chronic osteomyelitis. If this occurs, several strategies are available. Packing the hole with fresh autogenous cancellous bone graft has been successful in some cases. The graft is left open and kept moist with a physiologic irrigating solution. This technique is based on the excellent results obtained by filling skeletal defects with open cancellous bone grafts after saucerization or resection of an infected pseudarthrosis.

Another approach to the problem of chronic bone hole sepsis is to employ gentamicin-polymethyl methacrylate sticks developed in Germany.[161,162] These thin rods consist of hardened polymethyl methacrylate that is impregnated with gentamicin, a combination popularized by Buchholz[39] for use in total hip replacement. The rods are placed into the pin hole after removal of the infected pin or curettage of a ring sequestrum. Gentamicin leaches out of the rods, resulting in high local concentrations of the antibiotic. Immediately after insertion, the gentamicin concentrations exceed the minimum inhibitory concentration for most etiologically important bacteria by a hundredfold. Local concentration of gentamicin slowly decreases but remains greater than inhibitory levels for about eighty days. Traces of gentamicin can be detected in the *serum* for only a few days following implantation. If a saucerized cavity, instead of a pin hole, is present, gentamicin-polymethyl methacrylate beads—strung together like a pearl necklace—are available. These products are not presently marketed in the United States but will soon be released by the Food and Drug Administration.

PIN HOLE FRACTURE

Fractures through pin holes are rarely reported. Nevertheless, they do occur. Siris,[248] in 1944, noted one case of fracture through an osteomyelitic pin hole. The patient subsequently died of sepsis. Mears had a similar case, requiring extensive bone resection and reapplication of a Hoffmann fixator for control. At Rancho Los Amigos Hospital, we noted one pin hole fracture subsequent to curettage for a ring sequestrum. The fracture united in a plaster cast. Several reports describe isolated cases of pin hole fracture in osteoporotic bone.

Cooney[67] noted two (3.3%) nondisplaced pin hole fractures of the radius out of sixty patients treated with external fixation for displaced unstable Colles fractures. Burny's[37] 1,421 tibial fractures treated with external

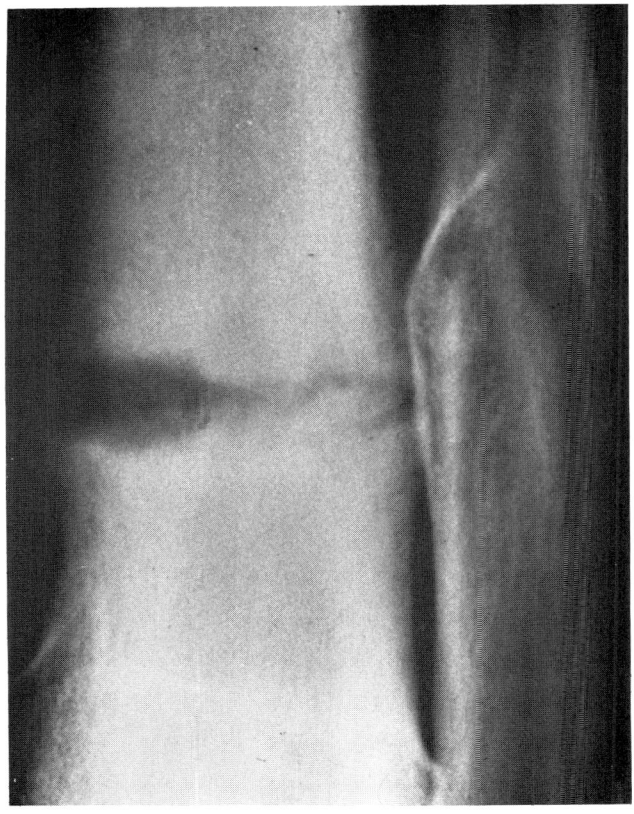

Figure 193: Stress fracture through a pin hole after debridement of ring sequestrum. The fracture united in a plaster cast.

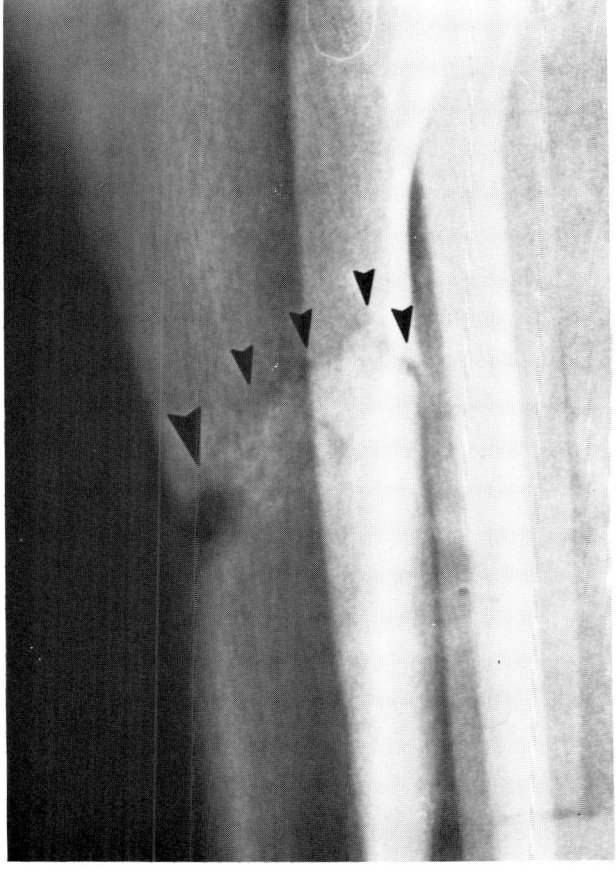

Figure 194: Fracture of tibia starting at bone hole in anterior crest of tibia (after skeletal traction).

fixation resulted in six (0.5%) patients with a fracture through a pin hole. Burny's experience probably represents the likelihood of developing a pin hole fracture in lower extremity applications of external fixation where a plaster or orthotic support is not applied following fixator removal.

Pin hole fractures occur because the bone hole acts as a local stress riser. Brooks et al.[28] have studied the problem of stress concentration and bone holes. They demonstrated that a small bone hole through both cortices, which is less than 30 percent of the diameter of bone, has a stress concentration factor of only 1.6. If the diameter of the hole approaches 50 percent of the diameter of the bone, the effect is much greater, approaching that of an open section. A recent development in the field of external fixation has been the introduction of fixators utilizing large (5.0 mm) pins. They were introduced to increase the stiffness of the fixator configuration which depends, to a considerable extent, on the stiffness of the transcutaneous pins. Time will tell whether this increase in pin diameter will be accompanied by an increase in the incidence of pin hole fracture.

The best way to avoid the prospect of pin hole fracture is to apply a cast or molded total contact orthosis for at least four weeks after fixator removal. During this time, progressive weight bearing can be expected to strengthen the bone around the pin hole. If a fracture does occur, it can be treated conservatively, provided no infection is present.

REFRACTURE

Refracture through a healed fracture is discouraging for the patient and his physician. It is a complication that is not frequently reported but that does occur nonetheless. Refracture is considered to be a fatigue or stress fracture due to incomplete healing.[152] The refracture usually occurs before the final remodeling phase

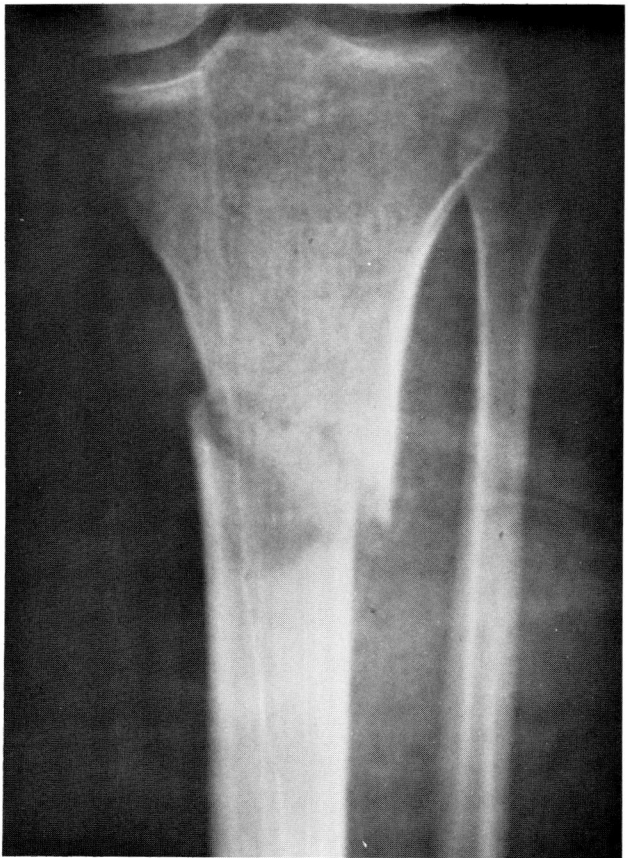

Figure 195: Anteroposterior projection of Figure 194.

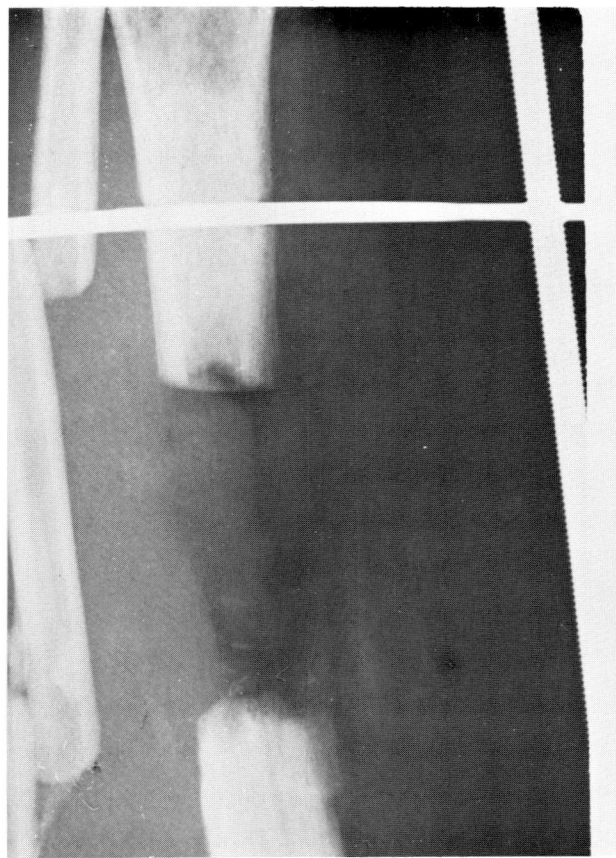

Figure 196A: Segmental resection of necrotic tibia for infected atrophic pseudarthrosis.

during which mature osteons cross the original fracture site. Fatigue fractures following internal fixation (and presumably external fixation) also occur through cortical bone that has become osteoporotic because weight-bearing stresses are detoured to the fixation device while it is in place. The likelihood of refracture can be reduced by gradually increasing the weight-bearing while protecting the limb with a snug fitting plaster of Paris cast or custom-molded total contact orthosis.

Incidence

Refractures occur following internal fixation, plaster treatment, and external fixation. The likelihood of refracture following compression plate fixation (employing ASIF techniques) for tibial fractures has been reported to be from 1.9 to 6 percent.[152] (Compare these figures with Chrisman and Snook's report of a 3 percent refracture rate following plaster cast treatment of tibial fractures.)[55]

The refracture rate following external fixation is comparable to the figures noted above. Burny,[39] provides the most reliable statistics in this regard. The overall incidence is 2.1 percent (25 of 1,421 tibial fractures). The incidence with closed (and open Type I) fractures is lower (1.8%). Recall that Burny's fixator management protocol encourages healing by fracture callus formation because his system is slightly flexible. Burny usually maintains the fixator in place until healing is complete (determined by strain gauge measurement) and does not follow external fixation with a cast or other support. Vizkelety,[277] of Hungary, reports eight refractures (2.7%) out of 292 cases treated with several different fixators, most commonly the Vidal-Adrey quadrilateral frame.

Refracture following bone grafting for septic nonunion is a more common problem. Klemm[161] reports

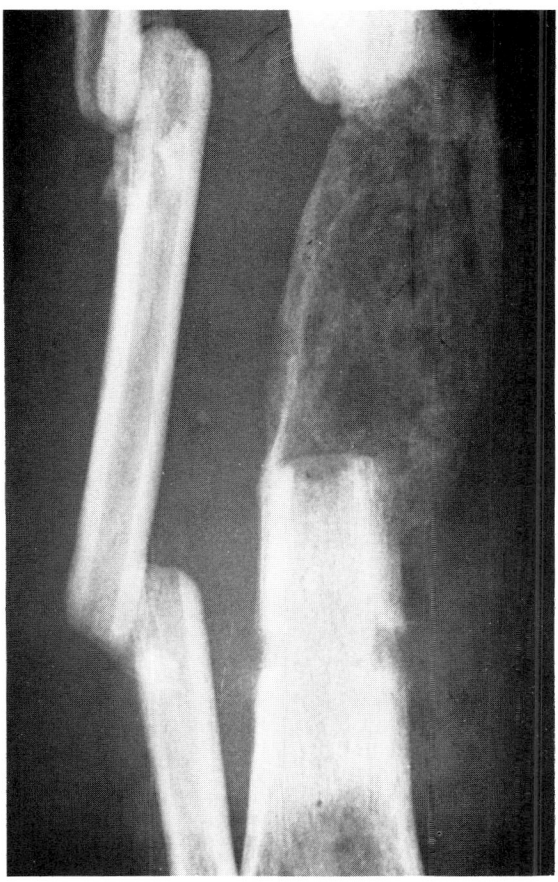

Figure 196B: Cancellous bone graft—note poor graft placement and failure to create tibia-fibula synostosis.

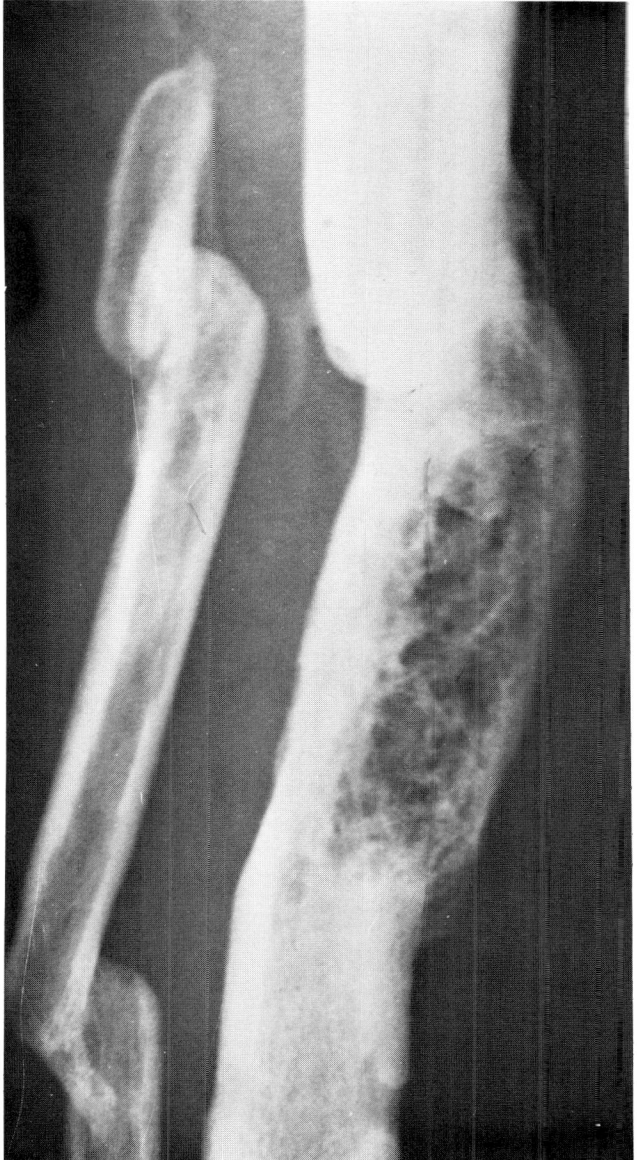

Figure 196C: Maturation of graft mass.

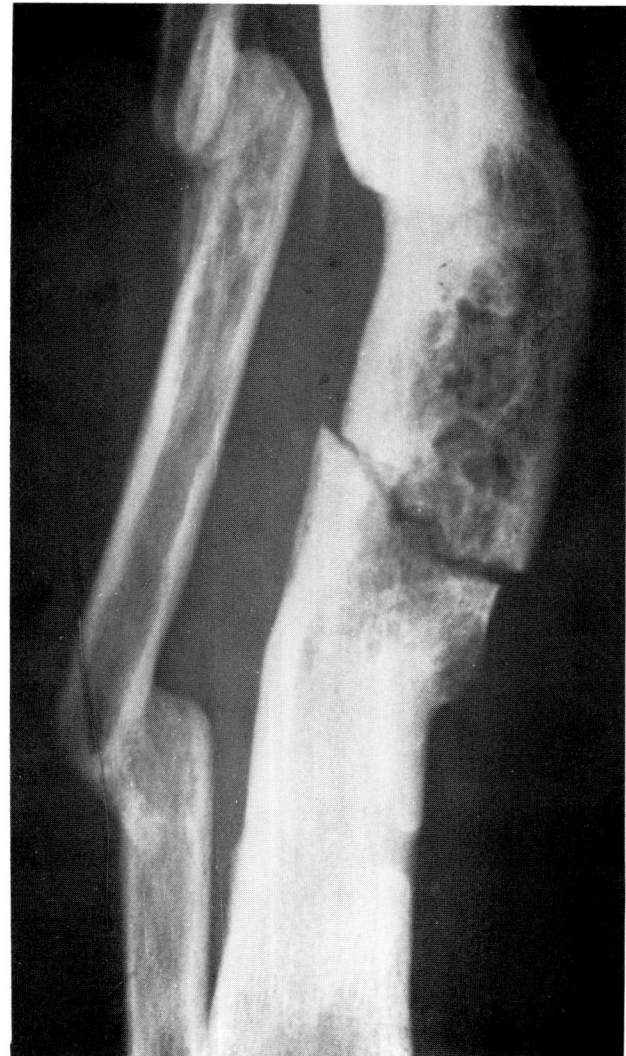

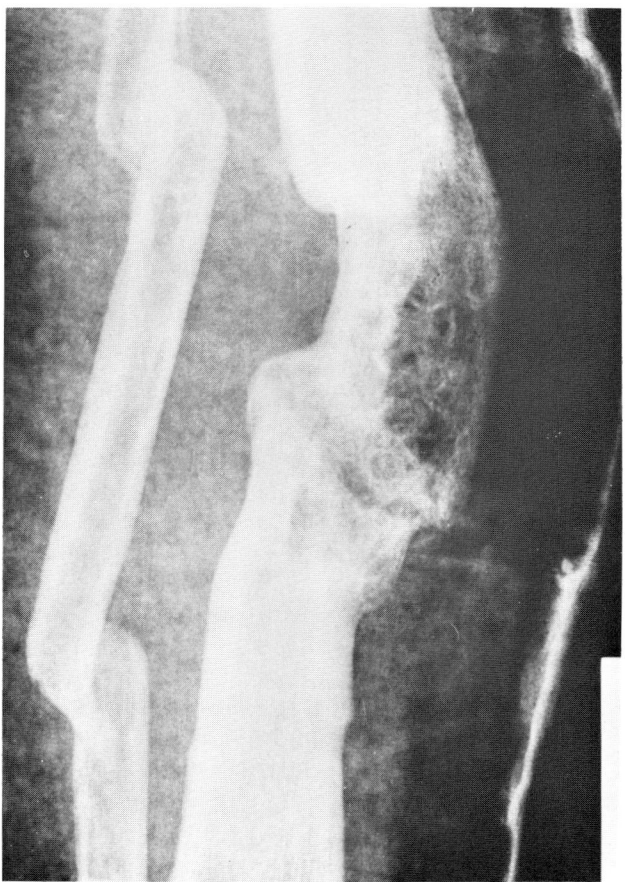

Figure 196E: Healing in plaster cast.

Figure 196D: Fracture of graft in barroom fight.

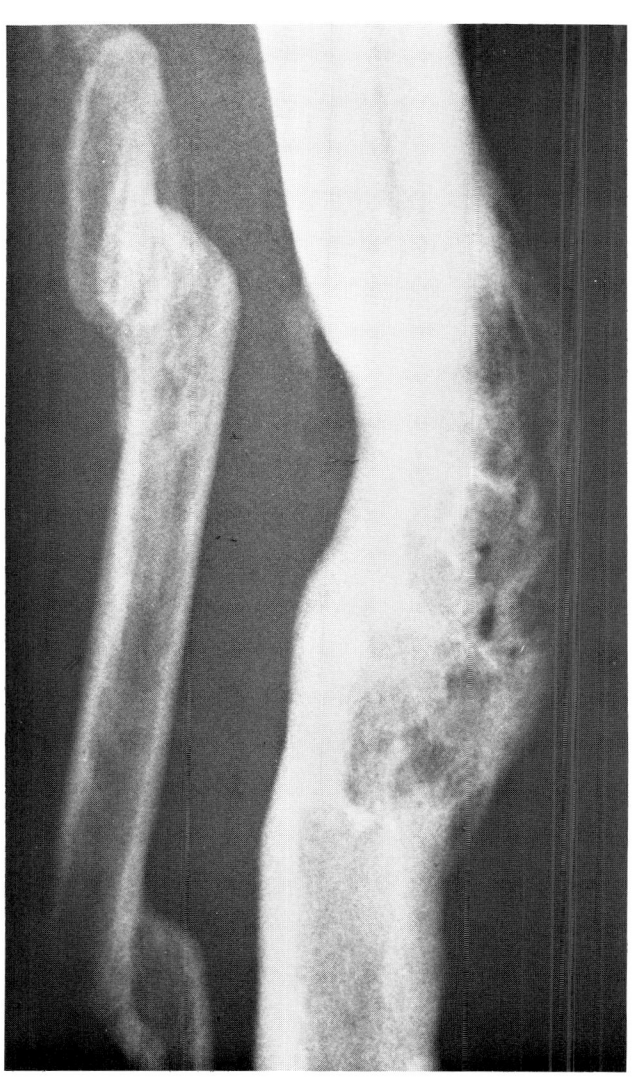

Figure 196F: Five year follow-up.

two (4%) refractures out of fifty septic nonunions treated with external fixation and gentamicin-PMMA beads. At Rancho,[116] we have had two refractures out of twenty-seven septic nonunions of the tibia treated with external fixation, debridement, and open bone grafting. One occurred following a bar room fight, the second when the patient fell from a short ladder. In spite of the apparent magnitude of the trauma, the patients insisted that the bone "snapped like a piece of chalk" with relatively little force.

It is apparent that protection of a recently healed fracture following removal of external skeletal fixation is important. The length of protection depends on the danger of refracture. Fractures that show adequate callus formation, especially in younger patients, may require as little as four weeks in a plaster cast or orthosis. On the other hand, if any of the factors known to contribute to delayed union or nonunion are present, the patient should be maintained in a snug fitting cast or molded orthosis for eight to twelve weeks. If a bone graft has been inserted to substitute for a segmental defect, the external support must be left on until cortical bone forms at the periphery of the graft mass. This may require two years or more in a removable molded polypropylene orthosis.

Chapter 10

SUMMARY AND CONCLUSIONS

The use of external fixation must be viewed as a race against time. Several biological processes are occurring simultaneously, some of which are detrimental to a successful application. For one thing, the incidence of pin tract sepsis rises slowly but steadily during the first 150 days (five months), and then begins to rise more rapidly thereafter.

At the same time, fracture callus formation may be somewhat suppressed by the absence of motion between the fracture fragments, while the fracture hematoma matures instead into collagenous scar tissue. Primary bone healing, a desirable goal of any rigid fixation system, proceeds while the frame is in place, but can only cross the fracture line if interfragmentary compression has virtually eliminated micromotion of the bone ends. Furthermore, primary healing of cortical bone is slow under the best of conditions.

Restriction of joint motion, caused by muscle impalement, may lead to permanent loss of mobility if the fixator is left in place too long.

With these considerations in mind, the surgeon should plan the fixator application with a strategy to deal with tardy bone union, should it occur. One can anticipate delayed bone healing if sequential roentgenograms (taken at three to four week intervals) do not show steady progression toward union, as evidenced by either a maturing external callus or obliteration of the fracture line.

A number of methods are available to decrease the likelihood of delayed union and nonunion. Anatomic reduction, compression when possible, and early bone grafting are the hallmarks of prudent fixator management. The danger of nonunion is greatest when suboptimal bone to bone contact occurs, a common characteristic of comminuted fractures. In this situation, early bone grafting and/or early removal of the fixator is important to prevent retarded bone healing. Any gaps between bone ends must be regarded as increasing the likelihood of delayed union and nonunion.

Sequential unloading of the fixator frame, flexible external fixation, and biocompression warrant further study as methods of transferring part of the mechanical load to bone during fracture healing. Periodic strain gauge or deflection gauge measurements during fracture healing deserves more widespread use. The information obtained with these techniques is invaluable in determining the progress of fracture healing while the patient is in a fixator.

Conversion from external fixation to internal fixation is a practice that is proven safe and effective, especially if an interval of two to four weeks is permitted to elapse before internal fixation. The risk of postoperative sepsis can be further reduced by utilizing plates rather than intramedullary rods.

Electrical stimulation of fracture healing, as an adjunct to external fixation, deserves evaluation to determine its place in the management of acute and chronic fracture problems.

The frame configuration should be appropriate to the problem being treated. A simple unilateral frame will suffice if the fractured bone makes a contribution to the configuration stability. This usually occurs with transverse fractures that are accurately reduced and with oblique fractures that are "back-cut." More rigid (and more complicated) frame assemblies are required where there is no intrinsic axial stability of the bone. Extremely rigid fixation is required when dealing with an established septic nonunion.

Pin technique is the key to a successful application of external skeletal fixation. The essential measures include the following:

1. Reduce the possibility of thermal necrosis of bone by the use of a hand drill.
2. Achieve a solid "screwed-in" fixation with multiple threaded pins.
3. Predrill the bone hole as a separate step, or use special pins to predrill the bone hole before tapping.
4. Eliminate tension at the pin-skin interface by aligning the fracture, and replacing all skin flaps to their original position prior to pin insertion.
5. Avoid pin insertion into the fracture site itself.

6. Transfix muscle without undue deep tissue tension.
7. Make adequate skin relaxing incisions after the pins are in place.
8. Release additional tension at the pin-skin interface if distraction or compression is employed.
9. Reduce skin motion at the pin by utilizing a bulky gauze wrap and requiring the patient to use crutches.

Chronic pin hole osteomyelitis is known to follow certain pin tract infections. It is important to warn patients that external fixation is associated with a 1-4 percent risk of chronic pin sepsis that will require additional surgical treatment.

Certain measures seem to reduce the likelihood of pin hole osteomyelitis:
1. Avoidance of pin hole insertion into the dense anterior crest of the tibia.
2. Removal of any pin emerging from a limb too hot to hold with the fingertips.
3. Use of a multiplanar fixator configuration for long-term application.
4. Early removal of loose pins, after which the pin hole should be curetted.
5. Aggressive treatment of pin tract infections.

Nerve and vessel injuries are largely avoidable. The zone system for pin placement described in this book has been designed for use in the operating room. The surgeon should mark the zones on the patient's skin for proper orientation. Dividing the selected zones for pin insertion into thirds will permit comparison with the atlas plates of this book.

The following neurovascular structures are most likely to be injured during pin insertion:
1. Radial nerve, as it winds around the humerus and the proximal radius.
2. Ulnar nerve, where it is adjacent to the medial cortex of the distal humerus and the proximal ulna.
3. Superficial femoral artery, which crosses the plane of the femur in the middle of the thigh.
4. Deep femoral artery, which lies first medial to the femur and then passes posterior to the bone.
5. Anterior tibial artery and deep peroneal nerve, which cross the lateral surface of the tibia in the distal leg, passing from the interosseous membrane to the front of the ankle.

Limb function, usually preserved with external fixation, may be restricted because of muscle impalement. Certain unilateral fixator configurations permit half-pin insertion into subcutaneous bone. Note, however, that the use of external skeletal fixation in this manner should be limited to short-term applications or employed in situations where the bone contributes significantly to the configuration stability.

Support of the foot and calf are important, especially if the limb is suspended with a fixator frame in place.

When planning to use an external skeletal fixator for joint arthrodesis, the key word to success is *exposure*. Adequate exposure permits careful bone carpentry, which will do much to prevent the common complications of arthrodesis surgery: failure to obtain union and unsatisfactory position at fusion. Supplemental cancellous bone graft should be utilized if there is deficiency of cancellous bone at the arthrodesis surfaces. Compression sufficient to deflect the pins two or three degrees should be applied if the arthrodesis is stable. Compression should not be applied, however, to unstable fusions and fractures, because the limb will certainly shorten, while union may not be achieved.

Pain after fixator application is a frequent occurrence, and usually consists of a dull ache. Undue or excessive pain requires evaluation and treatment. Early deep pin sepsis is a common cause, requiring release of tissue and removal of loose pins.

Bear in mind the stressful aspects of fixator applications to the patient's psyche. A clearly defined treatment program presented to the patient before fixator application will help ease the burden.

The application of a fixator to contaminated or infected fractures requires the same principles of wound management utilized with other treatment modalities. Attention to wound debridement, aggressive soft tissue management, and conversion to other methods of management, when possible, are important considerations. Skeletal defects involving more than one-third the diameter of the cortex of a long bone should be bone grafted, as should any site demonstrating tardy union—a feature common to severely traumatized limbs. The bone graft itself should consist of fresh autogenous bone. Cancellous bone should be used if there is any danger of graft infection; otherwise, cortico-cancellous strips will do.

It is obvious from the preceding chapters that complications continue to haunt external fixation in spite of the superbly designed frames and components currently available. For these reasons, fixators should be reserved for those situations where other treatment modalities are likely to fail.

As indicated in the preface, the relative merits of the various strategies available to deal with complex musculoskeletal injuries have not, as yet, been fully defined. Hopefully, the decade of the 1980s will more clearly delineate the rightful place of external skeletal fixation in the armamentarium of the orthopaedic surgeon.

BIBLIOGRAPHY

1. Adams, J. C.: Arthrodesis of the Ankle Joint. Experiences with the Transfibular Approach, *J. Bone Joint Surg.*, 30B:506, 1948.
2. Adrey, J.: Le *Fixateur Externe d'Hoffmann Couplé en Cadre. Etude Biomecanique dans les Fractures de Jambe.* Paris, Gead, 1970.
3. Ahlback, S. O. and Lindahl, O.: Hip Arthrodesis—The Connection Between Function and Position. *Acta Orthop. Scand.*, 37:77, 1966.
4. Ahmadi, B., Akbarnca, B. A., Ghobodi, F., Ganjavian, M-S. and Nasseri, D.: Experience with 141 Tibial Lengthenings in Poliomyelitis and Comparison of 3 Different Methods. *Clin. Orthop.*, 141:150, 1979.
5. Akeson, W., Woo, S., Amiel, D., and Doty, D.: Rapid Recovery from Contracture in Rabbit Hind Limbs. *Clin. Orthop.*, 122:359, 1977.
6. Anderson, J. T. and Gustilo, R. B.: Immediate Internal Fixation in Open Fractures. *Orthop. Clin. North Am.* 11:569, 1972.
7. Anderson, L. D. and Hutchins, W. C.: Fractures of the Tibia and Fibula Treated with Casts and Transfixing Pins. *South. Med. J.* 59:1026, 1966.
8. Anderson, R.: An Automatic Method of Treatment for Fractures of the Tibia and the Fibula. *Surg. Gynecol. Obstet.* 58:639, 1934.
9. Anderson, R.: An Ambulatory Method of Treating Fractures of the Shaft of the Femur. *Surg. Gynecol. Obstet.*, 62:865, 1936.
10. Anderson, W. V.: Leg Lengthening. *J. Bone Joint Surg.* 34B:150, 1952.
11. Arcq, M.: Le Risque Infectieux dans l'encouage Centromedullaire Secondaire a' un Traitement Par le Fixateur d'Hoffmann. *Acta Orthop. Belg.* 39:710, 1973.
12. Bargar, W. L., Cracchiolo, A., III, and Amstutz, H. C.: Results with the Constrained Total Knee Prosthesis in Treating Severely Disabled Patients and Patients with Failed Total Knee Replacements. *J. Bone Joint Surg.* 62A:504, 1980.
13. Barr, J. S., Freiberg, J. A., Colonna, P. C., and Pemberton, P. A.: Report of the Research Committee of the American Orthopaedic Association. A Survey of End Results on Stabilization of the Paralytic Shoulder. *J. Bone Joint Surg.* 24:699, 1942.
14. Bassett, C. A. L.: Current Concepts of Bone Formation. *J. Bone Joint Surg.* 44A:1217, 1962
15. Bassett, C. A. L.: Clinical Implications of Cell Function in Bone Grafting. *Clin. Orthop.* 87:49, 1972.
16. Bassett, C. A. L., Pilla, A. A., and Pawluk, R. J.: A Non-Operative Salvage of Surgically Resistant Pseudoarthroses and Non-Unions by Pulsating Electromagnetic Fields. *Clin. Orthop.* 124: 128, 1977.
17. Bonfield, W. and Li, C. H.: The Temperature Dependence of the Deformation of Bone. *J. Biomech.* 1:323, 1968.
18. Batten, R. L.: The Place of Compression Techniques in the Management of Long Bone Fractures in an Industrial City. *J. Bone Joint Surg.* 51B:177, 1969.
19. Bowers, W. H., Wilson, F. C., and Greene, W. B.: Antibiotic Prophylaxis in Experimental Bone Infections. *J. Bone Joint Surg.* 55A:795, 1973.
20. Boyd, H. B., Lipinski, S. W., and Wiley, J. H.: Observations on Non-Union of the Shafts of the Long Bones: With a Statistical Analysis of 842 Patients. *J. Bone Joint Surg.*, 43A:159, 1961.
21. Bradford, C. and Wilson, P. D.: Mechanical Skeletal Fixation in War Surgery. *Surg. Gynecol. Obstet.* 75:468, 1942.
22. Bradley, G. W., McKenna G. B., Dunn, H. K., Daniels, A. U., and Statton, W. O.: Effects of Flexural Rigidity of Plates on Bone Healing. *J. Bone and Joint Surg.* 61A:866, 1979.
23. Brighton, C. T.: Pathophysiology and Diagnosis of Non-Union. *American Academy of Orthopedic Surgeons.* Audio Cassette: 00S-9, 1980.
24. Brighton, C. T., Friedenberg, Z. B., Mitchell, E. I, and Booth, R. E.: Treatment of Non-Union with Constant Direct Current. *Clin. Orthop.* 124:106, 1977.
25. Brodersen, M. P., Fitzgerald, R. H., Peterson, L. F. A., Conventry, M. B., and Bryan, R. S.: Arthrodesis of the Knee following Failed Total Knee Arthroplasty. *J. Bone Joint Surg.* 61A:181, 1979.
26. Brooker, A.: Use of External Fixation in the Treatment of Burn Patients with Fractures. In Vidal, J. (Ed.): *Proceedings of the 7th International Conference on Hoffmann External Fixation.* Geneva, Diffince, 1979.
27. Brooks, A. L. and Saunders, E. A.: Fusion of the Ankle

in Denervated Extremities. *South. Med. J.* 60:30, 1967.
28. Brooks, D. B., Burnstein, A. H., and Frankel, V. H.: The Biomechanics of Torsional Fractures. The Stress Contraction Effect of a Drill Hole. *J. Bone Joint Surg.* 52A:507, 1970.
29. Brown, P. W.: The Prevention of Infection in Open Wound. *Clin. Orthop.*, 96:42, 1973.
30. Brown, P. W. and Urban, J. G.: Early Weight-Bearing Treatment of Open Fractures of the Tibia. An End-Result Study of Sixty-Three Cases. *J. Bone Joint Surg.*, 51A:59, 1969.
31. Buchholz, H. W. and Gartmann, H. D.: Infektionsprohylaxe und Operative Behandlung der Schleichenden Tiefen Infektion bie der Totalen Endoprosthese. *Der Chirurg.* 43:446, 1972.
32. Buncke, H. J. and Furnas, D. W.: Free Compound Bone-Skin Flap. *In* Daniel, R. K. and Terzis, J. K. (Eds.): *Reconstructive Microsurgery*. Boston, Little Brown, 1977.
33. Burkhalter, W. E., Butler, B., Metz, W., and Omer, G.: Experiences with Delayed Primary Closure of War Wounds of the Hand in Viet Nam. *J. Bone Joint Surg.*, 50A:945, 1968.
34. Burny, F.: Etude par Strain Gauges de la Consolidation des Fractures en Clinique. *Acta. Orthop. Belg.,* 34:917, 1968.
35. Burny, F.: Traitement par Osteotaxis des Fractures Diaphysaires du Tibia. Etude de 115 Cas. *Acta. Orthop. Belg.,* 38:280, 1972.
36. Burny, F.: Complications Liees à l'Utilisation de l'Ostéotaxis. *Acta. Orthop. Belg.,* 41:103, 1975.
37. Burny, F.: Elastic External Fixation of Tibial Fractures, Study of 1421 Cases. *In* Brooker, A. F., Jr. and Edwards, C. C. (Eds.): *External Fixation, The Current State of the Art.* Baltimore, Williams and Wilkins, 1979.
38. Burny, F.: Strain in Gauge Measurement of Fracture Healing. *In* Brooker, A. F., Jr., and Edwards, C. C. (Eds.): *External Fixation, The Current State of the Art.* Baltimore, Williams and Wilkins, 1979.
39. Burny, F. and Bourgois, R.: Etude Biomécanique du Fixateur Externe d'Hoffmann. *Acta. Orthop. Belg.* 38:265, 1972.
40. Burny, F., Elbana, S., Evrard, M., Van der Ghinst, M., Degeeter, L., Peeters, F., Verdonck, R., and Desmet, C.: Les Fractures Simples du Tibia. Traitement par Fixation Externe Elastique. *In* Vidal, J. (Ed.): *Proceedings of the 7th International Conference on Hoffman External Fixation.* Geneva, Diffinco, 1979.
41. Burny, F., Hinsenkemp, M., and Donkerwolcke, M.: External Fixation of the Humerus. Analysis of 100 Cases. *In* Vidal. J. (Ed.): *Proceedings of the 7th International Conference on Hoffmann External Fixation.* Geneva, Diffinco, 1979.
42. Burri, C.: *Post-Traumatic Osteomyelitis.* Bern, Huber, 1975.

43. Burwell, H. N.: Plate Fixation of Tibial Shaft Fractures A Survey of 181 Injuries. *J. Bone Joint Surg.,* 53B:258, 1971.
44. Cabanela, M. E.: Complications of External Fixation. *In* Johnston, R. M. (Ed.): *Advances in External Fixation.* Miami, Symposia Specialists, 1980.
45. Carnesale, P. G. and Stewart, M. J.: Complications of Arthrodesis Surgery. *In* Epps, C. (Ed.): *Complications of Orthopedic Surgery*, Philadelphia, Lippincott, 1978.
46. Carpenter, E. B.: Management of Fractures of Shaft of the Tibia and Fibula. *J. Bone Joint Surg.* 48A:1640, 1966.
47. Carpenter, E. D., Dobbie, J. J., and Siever, C. F.: Fractures of the Shaft of the Tibia and Fibula. Comparative End Results from Various Types of Treatment in a Teaching Hospital. *Arch. Surg.* 64:443, 1952.
48. Chacha, P. S. and Chong, K. C.: Experience with Tibial Leg Lengthening. *Clin. Orthop.* 125:100, 1977.
49. Chao, E. Y. S., Briggs, B. T., and McCoy, M. T.: Theoretical and Experimental Analysis of Hoffmann-Vidal External Fixation System. *In* Brooker, A. F., Jr. and Edwards, C. C. (Eds.): *External Fixation. The Current State of the Art.* Baltimore, Williams and Wilkins, 1979.
50. Chapman, M. W. and Mahoney, M.: The Role of Early Internal Fixation in the Management of Open Fractures. *Clin. Orthop.* 138:120, 1979.
51. Chapman, M. W.: The Use of Immediate Internal Fixation in Open Fractures. *Orthop. Clin. North Am.* 11:579, 1980.
52. Charnley, J.: Positive Pressure in Arthrodesis of the Knee Joint. *J. Bone Joint Surg.,* 30B:478, 1948.
53. Charnley, J.: Compression Arthrodesis of the Ankle and Shoulder. *J. Bone Joint Surg.,* 33B:180, 1951.
54. Charnley, J.: *Compression Arthrodesis.* Edinburgh, Livingstone, 1953.
55. Chrisman, O. D. and Snook, G. A.: The Problem of Refracture of the Tibia. *Clin. Orthop.,* 60:217, 1968.
56. Christie, H. K.: A Simple Appliance for the Correction of Gross Displacement in Tibial Fractures. *J. Bone Joint Surg.* 23:955, 1941.
57. Claffey, T.: Open Fractures of the Tibia. *J. Bone Joint Surg.* 42B:407, 1960.
58. Clark, C. R., Morgan, B. S., Sonstegard, D. A., and Mathews, L. S.: The Effect of Biopsy-Hole Shape and Size on Bone Strength. *J. Bone Joint Surg.* 59A:213, 1977.
59. Clawson, R. S. and McKay, D. W.: Arthrodesis in the Presence of Infection. *Clin. Orthop.* 114:209, 1976.
60. Codivilla, A.: Means of Lengthening in Lower Limbs the Muscles and Tissues Which Are Shortened Through Deformity. *Am. J. Orth. Surg.,* 2:353, 1904.
61. Cohen, J. and Harris, W. H.: The Three-Dimensional Anatomy of Haversian Systems. *J. Bone Joint Surg.,* 40A:419, 1958.
62. Cole, J. M. and Obletz, B. E.: Comminuted Fractures

of the Distal End of the Radius Treated by Skeletal Transfixation in Plaster Cast. An End Result Study in Thirty-three Cases. *J. Bone Joint Surg.* 48A:931, 1966.
63. Coleman, S. S. and Noonan, T. D.: Anderson's Method of Tibial Lengthening by Percutaneous Osteotomy and Gradual Distraction. *J. Bone Joint Surg.* 53A:411, 1971.
64. Coyler, R.: Compression External Fixation after Biplane Femoral Trochanteric Osteotomy for Severe Slipped Capital Femoral Epiphysis. *J. Bone Joint Surg.* 62A:557, 1980.
65. Connes, H.: *Hoffmann's Double Frame External Anchorage—Methods, Applications and Results in 160 Observations.* Paris, GEAD, 1973.
66. Connes, H.: *Hoffmann's External Anchorage.* Paris, GEAD, 1977.
67. Cooney, W. P. III: Current Management of Fractures of the Distal Radius and Forearm: Experience with External Pin Fixation. *In* Brooker, A. F., Jr. and Edwards, C. C. (Eds.): *External Fixation, The Current State of the Art.* Baltimore, Williams and Wilkins, 1979.
68. Cooney, W. P., III, Linscheid, R. L., and Dobyns, J. H.: External Pin Fixation for Unstable Colles Fractures. *J. Bone and Joint Surg.* 61A:840, 1979.
69. Copeland, C. X., Jr. and Enneking, W. F.: The Incidence of Osteomyelitis in Compound Fractures. *Am. Surg.*, 31:156, 1965.
70. Cotton, R. L.: A Clinical Study of Tibial Fractures Using Hoffmann External Fixation. *In* Brooker, A. F., Jr. and Edwards, C. C. (Eds.): *External Fixation, The Current State of the Art.* Baltimore, Williams. and Wilkins, 1979.
71. Crile, D. W.: Fracture of the Femur—A Method of Holding the Fragments in Difficult Cases. *Brit. J. Surg.*, 4:458, 1919.
72. Cruess, R. L., and Dumont, J.: Healing of Bone, Tendon and Ligament. *In* Rockwood, C. A., Jr., and Green, D. P. (Eds.): *Fractures.* Philadelphia, Lippincott, 1975.
73. Daniel, R. K. and May, J. W., Jr.: Free Flips: An Overview. *Clin. Orthop.* 133:122, 1978.
74. Daniel, R. K. and Taylor, G. I.: Distal Transfer of an Island Flap by Microvascular Anastomosis. *Plast. Reconstr. Surg.* 52:111, 1973.
75. Danis, R.: *Theorie et Pratique de l'Osteosynthese.* Paris, Masson et Cie, 1949.
76. Davis, A. G.: Primary Closure of Compound-Fracture Wounds, with Immediate Internal Fixation, Immediate Skin Graft and Compression Dressings. *J. Bone Joint Surg.*, 30A:405, 1948.
77. Decoulx, P., Kazemon, J. P., and Capron, J. P.: L'Utilisation de Fixateur External d'Hoffmann dans le Plasties Croisées. *Ann. Chir. Plast.*, 12:247, 1967.
78. Dee, R.: The Case for Arthrodesis of the Knee. *Orthop. Clin. North Am.* 10:249, 1979.
79. Dehne, E.: Treatment of Fractures of the Tibial Shaft. *Clin. Orthop.* 66:159, 1969.
80. Dehne, E., Metz, C. W., Deffer, P. A., and Hall, R. M.: Nonoperative Treatment of the Fractured Tibia and Immediate Weight Bearing. *J. Trauma* 1:514, 1961.
81. Die Poli, N. and Fiandaca, A.: Physiopathology of Compensating Joints in Hip Arthrodesis. *Minerva Orthop.* 21:81, 1970.
82. Delbet, P.: Methode de Traitement des Fractures. *Rev. Chir.*, 50:249, 1914.
83. Delbet, P.: *Annales de la Clinique Chirurgicale de Professeur Pierre Delbet. Méthode de Traitement de Fractures.* Paris, Librairie Felix Aken, 1916.
84. Derman, G. H. and Schenck, R. R.: Microsurgical Technique Fundamentals of the Microsurgical Laboratory. *Orthop. Clin. North Am.* 8:229, 1977.
85. Drennan, D. B., Fahey, J. J., and Maylahn, D. J.: Important Factors in Achieving Arthrodesis of the Charcot Knee. *J. Bone Joint Surg.* 53A:1180, 1971.
86. Dwyer, N.: Preliminary Report Upon a New Fixation Device for Fractures of Long Bones. *Injury*, 5:141, 1973.
87. Edwards, C. C.: Management of the Polytrauma Patient in a Major U.S. Center. *In* Brooker, A. F., Jr. and Edwards, C. C. (Eds.): *External Fixation, The Current State of the Art.* Baltimore, Williams and Wilkins, 1979.
88. Edwards, C. C.: New Directions in Hoffman External Fixation: The Maryland Experience with Major Trauma. *In* Vidal, J. (Ed.): *Proceedings of the 7th International Conference on Hoffmann External Fixation.* Geneva, Diffinco, 1979.
89. Edwards, C. C., Jaworski, M., Solana, J., and Aronson, B.: Management of Compound Tibia Fractures in the Multiply Injured Patient using External Fixation. *Am. Surg.*, 45:190, 1979.
90. Edwards, P.: Fractures of the Shaft of the Tibia. 492 Consecutive Cases in Adults. *Acta Orthop. Scand.*, Suppl. 76, 1965.
91. Evans, F. G.: *Mechanical Properties of Bone.* Springfield Thomas, 1973.
92. Eycleshymer, A. C. and Schoemaker, D. M.: *A Cross-Section Anatomy.* New York, Appleton-Century, 1911.
93. Fellander, M.: Treatment of Fractures and Pseudoarthrosis of the Long Bones by Hoffman's Transfixation Method (Osteotaxis). *Acta Orthop. Scand.* 33:132, 1963.
94. Finsterbush, A. and Friedman, B.: Early Changes in Immobilized Rabbits Knee Joints: A Light and Electron Microscopic Study. *Clin. Orthop.* 92:305, 1973.
95. Fischer, D. A.: The Hoffmann External Fixator: Technique of Application. *In* Brooker, A. and Edwards, C. (Eds.): *External Fixation, The Current State of the Art.* Baltimore, Williams and Wilkins, 1979.
96. Folschereiller, J. and Jenny, G.: Le Traitement de la Pseudarthrose Infectée du Tibia par la Solidarisation Tibio-Péronière Supérieure et Inférieure. *Rev. Orthop.* 50:499, 1964.

97. Frankel, V. H. and Burstein, A. H.: *Orthopedic Biomechanics.* Philadelphia, Lea and Febiger, 1970.
98. Freeland, A. E. and Mutz, S. B.: *Posterior Bone-Grafting for Infected Ununited Fracture of the Tibia. J. Bone Joint Surg.* 58A:653, 1976.
99. Freeman, L.: The Application of Extension to Overlapping Fractures, Especially of the Tibia, by Means of Bone Screws and a Turnbuckle without Open Reduction. *Ann. Surg.*, 70:231, 1919.
100. Freeman, M. A. R.: The Day Frame. *In* Vidal, J. (Ed.): *Proceedings of the 7th International Conference on Hoffmann External Fixation.* Geneva, Diffinco, 1979.
101. Frymoyer, J. W. and Hoaglund, F. T.: The Role of Arthrodesis in Reconstruction of the Knee. *Clin. Orthop.* 101:82, 1974.
102. Fulkerson, J. P.: Arthrodesis for Disabling Hip Pain in Children and Adolescents. *Clin. Orthop.* 128:296, 1977.
103. Fuller, J. E., Rostrup, O., and Huckell, J. R.: Ankle Arthrodesis—A Clinical Review. *In* Proceedings of the Canadian Orthopedic Association. *J. Bone Joint Surg.*, 56B:587, 1974.
104. Gallinaro, P., Crova, M., and Denicolai, F.,: Complications in 64 Open Fractures of the Tibia. *Injury* 5:157, 1973.
105. Gamble, W. E., Clayton, M. L., Leidholt, J. D., and Cletscher, J. O.: Complications Following Treatment of Tibial Fractures with Weight Bearing. *J. Bone Joint Surg.*, 54A:1343, 1972.
106. Ger, R.: The Management of Open Fracture of the Tibia with Skin Loss. *J. Trauma*, 10:112, 1970.
107. Glimcher, M. J.: A. Basic Architectural Principle in the Organization of Mineralized Tissues. *Clin. Orthop.* 61:16, 1968.
108. Goosens, M. J.: Nouveau Matériel Pour Osteosynthèse, Le Fixateur Externe et Rotules. *Le Scalpel*, 85:149, 1932.
109. Grant, J. C. B. and Basmajian, J. V.: *Grant's Method of Anatomy (Seventh Edition)* Baltimore, Williams & Wilkins, 1965.
110. Gray, H.: *Anatomy of the Human Body, Twenty-seventh Edition.* Goss, C. M. (Ed.) Philadelphia, Lea & Febiger, 1959.
111. Green, D. P.: Pins and Plaster Treatment of Comminuted Fractures of the Distal End of the Radius. *J. Bone Surg.*, 57A:304, 1975.
112. Green, D. P., Parkes, J. C., and Stinchfield, F. E.: Arthrodesis of the Knee—a Follow-up Study. *J. Bone Joint Surg.* 49A:1065, 1967.
113. Green, S. A.: Non-Union of the Lower Extremity. *In* Johnston, R. (Ed.) *Advances in External Fixation.* Miami, Symposia Specialists. 1980.
114. Green, S. A.: Arthrodesis. *In* Johnston, R. (Ed.) *Advances in External Fixation.* Miami, Symposia Specialists, 1980.
115. Green, S. A.: Placement of the Double Frame. *In* Vidal, J. (Ed.): *Proceedings of the 7th International Conference on Hoffmann External Fixation.* Geneva, Diffinco, 1979.
116. Green, S. A., and Bergdorff, T.: External Fixation in Chronic Bone and Joint Infections: The Rancho Experience. *Orthop. Trans. 4:*337, 1980.
117. Grosse, A.: Stabilization of Pelvic Fractures with Hoffmann External Fixation: The French Experience. *In* Brooker, A. F., Jr and Edwards C. C. (Eds.): *External Fixation, The Current State of the Art.* Baltimore, Williams and Wilkins, 1979.
118. Gross, A., Cutright, D. E., and Bhaskar, S. N.: Effectiveness of Pulsating Water Jet Lavage in Treatment of Contaminated Crushed Wounds. *Am. J. Surg.*, 124: 373, 1972.
119. Gustilo, R. B. and Anderson, J. T.: Prevention of Infection in the Treatment of One Thousand and Twenty-Five Open Fractures of Long Bones. *J. Bone Joint Surg.*, 58A:453, 1976.
120. Haynes, H. H.: Treating Fractures by Skeletal Fixation of the Individual Bone. *South Med. J.*, 32:720, 1939.
121. Ham, A. W.: *Histology.* Philadelphia, Lippincott, 1974.
122. Harii, K., Ohmori, K. and Ohmori, S.: Free Deltopectoral Skin Flaps. *Br. J. Plastic Surg.*, 27:321, 1974.
123. Harmon, P. H.: A Simplified Surgical Approach to the Posterior Tibia for Bone-Grafting and Fibular Transference. *J. Bone Joint Surg.* 27:496, 1945.
124. Harvey, J. P., Jr.: Management of Open Tibial Fractures. *Clin. Orthop.* 105:154, 1974.
125. Hellinger, J. and Mayer, G.: Biomechanische Versuche, Tierexperimente und Klinische Erfahrungen mit der Externen Fixation. *In* Vidal, J. (Ed.): *Proceedings of the 7th International Conference on Hoffmann External Fixation.* Geneva, Diffinco, 1979.
126. Henry, A. K.: *Extensile Exposure.* Baltimore, Williams and Wilkins, 1957.
127. Hoaglund, F. T. and States, J. D.: Factors Influencing the Rate of Healing in Tibial Shaft Fractures. *Surg. Gynecol. Obstet.* 124:71, 1967.
128. Hoffmann, R.: Rotules à Os Pour la Reduction Dirigée non Sanglante, de Fractures (Osteotaxis). *Helv. Med. Acta*, 844, 1938.
129. Hoffmann, R.: Closed Osteosynthesis with Special References to War Surgery. *Acta Chir. Scand.* 86:255, 1942.
130. Hoffmann, R.: L'osteotaxis, ostéosynthèse par fiches transcutanees et rotules. *Helv. Chir. Acta.* 18:282, 1951.
131. Hoffmann, R.: *Osteotaxis: Transcutaneous Osteosynthesis by Means of Screws and Ball-and-Socket Joints.* Paris, Gead, 1953.
132. Hoffmann, R.: Osteotaxis, Osteosyntheses Externe par Fiches et Rotules. *Acta Chir. Scand.* 107:72, 1954.
133. Hoffmann, R.: Le Traitement Transcutané des Fractures ou Ostéotaxis. *Rev. Méd. Suisse Rom.*, 4:206, 1954.
134. Hoffmann, R.: Du Danger des Fixateurs Externes et des Moyens d'y Pallier. *Acta Chir. Belg.* 49:583, 1957.
135. Horwitz, T.: The Use of the Transfibular Approach in Arthrodesis of the Ankle Joint. *Am. J. Surg.* 55:550, 1942.

136. Horwitz, T.: Surgical Treatment of Chronic Osteomyelitis Complicating Fractures. A Study of 50 Patients. *Clin. Orthop.* 96:118, 1973.

137. Ilisarov, L.: Results of Clinical Tests and Experience Obtained from the Clinical Use of the Set of Ilsarov Compression-Distraction Apparatus. *Med. Export.*, Moscow, 1976.

138. Jenny, G., Kempf, I., Jaeger, J. H., and Konsbruck, R.: Utilisation de Billes de Ciment Acrylique à la Gentamicine dans le Traitement de l'Infection Osseuse. *Rev. Chir. Orthop.* 63:491, 1977.

139. Johnson, E. W. and Boseker, E. H.: Arthrodesis of the Ankle. *Arch. Surg.*, 97:766, 1968.

140. Johnson, H. F. and Stovall, S. L.: External Fixation of Fractures. *J. Bone Joint Surg.*, 32A:466, 1950.

141. Johnston, R.: Stabilization of Pelvic Fractures with Hoffmann External Fixation: The Colorado Experience. In Brooker, A. F., Jr. and Edwards, C. C. (Eds.): *External Fixation, The Current State of the Art.* Baltimore, Williams and Wilkins, 1979.

142. Jones, E. C., Insall, J. C., Ingles, A. E., and Ranawat, C. S.: Gueper Knee Arthroplasty. Results and Late Complications. *Clin. Orthop.* 140:145, 1979.

143. Jorgensen, T. E.: Measurements of Stability of Crural Fractures Treated with Hoffmann Osteotaxis. *Acta. Orthop. Scand.* 43:188, 1972.

144. Jorgensen, T. E.: The Effect of Electric Current on the Healing Time of Crural Fractures. *Acta. Orthop. Scand.* 43:421, 1972.

145. Jorgensen, T. E.: A Simple Mechanical Method of Assessing Fracture Healing. In Brooker, A. F., Jr. and Edwards, C. C. (Eds.): *External Fixation, The Current State of the Art.* Baltimore, Williams and Wilkins, 1979.

146. Judet, H.: Immobilisation par Fixateur Externe des Extrémités Osseuses après Résection du Genou. *46e Congrès Français de Chirurgie*, 1028, 1937.

147. Judet, H.: Instrumentation pour Ostéosynthèse à Tuteur Externe. *Soc. Chir.*, Paris, 1932.

148. Judet, H.: Nouvelle Instrumentation pour L'ostéosynthèse à Tuteur Externe. *Soc. Cir.* Paris, 1934.

149. Judet, R.: Le Fixateur à Geometrie. Son Utilisation au Niveau de la Hanche. In Vidal, J. (Ed.): *Proceedings of the 7th International Conference on Hoffmann External Fixation.* Geneva, Diffinco, 1979.

150. Judet, R. and Judet, J.: Remarquè à Propos des Fixateurs Externes dans le Traitement des Fractures Ouvertes de Jambe. *Mem. Acad. Chir.* 84:288, 1958.

151. Juvara, E.: Traitement Ostéosynthésique des Fractures de Diaphyse par le Fixateur Externe et la Ligature. *Bull. Soc. Chir.*, Paris, 24, 1922.

152. Karlstrom, G. and Olerud, S.: Fractures of the Tibial Shaft. A Critical Evaluation of Treatment Alternatives. *Clin, Orthop.*, 105:82, 1974.

153. Karlstrom, G. and Olerud, S.: Percutaneous Pin Fixation of Open Tibial Fractures. Double-Frame Anchorage Using the Vidal-Adrey Method. *J. Bone Joint Surg.* 57A:915, 1975.

154. Karlstrom, G., and Olerud, S.: Stable External Fixation of Open Fractures. A Report of Five Years Experience with the Vidal-Adrey Double-Frame Method. *Orthop. Rev.* 6:25, 1977.

155. Kawamura, B.: Limb Lengthening. *Ortnop. Clin. North Am.* 9:155, 1978.

156. Kennedy, J. C.: Arthrodesis of the Ankle with Particular Reference to the Gallie Procedure. A Review of Fifty Cases. *J. Bone Joint Surg.* 42A:1308, 1960

157. Kennedy, J. C.: Complete Dislocation of the Knee Joint. *J. Bone Joint Surg.* 45A:889, 1963.

158. Key, J. A.: Positive Pressure in Arthrodesis for Tuberculosis of the Knee Joint. *South Med. J.* 25, 1932.

159. King, J. B.: The Day Frame. In Vidal, J. (Ed.) *Proceedings of the 7th International Conference on Hoffmann External Fixation.* Geneva, Diffinco, 1979.

160. Klebanoff, S. J.: Myeloperoxidase—Halide-Hydrogenperoxide Antibacterial System. *J. Bact.* 95:2131, 1968.

161. Klemm, K.: Treatment of Infected Non-Unions with Gentamicin-PMMA Beads and External Fixation. In Brooker, A. F., Jr. and Edwards, C. C. (Eds.): *External Fixation, The Current State of the Art.* Baltimore, Williams and Wilkins, 1979.

162. Klemm, K. and Jenny, G.: Traitment des Pseudarthroses Infectées par Fixateur Externe et Gentabilles. In Vidal, J. (Ed.): *Proceedings of the 7th International Conference on Hoffmann External Fixation.* Geneva, Diffinco, 1979.

163. Kolata, G. B.: Dilemma in Cancer Treatment. *Science*, 209:792, 1980.

164. Krempen, J. F., Silver, R. A., and Sotelo, A.: The Use of the Vidal-Adrey External Fixation System, Part 1: The Treatment of Open Fractures. *Clin. Orthop.* 140:111, 1979.

165. Krempen, J. F., Silver, R. A., and Sotelo, A.: The Use of the Vidal-Adrey External Fixation System, Part 2: The Treatment of Infected and Previously Infected Pseudoarthrosis. *Clin. Orthop.* 140:122, 1979

166. Kretzler, H. H. and Scham, S. M.: Roger Anderson Skeletal Pin Fixation in Fracture Treatment. *Orthop. Rev.* 11:97, 1978.

167. Kronner, R.: *The Kronner Device.* Memphis, Richards Co., 1978.

168. Lambotte, A.: L'intervention Operatoire dans les Fractures. Brussels, Lamertin, 1907.

169. Lambotte, A.: *Chirurgie Operatoire des Fractures.* Brussels, Lamertin, 1913.

170. Lance, E. M., Paval, A., Fries, I., Larsen, I., and Patterson, R. L., Jr.: Arthrodesis of the Ankle Joint. A Follow-up Study. *Clin. Orthop.* 142:146, 1979.

171. Lawyer, R.: Treatment of Complex Tibial Fractures. In Brooker, A. F., Jr. and Edwards, C. C. (Eds.): *External Fixation, The Current State of tne Art.* Baltimore, Williams and Wilkins, 1979.

172. Lazarides, E. and Revel, J.: The Molecular Basis of Cell Movement. *Sci Am.*, 240:100, 1979.

173. Lewis, K. M., Breidenbach, L., and Stader, O.: The Stader Reduction Splint for Treating Fractures of

the Shafts of the Long Bones. *Ann. Surg.* 116:623, 1942.
174. Lindahl, O.: Determination of Hip Adduction Especially in Arthrodesis. *Acta Orthop. Scand.*, 36:280, 1965.
175. Linson, M. A. and Scott, R. A.: Thermal Burns Associated with High Speed Cortical Drilling. *Orthopedics*, 1:394, 1978.
176. Lipscomb, P. R. and McCaslin, F. E.: Arthrodesis of the Hip. *J. Bone Joint Surg.* 43A:923, 1961.
177. Lockhart, R. D., Hamilton, G. F., and Fyfe, F. W.: *Anatomy of the Human Body*. Philadelphia, Lippincott, 1959.
178. Lopez-Antunez, L. and Gasparo, L.: *Atlas of Human Anatomy*. Philadelphia, Saunders, 1971.
179. Lord, G. and Besse, J. P.: Co-aptation Trochanteroiliaque *In* Vidal, J. (Ed.): *Proceedings of the 7th International Conference on Hoffmann External Fixation*. Geneva, Diffinco, 1979.
180. Lottes, J. O.: Medullary Nailing of the Tibia with the Triflange Nail. *Clin. Orthop.* 105:253, 1974.
181. Lucas, D. B. and Murray, W. R.: Arthrodesis of the Knee by Double Plating. *J. Bone Joint Surg.* 43A:795, 1961.
182. Lundeen, M., Jones, R., Mooney, V., and Murray, W.: Complex Open Tibia Fractures Treated by External Fixation. A Comparison Study Using the Roger Anderson Device and the Murray External Fixation System. *Orthop. Trans.* 4:348, 1980.
183. MacKaness, G. B.: Cell Mediated Immunity to Infection. *In* Good, R. A. and Fisher, D. W. (Eds.): *Immunobiology*. Stanford, Sinauer, 1971.
184. Malgaigne, J. F.: *Traitè de Fractures et de Luxations*. Paris, 1847. (*Treatise on Fractures*. Philadelphia, Lippincott, 1859.)
185. Mathysen, A.: *Du Bandàge plâtré et de son Application dans le Traitment des Fractures*. Liege, Grandmant-Donders, 1854.
186. Matter, P., Rittmann, W. W.: *The Open Fracture*. Bern, Huber, 1978.
187. Matthews, L. and Hirsch, C.: Temperatures Measured in Human Cortical Bone when Drilling. *J. Bone Joint Surg.*, 54A:297, 1972.
188. May, V.: Shoulder Fusion. A Review of Fourteen Cases. *J. Bone Joint Surg.*, 44A:65, 1962.
189. Mazur, J. M., Schartz, E., and Simon, S. R.: Ankle Arthrodesis, Long-Term Follow-up with Gait Analysis. *J. Bone Joint Surg.* 61A:964, 1979.
190. Mears, D.: The Management of Complex Pelvic Fractures. *In* Brooker, A. F., Jr. and Edwards, C. C. (Eds.): *External Fixation, The Current State of the Art*. Baltimore, Williams and Wilkins, 1979.
191. Mears, D.: The Use of External Fixation in Arthrodesis. *In* Brooker, A. F., Jr. and Edwards, C. C. (Eds.): *External Fixation, The Current State of the Art*. Baltimore, Williams and Wilkins, 1979.
192. Mears, D. C.: *Materials and Orthopaedic Surgery*, Baltimore, Williams and Wilkins, 1979.
193. Mears, D. C.: Modern Concepts in the Treatment of Pelvic Ring Fractures. In Vidal, J. (Ed.): *Proceedings of the 7th International Conference on Hoffmann External Fixation*. Geneva, Diffinco, 1979.
194. Mears, D. C.: The Use of External Fixation in Arthrodesis. *In* Vidal, J. (Ed.): *Proceedings of the 7th International Conference on Hoffmann External Fixation*. Geneva, Diffinco, 1979.
195. Morrey, B. F. and Wiedeman, G. P.: Complications and Long-Term Results of Ankle Arthrodeses Following Trauma. *J. Bone Joint Surg.*, 62A:777, 1980.
196. Morris, H. O. and Mosiman, R. S.: Arthrodesis of the Knee. A Comparative Study of Methods of Fusion of the Knee in Ninety-three Cases. *J. Bone Joint Surg.* 33A:982, 1951.
197. Morrison, W. A., O'Brien, B. M., and MacLeod, A.: Clinical Experiences in Free Flap Transfer. *Clin. Orthop.* 133:132, 1978.
198. Mueller, M. E., Allgoewer, M., and Willenegger, H.: *Manual of Internal Fixation*. New York, Springer-Verlag, 1970.
199. Muller, M. E. and Boitzy, A.: Le Traitement des Pseudarthroses Fistulisées de Jambe. *Rev. Chir. Orthop.*, 54:139, 1968.
200. Murray, C. R.: An Improved Method in the Use of Steinmann Pins for Traction. *Surg. Clin. North Am.* 8:1099, 1928.
201. Murray, D. G.: In Defense of Becoming Unhinged. (Editorial) *J. Bone Joint Surg.* 62A:496, 1980.
202. Naden, J. R.: External Skeletal Fixation in the Treatment of Fractures of the Tibia. *J. Bone Joint Surg.*, 31A:586, 1949.
203. Nelson, C. L., and Evarts, C. M.: Arthroplasty and Arthrodesis of the Knee Joint. *Ortho. Clin. North Am.* 2:245, 1971.
204. Nicoll, E. A.: Fractures of the Tibial Shaft. A Survey of 705 Cases. *J. Bone and Joint Surg.* 46B:373, 1964.
205. Nicoll, E. A.: Closed and Open Management of Tibial Fractures. *Clin. Orthop.*, 105:144, 1974.
206. O'Brien, B. M.: *Microvascular Reconstructive Surgery*. Edinburgh, Churchill Livingston, 1977.
207. Olerud, S.: External Fixation by the Hoffmann Device in Arthrodesis of the Talocrural Joint. *In* Vidal, J. (Ed.): *Proceedings of the 7th International Conference on Hoffmann External Fixation*. Geneva, Diffinco, 1979.
208. Olerud, S.: Treatment of Fractures by the Vidal-Adrey Method. *Acta Orthop. Scand.* 44:516, 1973.
209. Olerud, S. and Karlstrom, G.: Tibial Fractures Treated by AO Compression Osteosynthesis. Experience from a Five Year Material. *Acta Orthop. Scand.* Supp. 140, 1970.
210. Ombredanne: Ostéosynthèse Externe Temporaire Chez 1 'Enfant. *Presse Med.* 37:845, 1929.
211. Orr, H. W.: Early and Complete Immobilization as a Factor in the Preservation of Joint Function in the Treatment of Fractures. *Am. J. Surg.* 35:146, 1921.

212. Orr, H. W.: Compound Fractures with Special Reference to the Lower Extremity. *Am. J. Surg.* 46:733, 1939.
213. Papineau, L-J.: L'excision-greffe avec Fermeture Retardée Deliberée dans l'Osteomyelites Chronique. *Nouv Presse Med.*, 2:2753, 1973.
214. Parkhill, C.: A New Apparatus for the Fixation of Bones after Resection and in Fractures with a Tendency to Displacement. *Trans. Am. Surg. Assoc.* 15:251, 1897.
215. Paterson, D. C., Hillier, T. M., Carter, R. F., Ludbrook, J., Maxwell, G. M., and Savage, J. P.: Experimental Delayed Union of the Dog Tibia and Its Use in Assessing the Effect of an Electrical Bone Growth Stimulator. *Clin. Orthop.* 148:129, 1980.
216. Paterson, D. C., Lewis, G. N., and Cass, C. A.: Treatment of Delayed Union and Non-union with an Implanted Direct Current Stimulator. *Clin. Orthop.* 148:117, 1980.
217. Patzakis, M. J., Harvey, J. P., Jr., and Ivler, D.: The Role of Antibiotics in the Management of Open Fractures. *J. Bone Joint Surg.* 56A:532, 1974.
218. Perren, S. M.: Physical and Biological Aspects of Fracture Healing with Special Reference to Internal Fixation. *Clin. Orthop.* 138:175, 1979.
219. Petty, W., Bryan, R. S., Coventry, M. B., and Peterson, L.F.A.: Infection After Total Knee Arthroplasty. *Ortho. Clin. North Am.* 6:1005, 1975.
220. Pick, Y.: Quadricepsplasty. A Review, Case Presentation and Discussion. *Clin. Orthop.*, 120:138, 1976.
221. Plato, G.: Behandlungsergebnisse von 1536 Unterschenkelschaftfrakturen. *Verh. Dtsch. Ges. Orthop. Traum.* 57:308, 1971.
222. Potter, T. A.: Fusion of the Destroyed Arthritic Knee: Compression Arthrodesis versus Intramedullary Rod Techniques. *Surg. Clin. North Am.* 49:939, 1969.
223. Raimbeau, G., Chevalier, J. M., and Raguin, J.: Les Risques Vasculaires du Fixateur en Cadre a la Jambe. *Rev. Chir. Orthop.*, Supp. 11., 65 77, 1979.
224. Ramsey, W. S.: Analysis of Individual Leukocyte Behavior during Chemotaxis. *Exp. Cell Res.* 70:129, 1972.
225. Ramsey, W. S.: Locomotion of Human Polymorphonuclear Leukocytes. *Exp. Cell Res.*, 72:489, 1972.
226. Ratliff, A.H.C.: Compression Arthrodesis of the Ankle. *J. Bone Joint Surg.* 41B:524, 1959.
227. Reckling, F. W. and Roberts, M. D.: Primary Closure of Open Fractures of the Tibia and Fibula by Fibular Fixation and Relaxing Incision. *J. Trauma*, 10:853, 1970.
228. Rezaian, S. M.: Tibial Lengthening Using a New Extension Device. *J. Bone Joint Surg.* 58A:239, 1976.
229. Rhinelander, F. W.: The Normal Microcirculation of Diaphyseal Cortex and Its Response to Fracture. *J. Bone Joint Surg.* 50A:784, 1968.
230. Rittmann, W. W. and Perren, S. M.: *Cortical Bone Healing After Internal Fixation and Infection*. New York, Springer-Verlag, 1974.
231. Rosenthal, R. E., MacPhail, J. A., and Ortiz, J. E.: Non-Union in Open Tibial Fractures. Analysis of Reasons for Failure of Treatment. *J. Bone Joint Surg.*, 59A:244, 1977.
232. Rouiller, C. and Majno, G.: Morphologische und Chemische Untersuchungen und Knochen nach Hitzeinwir Kung. *Beitr. Path. Anat.* 113:100, 1953.
233. Rowe, C. R.: Re-evaluation of the Position of the Arm in Arthrodesis of the Shoulder in the Adult. *J. Bone Joint Surg.* 56A:913, 1974.
234. Roy-Camille, R., Reignier, B., Saillant, G., and Berteaux, D.: Résultats de l'Intervention de Papineau. A propos de 46 Cas. *Rev. Cir. Orthop.* 62:347, 1976.
235. Sakellarides, H. T., Freeman, P. A., and Grant, B. D.: Delayed Union and Non-Union of Tibial-Shaft Fractures. A Review of 100 Cases. *J. Bone Joint Surg.* 46A:557, 1964.
236. Sarmiento, A.: A Functional Below-the-Knee Cast for Tibial Fractures. *J. Bone Joint Surg.* 49A 855, 1967.
237. Sarmiento, A.: Functional Bracing of Tibial Fractures. *Clin. Orthop.* 105:202, 1974.
238. Schatzker, J., Horne, J. G., and Sumner-Smith, G.: The Effects of Movement on the Holding Power of Screws in Bone. *Clin. Orthop.* 111:257, 1975.
239. Schatzker, J., Horne, J. G., and Sumner-Smith, G.: The Reaction of Cortical Bone to Compression by Screw Threads. *Clin. Orthop.* 111:263, 1975.
240. Schenck, M.: Long-Term Follow-up of Treatment of Comminuted Fractures of the Distal End of the Radius by Transfixation with Kirschner Wires and Cast. *J. Bone Joint Surg.* 44A:337, 1962.
241. Scudese, V. A., Birotte, A., and Gialanella, J.: Tibial Shaft Fractures: Percutaneous Multiple Pin Fixation, Short-leg Cast and Immediate Weight Bearing. *Clin. Orthop.* 72:271, 1970.
242. Seligson, D. and Harmon, K.: Negative Experiences with Pins in Plaster for Femoral Fractures. *Clin. Orthop.* 138:243, 1979.
243. Shaar, C. M.: Treatment of Fractures and Bone and Joint Surgery with the Stader Reduction and Fixation Splint. *Bull. Am. Coll. Surg.* 28:128, 1943.
244. Shaar, C. M. and Kreuz, F. P.: Treatment of Fractures and Bone and Joint Surgery with the Stader Reduction and Fixation Splint. *Surg. Clin. North Am.* 22:1537, 1942.
245. Shaar, C. M. and Kreuz, F. P.: *Manual of Fractures. Treatment by External Skeletal Fixation*. Philadelphia, Saunders, 1943.
246. Shaar, C. M., Kreuz, F. P., and Jones, D. T.: Fractures of the Tibia and Fibula. Treatment with the Stader Reduction and Fixation Splint. *Surg. Clin. North Am.* 23:599, 1943.
247. Shaar, C. M., Kreuz, F. P., Jr., and Jones, D. T.: End Results of Treatment of Fresh Fractures by the Use of the Stader Apparatus. *J. Bone Joint Surg.* 26A:471, 1944.
248. Siris, I.: External Pin Transfixion of Fractures. An Analysis of Eighty Cases. *Ann. Surg.*, 120:911, 1944.
249. Solheim, K: Fractures of the Lower Leg. Immediate Results of Treatment in a Series of 500 Cases of Frac-

tures of the Shaft of Tibia and Fibula Treated with Plaster, Traction-Plaster, and Internal Fixation, with and without Exercise Therapy. *Acta Chir. Scand.* 119:268, 1960.
250. Solheim, K.: Tibial Fractures Treated According to the AO Method. *Injury* 4:213, 1973.
251. Stader, O.: A Preliminary Announcement of a New Method of Treating Fractures. *North Am. Vet.* 18:37, 1937.
252. Stewart, M.: Arthrodesis. *In* Edmondson, A. S. and Crenshaw, A.H. (Eds.): *Campbell's Operative Orthopedics.* St. Louis, Mosby, 1980.
253. Stewart, M. J. and Bland, W. G.: Compression in Arthrodesis. A Comparative Study of Methods of Fusion of the Knee in Ninety-three Cases. *J. Bone Joint Surg.* 40A:585, 1958.
254. Sudmann, E.: Treatment of Chronic Osteomyelitis by Free Grafts of Cancellous Autologous Bone Tissue. *Acta. Orthop. Scand.* 50:145, 1979.
255. Taylor, G. I.: Free Bone Transfer. *In* Daniels, R. K. and Terzis, J. K. (Eds.): *Reconstructive Microsurgery.* Boston, Little Brown, 1977.
256. Taylor, G. I.: Microvascular Free Bone Transfer. *Ortho. Clin. North Am.* 8:425, 1977.
257. Trais, A. and Ferry, A.: Cortical Circulation of Long Bones. *J. Bone Joint Surg.* 61A:1052, 1979.
258. Trueta, J.: The Role of the Vessels in Osteogenesis. *J. Bone Joint Surg.* 45A:402, 1963.
259. Uhtoff, H. K. and Dubuc, F. L.: Bone Structure Changes in the Dog under Rigid Internal Fixation. *Clin. Orthop.* 81:165, 1971.
260. Van der Linden, W. and Larsson, K.: Plate Fixation versus Conservative Treatment of Tibial Shaft Fractures. *J. Bone Joint Surg.*, 61A:873, 1978.
261. Veliskakis, K. P.: Primary Internal Fixation in Open Fractures of the Tibial Shaft. The Problem of Wound Healing. *J. Bone Joint Surg.* 41B:342, 1959.
262. Verbrugge, J.: A Propos de L'emploi de la Externe en Chirurgie Osseuse. *J. Chir. (Brux.)* 4:100, 1931.
263. Verhelst, M. P., Mulier, J. C., Hoogmarten, M. J., and Spaas, F.: Arthrodesis of the Ankle with Complete Removal of the Distal Part of the Fibula. Experience with the Transfibular Approach and Three Different Types of Fixation. *Clin. Orthop.*, 118:93, 1976.
264. Vidal, J.: Historique et Perspectives d'Avenir de la Fixation Externe. *In* Vidal, J. (Ed.): *Proceedings of the 7th International Conference on Hoffmann External Fixation.* Geneva, Diffinco, 1979.
265. Vidal, J. and Adrey, J.: *Hoffmann External Anchorage.* Paris, GEAD, 1971.
266. Vidal, J and Buscayret, C.: Traitement Initial et Secondaire des Fractures Ouvertes et des Pseudarthrosies Suppurees de Jambe. *In* Vidal, J. (Ed.): *Proceedings of the 7th International Conference on Hoffmann External Fixation.* Geneva, Diffinco, 1979.
267. Vidal, J., Buscayret, C., and Connes, H.: Treatment of Articular Fractures by "Ligamentotaxis" with External Fixation. *In* Brooker, A. F. Jr., and Edwards, C. C. (Eds.): *External Fixation, The Current State of the Art.* Baltimore, Williams and Wilkins, 1979.
268. Vidal, J., Buscayret, C., Connes, H., and Melka, J.: Treatment of Open Fractures with a Loss of Osseous Substance: Examples from Clinical Cases. *In* Brooker, A. F., Jr. and Edwards, C. C. (Eds.): *External Fixation, The Current State of the Art.* Baltimore, Williams and Wilkins, 1979.
269. Vidal, J., Buscayret, C., Connes, H., Paran, M., and Allieu, Y.: Traitement des Fractures Ouvertes de Jambe par le Fixateur Externe en Double Cadre. *Rev. Chir. Orthop.* 62:433, 1976.
270. Vidal, J. Buscayret, C., Faran, M., Fischbach, C., and Brahin, B.: Utilisation de la Technique de Papineau dans le Traitement des Fractures Ouvertes. *Acta Orthop. Belg.* 42:49, 1976.
271. Vidal, J. and Connes, H.: Evolution Technique du Double Cadre. *In* Vidal, J. (Ed.): *Proceedings of the 7th International Conference on Hoffmann External Fixation.* Geneva, Diffinco, 1979.
272. Vidal, J., Connes, H., Buscayret, C., Fischbach, C., Brahin, B., and Paran, M.: Complications et Incidences Socio-Professionelles du Fixateur Externe. *In* Vidal, J. (Ed.): *Proceedings of the 7th International Conference on Hoffmann External Fixation.* Geneva, Diffinco, 1979.
273. Vidal, J., Connes, H., Buscayret, C., and Trouillas, J.: Treatment of Infected Non-Union by External Fixation. *In* Brooker, A. F., Jr. and Edwards, C. C. (Eds.): *External Fixation, The Current State of the Art.* Baltimore, Williams and Wilkins, 1979.
274. Vidal, J. and Melda, J.: Chirurgie Reparatrice des Traumatismes Graves de la Jambe — Methodes Conventionelles et Transferts Libres et Microchirurgie. *In* Vidal, J. (Ed.): *Proceedings of the 7th International conference on Hoffmann External Fixation.* Geneva, Diffinco, 1979.
275. Vidal, J., Pous, J. G., Allieu, Y., Adrey, J., and Goaland, C.: Notre Experience de l'Irrigation Continue dans le Traitement des Suppurations et des Fracas de Membres. *Montpellier Chir.* 16:481, 1970.
276. Vidal, J. M., Rabischong, P., Bonnel, F., and Adrey, J.: Etude Biomecanique du Fixateur Externe d'Hoffmann dans les Fractures de Jambe. *Société de Chirurgie de Montpellier. Séance du 20 Janvin:* 43, 1970.
277. Vizkelety, T. and Salace, T.: Analyse de 292 Cas Traites par Fixateur Externe Dans les Plus Grand Service de Traumatologie et d'Orthopedic de Hongrie. *In* Vidal, J. (Ed.): *Proceedings of the 7th International Conference on Hoffmann External Fixation.* Geneva, Diffinco, 1979.
278. Volkov, M. V. and Oganesian, O. V.: Restoration of Function in the Knee and Elbow with a Hinged-Distractor Apparatus. *J. Bone Joint Surg.* 57A:599, 1975.

279. Von Lanz, T. and Wachsmuth, W.: *Praktische Anatomie*. Berlin, Springer-Verleg, 1959.
280. Wade, P. A.: The Need for Conservatism in the Treatment of Fractures. *Am. Surg.* 33:843, 1967.
281. Wade, P. A. and Campbell, R. D.: Open Versus Closed Methods in Treating Fractures of the Leg. *Am. J. Surg.* 95:599, 1958.
282. Wadkins, C. L., Luben, R., Thomas, M., and Humphreys, R.: Physical Biochemistry of Calcification. *Clin. Orthop.* 99:246, 1974.
283. Wagner, H.: Operative Lengthening of the Femur. *Clin. Orthop.* 136:125, 1978.
284. Wagner, H.: Surgical Lengthening or Shortening of the Femur. *In* Gschwend, N. (Ed.): *Progress in Orthopaedic Surgery*. New York, Springer-Verlag, 1977.
285. Weber, B. G. and Čech, O.: *Pseudarthrosis*. New York, Grune and Stratton, 1976.
286. Weis, E. B., Roberts, J. B., and Curtiss, P. H.: Salvage of Complicated Open Fractures by Transfixation. *J. Trauma* 16:266, 1976.
287. Weissman, S. L., Herold, H. Z., and Engelberg, M.: Fractures of the Middle Two-thirds of the Tibial Shaft. Results of Treatment without Internal Fixation in One-Hundred-Forty Consecutive Cases. *J. Bone Joint Surg.* 48A:257, 1966.
288. White, A. A., III: A Precision Posterior Ankle Fusion. *Clin. Orthop.* 98:239, 1974.
289. Wilkinson, P. C.: *Chemotaxis and Inflammation*. Edinburgh, Churchill Livingstone, 1974.
290. Williams, R. C., Jr. and Fudenberg, H. H.: *Phagocytic Mechanisms in Health and Disease*. New York, Intercontinental, 1971.
291. Witschi, T. H. and Omer, G. E.: The Treatment of Open Tibial Shaft Fractures from Vietnam War. *J. Trauma* 10:105, 1970.
292. Wolff, J.: *Das Gesetz der Transformation der Knochen*. Berlin, Hirschwald, 1892.
293. Woo, S. L.-Y, Akeson, W. H., Coutts, R. D., Rutherford, L., Doty, D., Jemmott, G. F., and Amiel, D.: A Comparison of Cortical Bone Atrophy Secondary to Fixation with Plates with Large Differences in Bending Stiffness. *J. Bone Joint Surg.* 58A:190 1976.
294. Zbikowski, J. L.: Biocompression. *In* Vidal, J. (Ed.) *Proceedings of the 7th International Conference on Hoffmann External Fixation*. Geneva, Diffinco, 1979.
295. Zucman, J. and Maurer, P.: Primary Medullary Nailing of Tibia for Fractures of the Shaft in Adults. *Injury* 2:80, 1970.

INDEX

A

Abductor hallucis, 75
Achilles tendon, 55
Adductor magnus, 40
Adrey, J., 9 (see also Fixator, type)
Ahmadi, B., 13
American Academy of Orthopedic Surgeons
    fixator survey of, 5, 12, 84
Anterior compartment syndrome (see Compartment syndrome)
Anterior humeral circumflex artery, 56, 57
Anterior interosseous artery and vein, 67-69
Anterior tibial artery and vein, 31, 33, 48-55, Fig. 60, 157
Antibiotics
    during bone grafting, 94
    during open fractures, 124, 125
    for pin tract infection, 30, 140, 150 (see also Gentamicin PMMA beads)
Anderson, J., (see Gustelo, R.)
Anderson, L., 14, 81, 86, 94
Anderson, R., 3, 86 (see also Fixator type)
Anderson, W. V., 7
Ankle
    arthrodesis of, 109-113
    frame configuration for, 110
    position of, 110
    stiffness (see Joint Mobility)
AO group
    fixator developed by, 6
Arcq, M., 105
Arthrodesis, 107-123, 157 (see also specific joint)
    compression, history of, 6
    causes of failure, 107-108
    frame configuration for (see specific joint)
    indications for, 107
Articulations, fixator, 9
Axillary artery, 57
Axillary nerve, 56, 57

B

"Back cut"
    fracture site, 91, Fig. 92, 156
Ball joint
    development of, 5
Bar, fixator, 8
    removal of (see Unloading the frame)
Basilic vein, 59-71

Batten, R., 85
Biceps tendon, 57
Bilateral fixator (see Fixator, configuration)
Biocompression, 103, 156
Bland, W. (see Stewart, M.)
Bleeding, pin hole, 25 (see also Vessel injury)
Bone defect, 132-135
Bone graft (see Graft, bone)
Bone pain (see Pain, bone)
Bone scan (see Scan, bone)
Boyd, H., 89
Brace, hand (see Drill)
Brachial artery and vein, 59-65
Breidenbach, L., (see Lewis, K.)
Brighton, C., 84
Brooker, A., 29
Brooks, A., 151
Browner, B., 31
Burney, F., 7, 12, 29, 31, 79, 80, 81, 87, 102, 105, 125, 137, 145, 150, 152
Burri, C., 84
Burwell, H., 85

C

Cabenela, M., 33, 34
C-arm image intensifier (see Fluoroscopy)
Calcaneus, 112
Calf, support of, 126, 157
Callus (see Fracture healing)
Cancellous bone
    graft (see Graft, bone)
    pin tract infection in, 145, 149
Carpenter, E., 85
Casts, plaster, 6, 29, 85
    after fixator, 87, 91-94, 101, 155
    open fracture, infection following, 124
Cephalin vein, 57-71
Chao, E., 9, Fig. 10, 17
Chapman, M., 86
Charcot joint (see Neuropathic arthropathy)
Charnley, J., 6, 108, 113, 115, 137
Chemotaxis, 16-17
Chevalier, J., (see Raimbeau, G.)
Chrisman, O., 152
Chronic pin tract infection (see Infection, pin tract)
Circumflex scapular artery and vein, 57
Clawson, R., 107
Clearance, fixator

skin and soft tissue, 137
Closure, wound (see Wound)
Coleman, S., 7
Common interosseous artery and vein, 65
Common peroneal nerve, 46-49
Common plantar digital nerve, 75
Compartment, anterior
 muscle transfixion of, 78
 necrosis of muscle in, 15
 pin pressure in, 33
 syndrome, 32
Complement system
 role in chemotaxis of, 16
Components, fixator
 terminology of, 8-11
Composite tissue transfer, 132 (see also Flaps, free)
Compound Fractures (see Wound, Debridement, and Infection. See also specific bone)
Compression, fixator, 27, 91
 arthrodesis with, 81, 107, 108
 interfragmentary, 90-91
 limit of patient comfort, 27
Compression plate (see Internal fixation)
Configuration, fixator (see Fixator)
Cooney, W., 29, 79, 80, 140, 150
Corocoid process, 57
Creeping substitution
 replacement of necrotic bone by, 130

D

Danis, R., 87
Davis, A., 119
Day, W. (see Freeman, M. See also Fixator type)
Debridement
 bone, 125-126
 pin hole, 30, 150
 serial wound, 125-128
 soft tissue, 125-127
Deep brachial artery, 59
Deep femoral artery and vein, 25, 40-45, 157
 injury to, 31, 33
Deep peroneal nerve, 48-55, 157
Dehne, E., 84
Delayed union (see Non-union)
Delbet, P., 85
Dorsal cutaneous nerve of forearm, 65
Dorsal intermediate cutaneous nerve, 55
Draping for surgery, 96-97
Drill, hand
 selection of, 15, 20, 156
Drill bit, 18-20
Drill guide (see Guide, pin)
Drill sleeve, 15, 19
Drill, power, 20, 21
Dwyer, N., 31

E

Edwards, C., 13, 33, 87, 94, 125, 127
Elbow
 arthrodesis of, 121-122
 frame configuration for, 121

position of, 121
 pin placement, 62-65
Elastic external fixation, 102-103, 156
Electrical stimulation
 bone healing with, 104, 156
Enchondral ossification, 88
Endosteal callus, 87 (see also Fracture healing)
Engelberg, M. (see Weissman, S.)
Extensor carpi radialis longus, 69
Extensor hallucis longus
 transfixion of, 79
External fixators (see Fixator)

F

Fellander, M., 12, 31, 86, 145
Femoral artery
 deep (see Deep femoral artery)
 superficial (see Superficial femoral artery)
Femoral nerve, 40-47
Femur, 40-47
 fractures of, 13
Fiber bone
 fracture healing with, 88
Fibula, 48-55
 synostosis to, 99, Fig. 128
First dorsal metatarsal artery, 75
Fischer, D., 7, 14, 29 (see also Fixator type)
Fixation, rigid
 effect of, 89, 156
Fixator, external
 configuration of
  bilateral, 9, 10
  biplanar, 9, 10, 133, 134, 157
  quadrilateral, 7, 10, 16
  unilateral, 9, 10, 156, 157
 draping of, 96
 early removal of, 91, 156
 history of, 3-11
 indications for, 18, 156
 leg lengthening, 7
 pain caused by (see Pain)
 partial removal of, 100
 removal of, 91, 140
  cast or orthosis after, 87, 91-94, 101, 156
 stability of (see Stability)
 type of
  Anderson, 4, 9, 10, 12, 22, 33, 80, 86, 145
  ASIF, Fig. 9, 10, 20, 145
  Charnley, Fig. 6, 10, 107, 108, 115
  Day, 7, Fig. 9, 10, 22, 145
  Fischer, 7, 10, 20
  Haynes, 12, 145
  Hoffmann, Fig. 6, 9, 10, 12, 13, 20, 22, 80, 81
  Ilisarov, 7
  Judet, Fig. 7, 10, 20
  Kawamura, 10
  Kronner, 7, 115
  Lambotte, 3, Fig. 4, 9
  Malgaigne, Fig. 3
  Murray, 11
  Parkhill, 3, 9
  pins-in-plaster, 6, 14

Reduction-Retention, 20
Russian, 7, 10, 87
Stader, 4, 9, 10, 12, 18
Vidal-Adrey, 7, Fig. 10, 13, 16, 31, 86, 101, 103, 104, 115, 145, 152
Wagner, 8, 9, 13, 80
Flaps
 free, 131, 132
 muscle, 127, 131
  dangers of, 127
  gastrocnemius, 131
  soleus, Fig. 13C, 162
 skin, 94, 127, 130
  cross leg, 130, Fig. 159
Flexible external fixation (see Elastic external fixation and Biocompression)
Flexor hallucis longus, 75
Fluoroscopy, 25, 35
Foot
 pin placement in, 74-75, 112, 113
 support, 78, 79
Fracture
 alignment, 21, 89-91
 back cut, 91, 156
 reduction of, 89-91, 156
Fracture healing, 87-91, 156
 effect of motion on, 88, 89, 104
 physiology of, 87-89
  callus formation, 87, 88
  primary, 88, 156
 testing of, 101, 102, 103
Frame configuration (see Fixator, configuration and specific joints)
Free flaps (see Flaps)
Freeman, M., 7, 22
Full pin
 definition of, 8
Fusion, joint (see Arthrodesis)

G

Galvanic reaction
 pin and bone, 5, 12
Gentamicin - PMMA beads, 13, 150
Ger, R., 127
Goosens, M., 5
Graft
 bone, 79, 94-99, 107, 132, 156
  cancellous, 94, 113, 129, 132, 133, 134, 135, 157
  cortico-cancellous, 131, 134
  selection of, 132
  pseudarthrosis, of, 100
  technique, 95 (see also specific donor site)
  treatment of pin hole infections with, 148
 skin, 130, 94 (see also Flaps)
Granulation tissue
 pin hole, 148
 wound healing by, 130
Greater saphenous vein, 42-55
Greater trochanter
 bone graft from, 98
Green, D., 113
Guepar prosthesis
 arthrodesis after, 115
Guide, pin
 use of, 19, 20
Gustilo, R., 124, 125, 127

H

Half pin, 8
Half ring, 10
Hand
 pin placement, 72, 73
Harmon, P., 99
Haynes, H., 5
Haversian system, 87, 88
Heat
 effect on bone of, 15, 16, 19
Henry, A., 99
Herold, H. (see Weissman, S.)
Hoffmann, R., 5, 6, 102
Hip
 arthrodesis of, 117-119
  frame configuration, 118
  position of, 119
Hirsch, C., 15
Humerus, 56-63
 fractures of, 13
Hutchins, W., (see Anderson, L.)

I

Iliac crest
 bone graft from, 98
Iliacus muscle, 76
Ilio-trochanteric coaptation, 118
Ilium, 76, 77 (see also Pelvis)
Ilisarov, L., 9 (see also Fixator type)
Image intensifier (see Fluoroscopy)
Infection
 bone, 124-135
  pathophysiology of chronic, 129
 joint
  chronic, 107
 pin tract, 5, 12-30, 145-150, 156
  causes of, 14
  chronic, 145-150, 157 (see also Ring sequestrum)
  effect of motion on, 103
  frequency of, 12
  incidence, 12, 13, 14, 156
  pathophysiology of, 14-18
  prevention of, 18-25
  time in fixator, 13
 wound
  culture of, 125
  following arthrodesis, 108
  persistent, 124-135
Inferior gluteal artery and vein, 41
Internal fixation, 85, 89
 follow external fixation with, 105, 156
 infection following, 85, 124
Intraarticular fractures, 80
Ischium, 41

J

Johnson, H., 12, 84

Joint (see specific joint)
Joint mobility
  restriction of, 78, 79, 80, 81, 156
Joint sepsis
  chronic (see Infection, joint)
Jorgensen, T., 101
Judet, H., 6, 118 (see also Fixator type)
Juvara, E., 5

K

Karlstrom, G., 13, 85, 86, 124, 125
Kawamura, B., 7, 13, 29, 138 (see also Fixator type)
Key, J., 113
King, J., 94
Klemm, K., 13, 155
Knee
  arthrodesis of, 113-117
    frame configuration for, 115
    position of, 115
  motion in fixator of, 18
  pin placement in, 46-49
Krempen, J., 13, 29, 31, 79, 81, 105, 145
Kretzler, H., 14, 18
Kreuz, F. (see Shaar, C.)
Kronner, R., 7 (see also Fixator type)

L

Landmarks (see Pin, placement of)
Lambotte, A., 3, 102
Lance, E., 107, 108, 109
Larrson, K. (see Van der Linden, W.)
Lateral cutaneous nerve of forearm, 61, 63, 67
Lateral femoral circumflex artery, 41
Lateral femoral cutaneous nerve, 41-43
Lateral plantar nerve, 75
Lateral sural cutaneous nerve, 49, 51
Leg lengthening
  complications of, 7, 13
Lewis, K., 5, 12
Ligamentotaxis, 80
Lipinski, S. (see Boyd, H.)
Lipscomb, P., 119
Loosening, pin, 12, 17, 18, 30
Lottes, J., 85

M

Mahoney, M. (see Chapman, M.)
Malgaigne, J., Fig. 3
Mathews, L., 15
McCaslin, F. (see Lipscomb, P.)
McKay, D. (see Clawson, R.)
Mears, D., 29, 79, 94, 118, 145, 150
Medial brachial cutaneous nerve, 59
Medial cutaneous nerve of forearm, 61-67
Medial sural cutaneous nerve, 49-51
Median nerve, 59-73
Mesenchymal cells
  fate of
    fracture healing, 88
Microorganisms, wound, 125

pin tract, 30, 145-150
Mobility, joint (see Joint Mobility)
Morrey, B., 109
Motion
  effect on host defense of, 16
  pin-bone interface, 17
  pin-soft tissue, 17
  pin-skin, 17
Muscle impalement, 78, 79
Musculocutaneous nerve, 57-59

N

Naden, J., 12, 18, 29, 32, 81, 86, 94, 145
Nail, fixator (see Pin)
Necrosis
  bone
    due to heat (see Thermal necrosis)
    due to pressure, 16
  skin, 15, 21
Nerve, anatomy of (see specific nerve)
  injury, fixator related, 34
Neurogenic pain (see Pain)
Neuropathic arthropathy
  arthrodesis of, 107, 113
Neurovascular injury
  mechanism of, 59
Nicoll, E., 79
Non-unions, 84-106, 124-135, 156, 157
  definition, 84
  infected, 6, 10, 79, 124-135
Noonan, T. (see Coleman, S.)

O

Obturator nerve, 41
Olerud, S., 13, 105 (see also Karlstrom, G.)
Open fracture (see Wound, Debridement, and Infection)
Orthosis
  after fixator, 87, 91-94, 101, 155
  preventing refracture, 152
Osteoarthritis
  arthrodesis for, 107
Osteocytes
  thermal damage to, 15
Osteolysis (see loosening)
Osteomyelitis
  bone (see Infection, bone)
  pin tract (see Infection, pin tract)
Osteon (see Haverson system)

P

Pain
  bone
    after external fixation, 140
    excessive during fixation, 139
    following arthrodesis, 108
    pin insertion, 138
    post-operative, 138
    pressure on nerve, 140
    removal of pins causing, 140
    while in fixator, 139

Papineau, L. J., 99, 133
Parkhill, C., 3, 102 (*see also* Fixator type)
Patella, 47
 fractures of, 3
Patzakis, M., 123, 126
Perforating arteries, 41, 45
Peroneal artery and vein, 48-55
Perren, S., 17
Pelvis
 bone graft from, 98
 clearance for fixator from, 136
 pin placement, 76, 77, 119
 pin tract infection, 145
Phagocytosis, 16
Piezoelectric effect, 88, 104, 149
Pin, 8
 alignment guide (*see* Guide)
 Bonnel, 8, 18
 breakage of, 137
 care, 29-30
 Crowe, 19
 divergent, 22
 gripper, 8, 136
 infection (*see* Infection, pin tract)
 insertion technique, 21-29
 loosening of, 147, 148
 number
  effect on stability (*see* Stability)
 parallel, 22
 placement of
  lower extremity, 40-55, 74-75
  upper extremity, 56-73, 137
  Zone systems for, 38
 position within bone, 25, 34
 removal
  curretting after, 149
 smooth, 18
 Steinmann, 18, 137
 threaded, 17
 wrapping of, 29, 136
Pins-in-plaster (*see* Fixator, type)
Pin-skin interface (*see* Infection, pin tract. *See also* Motion)
Plantar arterial arch, 75
Plato, G., 85
Polycentric hinges
 joint mobilization with, 81
Popliteal artery and vein, 46-49
Posterior femoral cutaneous nerve, 41, 47
Posterior humeral circumflex artery, 56, 57
Posterior interosseous artery, 67, 69
Posterior interosseous nerve, 67
Posterior tibial artery and vein, 48-54
Predrilling, 18, 156
Pressure necrosis
 bone, 16, 145, 147
 skin, 15, 21
  from fixator, 136
Primary bone healing, 87, 88, 156
Primary wound closure, 126
Pseudarthrosis
 bone (*see* Non-union)
 bone graft (*see* Graft, bone)
Psychologic problems

 fixator related, 157

R

Radial artery and vein, 65-71
Radial collateral artery, 61
Radial nerve, 34, 57-71, 156
Radius, 64-71
 fracture of, 80
Raguin, J. (*see* Raimbeau, G.)
Raimbeau, G., 32, 33, 34
Ramsey, W., 16
Ratliff, A., 107, 112
Reckling, F., 127
Reduction, fracture
 fixator use for, 21
Refracture, 151-155
Rezaian, S., 13
Rigidity, fixator (*see* Stability)
Ring, fixator, 8, 10
Ring sequestrum, Fig. 185, 186, 187, 188, 189; 145 (*see also* Infection, pin tract)
Roberts, M. (*see* Reckling, F.)
Rosenthal, R., 85, 86, 124
Roy-Camille, R., 133, 134

S

Sakellandes, H., 85, 86
Saphenous nerve, 42-55
Sarmiento, A., 85
Scan, bone, 132
Scapula, 57
Scar tissue
 fracture healing by, 94
Schatzker, J., 17, 145
Sciatic nerve, 40-47
Screw, fixator (*see* Pin)
Secondary intention
 wound healing by, 128
Segmental bone defect, 94, 100
Seligson, D., 31
Sepsis, wound (*see* Infection, wound)
Sequestrum (*see* Infection, bone)
 ring (*see* Ring sequestrum)
Shaar, C., 5, 86
Shiers prosthesis
 arthrodesis after, 115
Shoulder
 arthrodesis of, 119-121
  frame configuration of, 120
  position of, 120
 pin placement in, 56-57
Silver, R., 13
Sinus tract
 formation of, 129
Siris, I., 29, 86, 94, 145, 150
Skin
 effects of motion on (*see* Motion)
 incision for pins, 22
 tension of, 15, 21
Skin flaps (*see* Flaps, skin)
Skin graft (*see* Graft, skin)

Skin necrosis
    fixator causing, 136
Sling, calf, 126
Snook, G. (*see* Chrisman, O.)
Soft tissue coverage, 128 (*see also* Flaps)
Sotelo, A. (*see* Krempen, J.)
Soudure Autogen, 87
Stability, fixator, 17, 18, 26, 27, 30, 91, 100, 102, 103, 115, 134, 156
Stader, O., 4, 12
Stewart, M., 113, 119
Stiffness, joint (*see* Joint mobility)
Stiffness, fixator (*see* Stability)
Stovall, S. (*see* Johnson, H.)
Strain (local deformation)
    effect of, 148 (*see also* Piezoelectric effect)
Strain gauge
    fracture testing with, 101-102, 103, 156
Superficial femoral artery, 40-47
    injury to, 31, 32
Superior ulnar collateral artery, 61
Suprascapular nerve, 57
Sural nerve, 53-55
Synovium
    transfixion of, 78

T

Talus, 112
Tap, drill, 20
Thermal necrosis, 15, 16, 19
Tibia
    bone graft from, 99
    fracture of, 6, 13, 84, 85, 124, 152
    pin placement, 48-55
    treatment of
        cast, 29, 85, 87, 91-94, 101
        internal fixation, 85, 89, 105, 124, 156
Tibial artery, anterior (*see* Anterior tibial artery)
Tibial nerve, 46, 47
Total joint replacement
    arthrodesis following, 107, 113-114, 118
Transverse scapular artery and veins, 57
Triceps, 59
    transfixion of, 79

U

Ulna
    pin placement in, 64-71

Ulnar nerve, 34, 57-73, 157
Unilateral fixator (*see* Fixator, configuration)
Unloading the frame, 100, 156

V

Van der Linden, W., 86, 124
Vastus lateralis
    transfixion of, 78
Vastus medialis
    transfixion of, 78
Verbrugge, J., 5
Verhelst, M., 109
Vessel injury, 31-33 (*see also* specific vessel)
Vidal, J., 6, 10, 17, 32, 34, 79, 118 (*see also* Fixator type)
Vizkelety, T., 87, 152

W

Wagner, H., 7, 13
Weis, E., 13
Weissman, S., 85
Weidemann, G. (*see* Morrey, B.)
Wiley, J. (*see* Boyd, H.)
Wrist
    arthrodesis of, 122-123
        frame configuration for, 122
        position of, 122 (*see also* ligamentotaxis)
    fracture, 6, 150 (*see also* ligamentotaxis)
    pin placement in, 70-73
Wolff, J., 88
World War II
    fixator use during, 5
Wound
    classification of, 124
    closure, 125, 126
        delayed, 126
        primary, 126
    culture, 125
    infection, 108, 124-135
    debridement, 127
Woven bone (*see* Fiber bone)

Z

Zbikowski, J., 103
Zone system (*see* Pin, placement)